Graphene and Other 2D Layered Nanomaterials and Hybrid Structures: Synthesis, Properties and Applications

Graphene and Other 2D Layered Nanomaterials and Hybrid Structures: Synthesis, Properties and Applications

Editors

Federico Cesano
Domenica Scarano

MDPI • Basel • Beijing • Wuhan • Barcelona • Belgrade • Manchester • Tokyo • Cluj • Tianjin

Editors
Federico Cesano
Department of Chemistry
University of Torino
Torino
Italy

Domenica Scarano
Department of Chemistry
University of Torino
Torino
Italy

Editorial Office
MDPI
St. Alban-Anlage 66
4052 Basel, Switzerland

This is a reprint of articles from the Special Issue published online in the open access journal *Materials* (ISSN 1996-1944) (available at: www.mdpi.com/journal/materials/special_issues/ graphene_nanomat_struct).

For citation purposes, cite each article independently as indicated on the article page online and as indicated below:

LastName, A.A.; LastName, B.B.; LastName, C.C. Article Title. *Journal Name* **Year**, *Volume Number*, Page Range.

ISBN 978-3-0365-3181-6 (Hbk)
ISBN 978-3-0365-3180-9 (PDF)

© 2022 by the authors. Articles in this book are Open Access and distributed under the Creative Commons Attribution (CC BY) license, which allows users to download, copy and build upon published articles, as long as the author and publisher are properly credited, which ensures maximum dissemination and a wider impact of our publications.

The book as a whole is distributed by MDPI under the terms and conditions of the Creative Commons license CC BY-NC-ND.

Contents

About the Editors vii

Preface to "Graphene and Other 2D Layered Nanomaterials and Hybrid Structures: Synthesis, Properties and Applications" ix

Domenica Scarano and Federico Cesano
Graphene and Other 2D Layered Nanomaterials and Hybrid Structures: Synthesis, Properties and Applications
Reprinted from: *Materials* **2021**, *14*, 7108, doi:10.3390/ma14237108 1

Tian Yu and Carmel B. Breslin
Graphene-Modified Composites and Electrodes and Their Potential Applications in the Electro-Fenton Process
Reprinted from: *Materials* **2020**, *13*, 2254, doi:10.3390/ma13102254 7

Daniele C. da Silva Alves, Bronach Healy, Tian Yu and Carmel B. Breslin
Graphene-Based Materials Immobilized within Chitosan: Applications as Adsorbents for the Removal of Aquatic Pollutants
Reprinted from: *Materials* **2021**, *14*, 3655, doi:10.3390/ma14133655 31

Hai Tan, Deguo Wang and Yanbao Guo
A Strategy to Synthesize Multilayer Graphene in Arc-Discharge Plasma in a Semi-Opened Environment
Reprinted from: *Materials* **2019**, *12*, 2279, doi:10.3390/ma12142279 61

Rimantas Gudaitis, Algirdas Lazauskas, Šarūnas Jankauskas and Šarūnas Meškinis
Catalyst-Less and Transfer-Less Synthesis of Graphene on Si(100) Using Direct Microwave Plasma Enhanced Chemical Vapor Deposition and Protective Enclosures
Reprinted from: *Materials* **2020**, *13*, 5630, doi:10.3390/ma13245630 73

Farzaneh Farivar, Pei Lay Yap, Tran Thanh Tung and Dusan Losic
Highly Water Dispersible Functionalized Graphene by Thermal Thiol-Ene Click Chemistry
Reprinted from: *Materials* **2021**, *14*, 2830, doi:10.3390/ma14112830 89

Rosalía Poyato, Reyes Verdugo, Carmen Muñoz-Ferreiro and Ángela Gallardo-López
Electrochemically Exfoliated Graphene-Like Nanosheets for Use in Ceramic Nanocomposites
Reprinted from: *Materials* **2020**, *13*, 2656, doi:10.3390/ma13112656 101

Michael M. Slepchenkov, Igor S. Nefedov and Olga E. Glukhova
Controlling the Electronic Properties of a Nanoporous Carbon Surface by Modifying the Pores with Alkali Metal Atoms
Reprinted from: *Materials* **2020**, *13*, 610, doi:10.3390/ma13030610 115

Zhuhua Xu, Yanfei Lv, Feng Huang, Cong Zhao, Shichao Zhao and Guodan Wei
ZnO-Controlled Growth of Monolayer WS_2 through Chemical Vapor Deposition
Reprinted from: *Materials* **2019**, *12*, 1883, doi:10.3390/ma12121883 127

Nuria Jiménez-Arévalo, Eduardo Flores, Alessio Giampietri, Marco Sbroscia, Maria Grazia Betti, Carlo Mariani, José R. Ares, Isabel J. Ferrer and Fabrice Leardini
Borocarbonitride Layers on Titanium Dioxide Nanoribbons for Efficient Photoelectrocatalytic Water Splitting
Reprinted from: *Materials* **2021**, *14*, 5490, doi:10.3390/ma14195490 135

Nawal Drici-Setti, Paolo Lelli and Noureddine Jouini
LDH-Co-Fe-Acetate: A New Efficient Sorbent for Azoic Dye Removal and Elaboration by Hydrolysis in Polyol, Characterization, Adsorption, and Anionic Exchange of Direct Red 2 as a Model Anionic Dye
Reprinted from: *Materials* **2020**, *13*, 3183, doi:10.3390/ma13143183 . . . . **147**

About the Editors

Federico Cesano received his Degree in Chemistry in 1999 at the University of Torino. After two years spent at the Italian National Research Council (2000-2002), in 2005 he completed his PhD in Materials Science. Since 2006, he has been working at the Department of Chemistry of the University of Turin. He supervised several undergraduate students from his research group. He is the co-author of more than 80 ISI publications and several chapter books published in the journals of *Chemistry* and *Materials Science*. His main research interests are 1D, 2D and 3D nanostructured materials (including oxides, carbon nanomaterials, transition metal dichalcogenides, polymers), either alone or combined to form hybrid structures and composites.

Domenica Scarano previously taught Electrochemistry (1st level in Chemistry Degree), Spectroscopic Methods and Microscopy (1st level in Material Science Degree), and Interaction and Molecular Recognition (2nd level in Chemistry Degree). Now she teaches Physical Chemistry (1st level in Chemistry Degree), Physical Chemistry of Materials (1st level in Material Science and Technology Degree), Spectroscopic Methods and Microscopy (1st level in Material Science and Technology Degree), Materials Today (1st level in Material Science and Technology Degree). She has been a tutor of many laurea theses in Chemistry and Material Science.

Her research activity is focussed on oxide-based materials and, more recently, either on carbon-based systems and on ibrid systems, within nationally and internationally coordinated projects, as documented by more than 140 ISI publications. The activity, focused on surfaces defectivity and on morphological/structural characterizations of thin-film oxides, as well as polycrystalline materials, was developed with the joint use of FTIR, UV–vis, and Raman spectroscopy, together with HRTEM, SEM and AFM microscopy. For more than ten years, she has focused on themes concerning carbon fibers, CNT/GRM-based composites, and hybrid carbon-oxide composites.

She has also been the coordinator of many national and regional research projects. Since 2005, she has been responsible for the National Project "Scientific Degrees" in Materials Science for the Torino Unit. Today, she is referent for Physical Chemistry Section in the Didactic Commission of the Chemistry Department.

Preface to "Graphene and Other 2D Layered Nanomaterials and Hybrid Structures: Synthesis, Properties and Applications"

Graphene, one of the most interesting and versatile materials of modern world, is recognized for its unique properties, which are strongly different from its bulk counterpart. This discovery has recently stimulated research on other two-dimensional (2D) systems, all consisting of a single layer of atoms.

Two-dimensional materials have also emerged as major candidates for use in next-generation applications as a result of the rapid discovery of their properties. In this Special Issue, we have collected a few recent studies that examine some of these new areas of work in the field of 2D materials.

Federico Cesano, Domenica Scarano
Editors

Editorial

Graphene and Other 2D Layered Nanomaterials and Hybrid Structures: Synthesis, Properties and Applications

Domenica Scarano and Federico Cesano *

Department of Chemistry and NIS Interdepartmental Centre, University of Torino, Via P. Giuria 7, 10125 Torino, Italy; domenica.scarano@unito.it
* Correspondence: federico.cesano@unito.it; Tel.: +39-011-6707548

Citation: Scarano, D.; Cesano, F. Graphene and Other 2D Layered Nanomaterials and Hybrid Structures: Synthesis, Properties and Applications. *Materials* **2021**, *14*, 7108. https://doi.org/10.3390/ma14237108

Received: 14 October 2021
Accepted: 15 November 2021
Published: 23 November 2021

Publisher's Note: MDPI stays neutral with regard to jurisdictional claims in published maps and institutional affiliations.

Copyright: © 2021 by the authors. Licensee MDPI, Basel, Switzerland. This article is an open access article distributed under the terms and conditions of the Creative Commons Attribution (CC BY) license (https://creativecommons.org/licenses/by/4.0/).

The field of two-dimensional (2D) layered nanomaterials, their hybrid structures, and composite materials has been suddenly increasing since 2004, when graphene—almost certainly the most known 2D material—was successfully obtained from graphite via mechanical exfoliation [1]. Since then, 2D crystals and layered nanomaterials have been more actively and widely involved in research, as shown by the growing number of scientific contributions (Figure 1a). From a geographical point of view, the majority of such contributions are from China (c.a. 29%) and the United States (18%), followed by South Korea, Germany and India (4%), United Kingdom, Singapore and Japan (3%). These scientific contributions categorized by subject area suggest a clear direction towards the application of these materials in various sectors, including engineering, energy, biochemistry-genetics-molecular biology, and computer science (Figure 1b). In this regard, there are actually emerging applications in many fields, including electronics, sensing, spintronics, plasmonics, photodetectors, ultrafast lasers, batteries, supercapacitors, piezoelectrics, thermoelectrics and catalytic applications [2–6]. Such applications are the results of the extremely wide variety of 2D materials that have been fabricated and studied, including inorganic, organic, hybrid compounds and heterostructures according to a traditional definition (Figure 2). Hundreds of compounds potentially stable at the atom-thin layer have been identified [7] and expand the actual group of 2D materials. As for the 2D material family, graphene is undoubtedly the most studied material, followed by MoS_2, which represents the transition metal dichalcogenide compounds (MX_2 type, in which M is a transition metal, such as Mo or W and X a chalcogen atom, such as S, Se, or Te). Single-element 2D materials (i.e., borophene, phosphorene, etc.), with rare exceptions, redefine the physics and chemistry of the elements [8], while ternary 2D materials have additional freedom degrees for tailoring their band gaps and physicochemical properties via stoichiometric engineering [9].

The assembly of 2D materials directly onto the surface of solids (such as in-situ fabrication of photosensitizers at TiO_2 surface [10]) is still an evolving field, but the engineered interface can improve some properties with respect to heterostructures with weakly bonded van der Waals interactions.

As reported in the previous special issue [11], the current one highlights a few achievements, past/present developments, and future perspectives in the 2D layered nanomaterials and the related hybrid structures fields. It includes two reviews and height research articles.

Yu et al. [12] reviewed the subject of graphene-based materials (i.e., graphene, graphene oxide or reduced graphene oxide) as emerging new electrode material in electro-Fenton reactions to be adopted in the contaminant removal from water, thus preserving valuable water resources. Notably, when graphene or graphene analogues are combined and supported with other carbon materials, such as carbon fiber felts or CNTs, and with Fe or other metal oxide catalysts have the potential to provide true Fe- and metal-free E-Fenton catalysts. For instance, reduced-graphene oxide may be on carbon felts or graphite electrodes and combined with CNTs, to be used in gas diffusion electrodes and when doped with N and other elements, the N-doping is appearing to be the best option in E-Fenton.

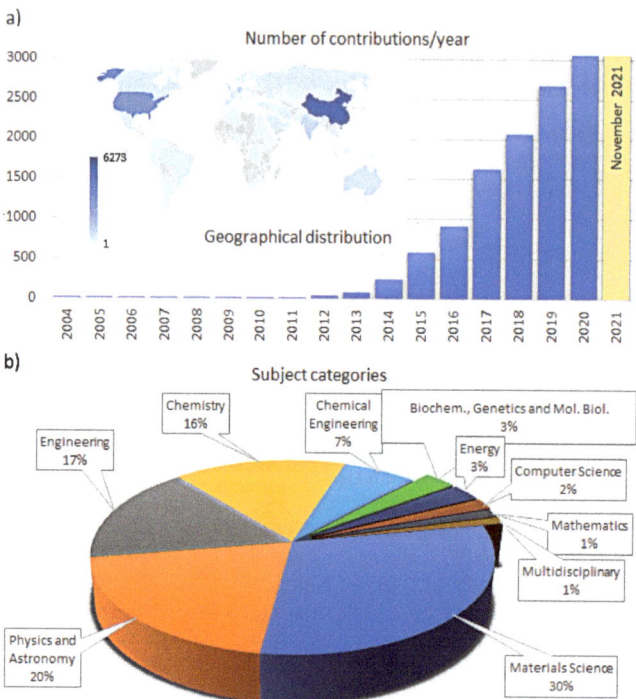

Figure 1. Scientific contributions dedicated to 2D materials (**a**) numbers of document distributed in the last 20 years and their geographical distribution; (**b**) subject areas. Keyword: "2D Materials" (source: Scopus).

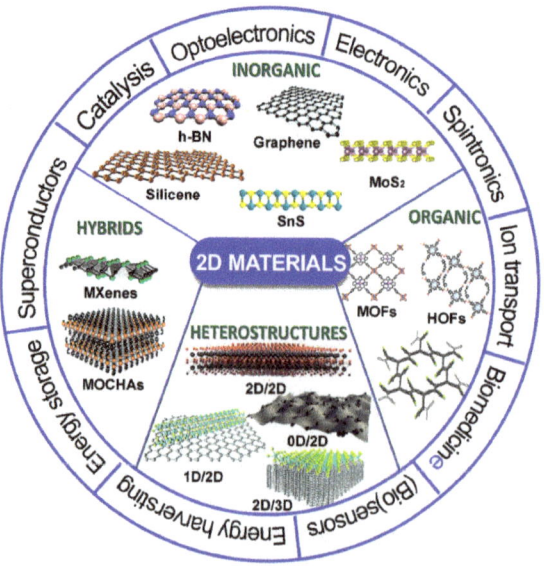

Figure 2. Classification of 2D materials and their most relevant applications.

In another review, Alves et al. [13] reviewed the field of graphene and graphene derivatives to be employed with bio-adsorbents. Among all, GO combined with chitosan has the potential to remove organic pollutants and metal cations that escaped into the water environment. The mutual role played by graphene (mechanical properties) and chitosan hydrogel (i.e., immobilization matrix) of the composite materials can increase sorption capabilities and performances as compared to graphene or chitosan alone as independent sorbent materials. The authors have also shown that additional components (including magnetic iron oxides, chelating agents, cyclodextrins, additional adsorbents and polymeric blends) can be positively added. In this regard, the performances of these materials in the removal of organic molecules, dyes and heavy metal ions are discussed together with regeneration strategies, selectivity in the adsorption process and involved costs.

The topic of graphene synthesis is still relevant for the exploration of more favorable conditions and for mass graphene production. Tan et al. [14] reported a method to obtain few-layer graphene under semi-open environmental conditions by introducing arc-discharge plasma technology. Compared with other fabrication technologies (i.e., chemical vapor deposition, mechanical/chemical exfoliation and chemical reduction of GO), no toxic gases and hazardous chemicals are generated. In the plasma discharge process, when the gas between the cathode and the anode is activated by the arc-discharge, a hexagonal arrangement of carbon atoms is observed, resulting from the nucleation and rearrangement of individual carbon atoms that have evaporated at the anode and cathode. Gudaitis et al. [15] demonstrated that graphene can be grown on the Si (100) substrates. More in detail, few-layer graphene was synthesized from CH_4 and H_2 on the Si (100) substrates without the use of catalysts via the use of direct microwave plasma-enhanced chemical vapor deposition (PECVD). N-type self-doping graphene is obtained in the process, due to the charge transfer from the Si (100) substrate. The authors observed large graphene sheets with the occurrence of compressive stresses, presumably arising from thermal stress due to the huge lattice mismatch between the Si (100) substrate and the growing graphene.

The theme of graphene functionalization/dispersion is still very active, particularly for environmentally friendly and effective methods that do not require the use of strong acids or oxidants. Farivar et al. [16] reported a simple and green modification method for preparing highly dispersible functionalized graphene via thermal thiol-ene click reactions. The method provides specific chemical functionalities such as –COO, –NH_2 and –S to the graphene sheets by using modified L-cysteine ethyl ester. The direct attachment of specific functional groups on the surface of graphene is obviously highly demanded towards the application of such materials, including ink formulations, coatings, adsorbents, sensors and supercapacitors.

Poyato et al. [17] reported the preparation of graphene nanosheets via the electrochemical exfoliation method and fabricated Yttria tetragonal zirconia (3YTZP)-based composites. The authors investigated the morphology, structure and surface properties of the composites, which were found to be formed of a graphene layer with stacking number n < 10, containing amorphous carbon and vacancy-type defects. Finally, the authors verified their Vickers hardness, which was compared with that of sintered monolithic 3YTZP ceramics. As for defectivity, there is a growing number of theoretical reports highlighting the role played by the different types of structural defects and properties.

Along this research line, Slepchenkov et al. [18] investigated a method for controlling the electronic properties of nanoporous carbon glass-like surfaces, when the pores are filled with K atoms. The presence of surface impurities, such as chemically adsorbed H and O atoms, and -OH groups, was investigated. The authors showed the calculated work function in the presence of impurities on the carbon nanoporous surface. Furthermore, the state of K atoms was shown, providing insights for the effective control of the properties, such as electronic structure and emission.

Xu et al. [19] reported the growth of monolayer WS_2 on SiO_2 substrate via the chemical vapor deposition (CVD) process in the presence of ZnO crystalline whiskers as growth promoters. As monolayer WS_2 was found on both sides of ZnO crystal whiskers, the authors discussed the monolayer growth mechanisms by approaching a concentration distribution model. According to this growth model, S and W volatile compounds and their concentrations were suggested to play a role in the WS_2 sheet thickness. Arevalo et al. [20] have recently shown the preparation of heterostructures formed by thin borocarbonitride (BCN) layers grown on TiO_2 nanoribbons. Such nanoribbons were first obtained by thermal oxidation of TiS_3 samples. Then, BCN layers were successfully grown by PECVD. The obtained TiO_2-BCN heterostructures were successfully employed in a photoelectrochemical cell, showing a boosted current density under dark conditions and higher photocurrents when compared with the bare TiO_2. The excellent photo-electrocatalytic properties of BCN suggest its use as a metal-free material in water-splitting devices.

Drici-Setti et al. [21] prepared layered double hydroxides (LDHs) based on Co, Fe and acetate ions by forced hydrolysis in a polyol medium. The synthesized Co-Fe-acetate LDH exhibited anion exchange properties and the acetate interlayer species were successfully exchanged by carbonate anions with a topotactic reaction. The exchange reactions are also favored by the high interlamellar distance. The authors have also tested LDH-Co-Fe-acetate system in sorption experiments of azoic anionic dyes from wastewater and high dye uptake was observed due to both physisorption and chemical sorption processes.

We truly hope that the reviews and research articles collected in this special issue may benefit readers and researchers in diverse fields for rising their knowledge in the fields of 2D layered nanomaterials and of the related hybrid structures, thus motivating and giving motivation for new relevant studies. We also express our sincere gratitude toward the authors, referees and the editorial staff for their valuable contributions, appropriate and insightful comments, and for the rapid and constant support.

Author Contributions: Editorial was written through the contributions of all authors. All authors have read and agreed to the published version of the manuscript.

Funding: This research received no external funding.

Acknowledgments: This work was supported by MIUR (Ministero dell'Istruzione, dell'Università e della Ricerca), INSTM Consorzio and NIS (Nanostructured Interfaces and Surfaces) Inter-Departmental Centre of University of Torino. The Guest Editors express their gratitude to the valuable contributions by all authors, referees, the editorial team of Materials, and especially to Floria Liu for helping me in managing this Special Issue.

Conflicts of Interest: The authors declare no conflict of interest.

References

1. Novoselov, K.S.; Geim, A.K.; Morozov, S.V.; Jiang, D.; Zhang, Y.; Dubonos, S.V.; Grigorieva, I.V.; Firsov, A.A. Electric Field Effect in Atomically Thin Carbon Films. *Science* **2004**, *306*, 666–669. [CrossRef]
2. Glavin, N.R.; Rao, R.; Varshney, V.; Bianco, E.; Apte, A.; Roy, A.; Ringe, E.; Ajayan, P.M. Emerging Applications of Elemental 2D Materials. *Adv. Mater.* **2020**, *32*, 1904302. [CrossRef]
3. Khan, K.; Tareen, A.K.; Aslam, M.; Wang, R.; Zhang, Y.; Mahmood, A.; Ouyang, Z.; Zhang, H.; Guo, Z. Recent developments in emerging two-dimensional materials and their applications. *JPCC* **2020**, *8*, 387–440. [CrossRef]
4. Ares, P.; Novoselov, K.S. Recent advances in graphene and other 2D materials. *Nano Mater. Sci.* **2021**. In Press. [CrossRef]
5. Zeng, M.; Xiao, Y.; Liu, J.; Yang, K.; Fu, L. Exploring Two-Dimensional Materials toward the Next-Generation Circuits: From Monomer Design to Assembly Control. *Chem. Rev.* **2018**, *118*, 6236–6296. [CrossRef] [PubMed]
6. Cui, C.; Xue, F.; Hu, W.-J.; Li, L.-J. Two-dimensional materials with piezoelectric and ferroelectric functionalities. *NPJ 2D Mater. Applic.* **2018**, *2*, 18. [CrossRef]
7. Mounet, N.; Gibertini, M.; Schwaller, P.; Campi, D.; Merkys, A.; Marrazzo, A.; Sohier, T.; Castelli, I.E.; Cepellotti, A.; Pizzi, G.; et al. Two-dimensional materials from high-throughput computational exfoliation of experimentally known compounds. *Nat. Nanotechnol.* **2018**, *13*, 246–252. [CrossRef] [PubMed]
8. Grazianetti, C.; Martella, C.; Molle, A. The Xenes Generations: A Taxonomy of Epitaxial Single-Element 2D Materials. *Phys. Stat. Sol. Rap. Res. Lett.* **2020**, *14*, 1900439. [CrossRef]
9. Wang, L.; Hu, P.; Long, Y.; Liu, Z.; He, X. Recent advances in ternary two-dimensional materials: Synthesis, properties and applications. *J. Mater. Chem. A* **2017**, *5*, 22855–22876. [CrossRef]
10. Cravanzola, S.; Cesano, F.; Gaziano, F.; Scarano, D. Carbon domains on MoS_2/TiO_2 system via acetylene polymerization: Synthesis, structure and surface properties. *Front. Chem.* **2017**, *5*, 91. [CrossRef]

11. Cesano, F.; Scarano, D. Graphene and Other 2D Layered Hybrid Nanomaterial-Based Films: Synthesis, Properties, and Applications. *Coatings* **2018**, *8*, 419. [CrossRef]
12. Yu, T.; Breslin, C.B. Graphene-Modified Composites and Electrodes and Their Potential Applications in the Electro-Fenton Process. *Materials* **2020**, *13*, 2254. [CrossRef] [PubMed]
13. Alves, D.C.D.S.; Healy, B.; Yu, T.; Breslin, C.B. Graphene-Based Materials Immobilized within Chitosan: Applications as Adsorbents for the Removal of Aquatic Pollutants. *Materials* **2021**, *14*, 3655. [CrossRef]
14. Tan, H.; Wang, D.; Guo, Y. A Strategy to Synthesize Multilayer Graphene in Arc-Discharge Plasma in a Semi-Opened Environment. *Materials* **2019**, *12*, 2279. [CrossRef] [PubMed]
15. Gudaitis, R.; Lazauskas, A.; Jankauskas, Š.; Meškinis, Š. Catalyst-Less and Transfer-Less Synthesis of Graphene on Si(100) Using Direct Microwave Plasma Enhanced Chemical Vapor Deposition and Protective Enclosures. *Materials* **2020**, *13*, 5630. [CrossRef]
16. Farivar, F.; Yap, P.L.; Tung, T.T.; Losic, D. Highly Water Dispersible Functionalized Graphene by Thermal Thiol-Ene Click Chemistry. *Materials* **2021**, *14*, 2830. [CrossRef]
17. Poyato, R.; Verdugo, R.; Muoz-Ferreiro, C.; Gallardo-Lpez, N. Electrochemically Exfoliated Graphene-Like Nanosheets for Use in Ceramic Nanocomposites. *Materials* **2020**, *13*, 2656. [CrossRef] [PubMed]
18. Slepchenkov, M.M.; Nefedov, I.S.; Glukhova, O.E. Controlling the Electronic Properties of a Nanoporous Carbon Surface by Modifying the Pores with Alkali Metal Atoms. *Materials* **2020**, *13*, 610. [CrossRef]
19. Xu, Z.; Lv, Y.; Huang, F.; Zhao, C.; Zhao, S.; Wei, G. ZnO-Controlled Growth of Monolayer WS2 through Chemical Vapor Deposition. *Materials* **2019**, *12*, 1883. [CrossRef]
20. Jiménez-Arévalo, N.; Flores, E.; Giampietri, A.; Sbroscia, M.; Betti, M.G.; Mariani, C.; Ares, J.R.; Ferrer, I.J.; Leardini, F. Borocarbonitride Layers on Titanium Dioxide Nanoribbons for Efficient Photoelectrocatalytic Water Splitting. *Materials* **2021**, *14*, 5490. [CrossRef]
21. Drici-Setti, N.; Lelli, P.; Jouini, N. LDH-Co-Fe-Acetate: A New Efficient Sorbent for Azoic Dye Removal and Elaboration by Hydrolysis in Polyol, Characterization, Adsorption, and Anionic Exchange of Direct Red 2 as a Model Anionic Dye. *Materials* **2020**, *13*, 3183. [CrossRef] [PubMed]

Review

Graphene-Modified Composites and Electrodes and Their Potential Applications in the Electro-Fenton Process

Tian Yu and Carmel B. Breslin *

Department of Chemistry, Maynooth University, Maynooth, Co. Kildare, Ireland; Tian.Yu.2020@mumail.ie
* Correspondence: Carmel.Breslin@mu.ie

Received: 25 April 2020; Accepted: 11 May 2020; Published: 14 May 2020

Abstract: In recent years, graphene-based materials have been identified as an emerging and promising new material in electro-Fenton, with the potential to form highly efficient metal-free catalysts that can be employed in the removal of contaminants from water, conserving precious water resources. In this review, the recent applications of graphene-based materials in electro-Fenton are described and discussed. Initially, homogenous and heterogenous electro-Fenton methods are briefly introduced, highlighting the importance of the generation of H_2O_2 from the two-electron reduction of dissolved oxygen and its catalysed decomposition to produce reactive and oxidising hydroxy radicals. Next, the promising applications of graphene-based electrodes in promoting this two-electron oxygen reduction reaction are considered and this is followed by an account of the various graphene-based materials that have been used successfully to give highly efficient graphene-based cathodes in electro-Fenton. In particular, graphene-based composites that have been combined with other carbonaceous materials, doped with nitrogen, formed as highly porous aerogels, three-dimensional materials and porous gas diffusion electrodes, used as supports for iron oxides and functionalised with ferrocene and employed in the more effective heterogeneous electro-Fenton, are all reviewed. It is perfectly clear that graphene-based materials have the potential to degrade and mineralise dyes, pharmaceutical compounds, antibiotics, phenolic compounds and show tremendous potential in electro-Fenton and other advanced oxidation processes.

Keywords: electro-Fenton; graphene; oxygen reduction reaction; advanced oxidation; hydrogen peroxide

1. Introduction

As the quality of water continues to decrease, there has been an ever-increasing interest in advanced oxidation processes (AOPs) that are capable of mineralising organic pollutants to CO_2, H_2O and inorganic ions, or at least to harmless products [1]. These organic pollutants, which include pesticides, herbicides, dye molecules, phenolic compounds, antibiotics, pharmaceuticals and surfactants, are normally very difficult to degrade [2,3]. Conventional water treatment plants are not always capable of removing these emerging contaminants. Although they are present in water at relatively low concentrations, their presence and ability to produce even more harmful metabolites have led to the development of AOPs. These processes range from UV irradiation and ozonation to electrochemical oxidation [1,4,5]. The electro-Fenton (E-Fenton) process has been identified as a particularly attractive technology, as it is clean and can be used to generate reasonably high concentrations of hydroxy radicals, OH•, that can be employed in the oxidation, degradation and mineralisation of various organic compounds and is considered to be one of the more promising and emerging AOPs [1,6,7].

The main principle of the E-Fenton process is summarised in Equation (1), where the oxidation of Fe^{2+} to Fe^{3+} facilitates the conversion of H_2O_2 to the highly oxidising OH•, with a standard reduction potential of 2.56 V vs. SCE (saturated calomel electrode). Consequently, these OH• radicals can be employed to mineralise a large number of organic contaminants. The sustained production of OH• requires both Fe^{2+} and H_2O_2. The Fe^{2+} ions can be regenerated through the reduction of Fe^{3+} at the cathode (Equation (2)), provided the Fe^{3+} ions do not form hydroxide precipitates in the solution phase, to give a near-continuous supply of Fe^{2+}. The H_2O_2 is produced by the two-electron reduction of dissolved oxygen; see Equation (3).

$$Fe^{2+} + H_2O_2 + H^+ \rightarrow Fe^{3+} + OH^\bullet + H_2O \tag{1}$$

$$Fe^{3+} + e^- \rightarrow Fe^{2+} \tag{2}$$

$$O_2 + 2H^+ + 2e^- \rightarrow H_2O_2 \tag{3}$$

However, the classical homogeneous E-Fenton process suffers from a number of limitations. The generation of secondary sludge, as ferric and ferrous ions in the treated wastewater, gives hydroxide precipitates and these must be removed, adding cost and reducing the overall efficiency. To limit the formation of solid $Fe(OH)_2$ and $Fe(OH)_3$, the system must be operated under stringent pH control as these hydroxide species are only soluble at pH values lower than about 4.0. Therefore, the pH of water samples or effluents must be acidified and brought to an acidic pH value of approximately 2.0 to 3.0. Moreover, the OH• radicals are not continuously generated and require a supply of pure oxygen.

In more recent years, heterogeneous E-Fenton has emerged as a solution to the issues with iron hydroxide precipitation [8–10]. In this case, iron catalysts are incorporated as solids, usually oxides, such as Fe_3O_4, into a suitable electrode material. In Figure 1, a schematic is provided, illustrating homogenous and heterogeneous E-Fenton. As shown in Figure 1a for homogeneous E-Fenton, Fe^{2+} is generated from a sacrificial anode and reacts with the H_2O_2 in the bulk solution. An iron salt can also be added to the cell to facilitate this reaction. In contrast, the aim in heterogeneous E-Fenton (Figure 1b) is to maintain the Fe^{2+}/Fe^{3+} couple in the solid state [11] and, provided the cathode promotes the two-electron reduction reaction to give H_2O_2, these coupled reactions can be sustained.

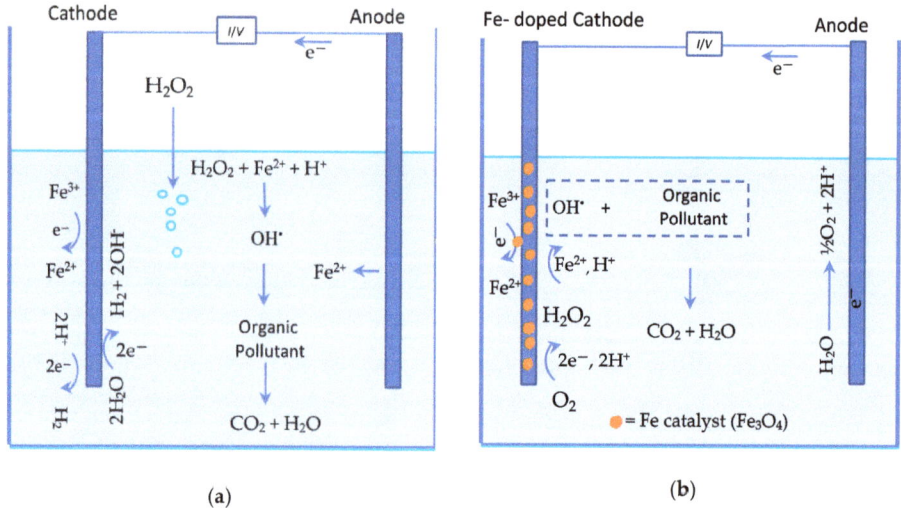

Figure 1. Schematic of (**a**) homogeneous and (**b**) heterogeneous electro-Fenton (E-Fenton).

Although heterogeneous E-Fenton can be employed over a wide pH range, issues still remain with the sluggish reduction of Fe^{3+} to Fe^{2+} within the iron-containing catalysts and long-term instability of the catalysts. Furthermore, high and sustainable amounts of H_2O_2 are required to provide efficient levels of OH• and this can be difficult as the two-electron oxygen reduction reaction is often complicated by the competing four-electron transfer reaction. Therefore, it is no surprise that considerable effort has been devoted to the design and production of cathode materials that facilitate the formation of high yields of H_2O_2. Metals such as Au, Pt, Pd and Ru [12,13] are effective catalysts for the production of H_2O_2, with relatively low overpotentials and very good conductivity. However, these are not cost-effective. Recently, carbon-based electrodes and, in particular, graphene-based materials are beginning to be employed in E-Fenton cells and in E-Fenton technologies. This developing interest in the use of graphene-based materials in E-Fenton can be seen clearly in Figure 2, where the number of publications is shown as a function of the year of publication for Fenton, which covers the classical Fenton reagents, E-Fenton, and graphene-based materials coupled with E-Fenton. The publications assigned to Fenton include E-Fenton and this comparison highlights the rise in the popularity of E-Fenton over recent years. It is also very evident from this analysis that graphene-based materials are being increasingly considered as electrodes in E-Fenton cells and are likely to make a more significant impact in the near future.

In this review, graphene-modified electrodes and composites and their applications in E-Fenton are reviewed and discussed. There are a number of review papers devoted to AOPs and E-Fenton technologies [14–17], for example, Brillas and Martinez-Huitle [15] have reviewed various electrochemical treatments, including OH• radicals as oxidants; Nidheesh et al. [18] have reviewed various electrochemical advanced oxidation processes for the removal of dye molecules, while Bechelany and co-workers [19] have considered a number of carbonaceous materials for energy and environmental applications, such as E-Fenton. Likewise, there are some excellent reviews published on graphene/rGO and its applications [20,21]. Nag et al. [20] have described the applications of graphene/rGO in sensors, Chang et al. [21] have reviewed the use of graphene-based materials as anodes in batteries, while the applications of graphene-based composites to electrocatalysis [22], energy storage [23] and flexible electronics [24] have all been described and reviewed. However, to the best of our knowledge, there is only one mini-review that considers graphene-based cathodes in E-Fenton [25]. In this present review, we consider a more extensive variety of graphene-based composites and electrodes and discuss their emerging applications in E-Fenton. The methods employed in forming these composites and their subsequent performance in catalysing the selective two-electron oxygen reduction reaction and the removal of a number of micropollutants and organic contaminants in E-Fenton are reviewed and discussed.

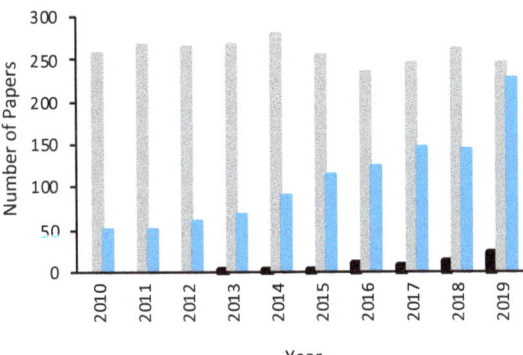

Figure 2. Number of publications shown as a function of the year of publication, taken from Scopus, for Fenton (grey), E-Fenton (blue) and E-Fenton coupled with graphene-based materials (black).

2. Oxygen Reduction Reaction and Graphene-Based Electrodes

The oxygen reduction reaction is important in several applications [26,27] and occurs by two different pathways, depending on the solution pH and cathode material. The two-electron reduction reaction is illustrated in Equation (3), while the competing four-electron reduction is given in Equation (4).

$$O_2 + 4H^+ + 4e^- \rightarrow 2H_2O \tag{4}$$

As high yields of H_2O_2 are needed to provide the essential $OH^\bullet$ species, the two-electron reduction reaction is the preferred reaction in the E-Fenton cell and therefore new catalysts are required to give high selectivity for the two-electron over the four-electron oxygen reduction reaction. Not only is this reaction highly relevant to E-Fenton, but H_2O_2 is used in a number of other applications and it is currently formed using an energy-demanding anthraquinone oxidation process [28]. Consequently, it is no surprise that, in the past decade, considerable attention has been devoted to designing new materials that can be employed to generate H_2O_2 through the two-electron reduction of dissolved oxygen [29,30]. Despite this considerable interest, the origin of the selectivity is still poorly understood. The four-electron transfer reaction is favoured by thermodynamics, which may indicate that selectivity for the two-electron transfer reaction originates in kinetics [31]. In addition to selectivity issues, there are a number of parasitic reactions that can occur at the cathode or in the bulk solution [29]. The reduction of H_2O_2 to H_2O may take place at the cathode–solution interface [32] (Equation (5)), or it may give O_2 and H_2O through a disproportionation reaction [33] (Equation (6)), or be oxidised at the anode in the cell, or through the generation of the $HO_2^\bullet$ intermediate (Equation (7)) [34]. It is generally accepted that the highest yield of H_2O_2 is achieved at pH values between 2.0 and 3.0 [35], with more acidic conditions favouring the reduction of H^+, while the lack of protons with increasing pH reduces the rate of the reaction.

$$H_2O_2 + 2e^- + 2H^+ \rightarrow 2H_2O \tag{5}$$

$$H_2O_2 \rightarrow O_2 + 2H_2O \tag{6}$$

$$H_2O_2 \rightarrow HO_2^\bullet + H^+ + e^- \rightarrow O_2 + 2H^+ + 2e^- \tag{7}$$

In an attempt to overcome these complex issues, various carbon-based materials have been considered, as these are generally cost effective and facilitate the two-electron reduction of oxygen. In particular, carbon in different forms, such as amorphous carbon, glassy carbon, graphite, fullerenes and carbon nanotubes have all been evaluated [29]. However, research is increasingly focused on the graphene family, as it presents a genuine alternative, with high surface area, moderate to good conductivity, excellent stability and its properties can be tailored using heteroatom doping and it can be easily combined with other materials [36–39]. For example, Yang et al. [40] employed a graphite felt electrode modified with electrochemically generated exfoliated graphene and carbon black and reported a H_2O_2 production rate of 7.7 mg h^{-1} cm^{-2}.

Graphene is a two-dimensional material with sp^2-hybridised carbon atoms arranged in a two-dimensional honeycomb monolayer to produce a one-atom-thick sheet. Graphene can be produced using a variety of both top-down and bottom-up approaches, including exfoliation, sonication, ball milling, chemical vapour deposition and epitaxial growth [41]. However, the production of pristine graphene with minimum defects is still challenging and while mechanical cleavage of graphite results in high quality graphene flakes, the yield is low, making the mass production of graphene demanding and time consuming [41]. One of the more common routes for the production of graphene-based materials is the formation of graphene oxide (GO) followed by its reduction to reduced GO (rGO) [42–44]. GO is typically synthesised by oxidising graphite in a mixture of $NaNO_3$, $KMnO_4$ and H_2SO_4, which is widely known as the modified Hummers method [42–46]. Once the graphite is oxidised, the interlayer spacing increases and the resulting expanded interlayer, combined with ultrasonication, allows for liquid-phase exfoliation in order to produce the GO sheets.

This oxidation process introduces a number of negatively charged oxygen-containing groups, such as hydroxy and carboxy groups, to the GO sheets and as a result, the GO sheets have good hydrophilic properties. GO exhibits good stability as a colloidal solution and this is very useful for solution processing approaches. The GO sheets can be subsequently reduced to give rGO, which has a much higher conductivity, and it is this member of the graphene family that is most suitable for electrochemical applications, such as the reduction of dissolved oxygen to produce H_2O_2 in E-Fenton applications. The reduction methods employed are generally thermal or chemical reduction processes [47]. The thermal reduction involves heating the GO to temperatures in the vicinity of 400 to 1100 °C, where most of the oxygen-containing groups are transformed to gaseous CO or CO_2, giving a reduction of GO and the formation of rGO. Various reducing agents, such as borohydride, hydrazine or hydrogen iodide can be employed at room temperature or with mild heating to give rGO. However, some of these chemicals have environmental concerns and they can introduce impurities to the carbon matrix. These health and environmental issues can be overcome by using ascorbic acid as the reducing agent. This has been successfully employed to reduce GO and the resulting rGO was shown to have good conductivity [48]. The electrochemical reduction of GO is another very effective and simple route to obtain rGO [49–51]. For example, Guo et al. [50] used a potential of −1.5 V to reduce GO, giving a green and fast process with no evidence of contamination of the rGO sheets. However, the rGO produced from all these approaches contains some oxygen-containing functional groups.

Different approaches have been used in forming graphene-based cathodes in an attempt to give enhanced and selective production of H_2O_2. These approaches include various electrophoretic and electrodeposition routines to generate graphene-based cathodes, graphene-based materials deposited from slurries or suspensions at graphite or carbon felt electrodes which act as supports, graphene-based inks, porous graphene-based aerogels, three-dimensional graphene-modified electrodes, heteroatom doped graphene composites, graphene composites combined with iron oxides and graphene-based diffusion electrodes. These are now described and discussed in the following sections.

2.1. Graphene Modified Carbon/Graphite Felt Electrodes and Other Supports

One of the most common approaches is to use carbon or graphite felt as a support for GO or rGO as these three-dimensional felt cathodes have low cost, high surface area, excellent conductivity and high porosity. The exfoliated GO/rGO can be deposited at the carbon or graphite electrodes using electrophoretic deposition [52], coated from a liquid solution phase with different additives, such as polytetrafluoretyhylene (PTFE) and carbon black to form a slurry, followed by a heating or an annealing process of the electrode [39,53,54]. Various additives have been employed in formulating these graphene-containing suspensions or slurries. In several papers polytetrafluroethylene (PTFE), a synthetic fluropolymer with good adhesion and lubricating properties, has been used successfully [55,56]. Spin coating has also been employed to deposit GO from slurries [57], while in some cases, the treated carbon or graphite felt is immersed in the rGO containing suspension or slurry [4], or dip-coated [58,59]. The solution or slurry coating approaches enable the addition of various additives that have the potential to enhance the removal of micropollutants using E-Fenton. However, the addition of binders, such as PTFE, can reduce the electric conductivity and increase the impedance of the final composite [60], while hindering ion permeability at the electrode-solution interface [61]. The electrophoretic deposition routines can be easily coupled with the electrochemical reduction of GO to form rGO without the need for reducing agents and binders that are toxic in many cases. Interestingly, in a comparison of the reduction of GO to rGO using a constant potential reduction, chemical reduction and thermal reduction, it was concluded that the electrochemical reduction was the best option, in terms of simplicity, cost, ecology and performance in the mineralisation of an azo dye [62].

Indeed, these electrochemical approaches, employing combinations of electrochemical exfoliation and electrophoretic deposition, followed by the reduction of GO to rGO, have been used to coat carbon felt electrodes [52], while in some cases the electrochemically exfoliated GO is combined with carbon black before being deposited at the felt electrodes [3,40]. These graphene-modified electrodes have been shown to faciliate the oxygen reduction reaction, producing H_2O_2 with low energy consumptions of 9.7 kWh kg^{-1} [39] and 3.08 kWh kg^{-1} [3] and have been employed successfully to remove acid orange [52], sulfadiazine [3] and imatinib [55].

On the other hand, graphene-containing slurries can be readily formed using exfoliated or electrochemically exfoliated GO and these have been combined with quinones (AQ), which have the ability to generate H_2O_2 (Equations (8) and (9)). For example, Gao et al. [53] used GO, PTFE and anthraquinone sulfonate to modify carbon felt to fabricate a hybrid electrode. The catalytic oxygen reduction reaction was greatly enhanced in the presence of the quinone, resulting in the efficient removal of Rhodamine B. The authors proposed a mechanism whereby a semi-quinone (s-Q) anion radical is formed at the cathode (Equation (10)). This is followed by a catalysed reduction of dissolved oxygen to generate the oxygen radical anion (Equation (11)), which then combines with protons to generate H_2O_2; see Equation (12).

$$AQ + 2H^+ + 2e^- \rightarrow AQH_2 \qquad E^0 = 0.43 \text{ V vs. SCE} \tag{8}$$

$$AQH_2 + O_2 \rightarrow AQ + H_2O_2 \tag{9}$$

$$AQ + e^- \rightarrow s\text{-}Q^{\bullet-} \tag{10}$$

$$O_2 + s\text{-}Q^{\bullet-} \rightarrow O_2^{\bullet-} + s\text{-}Q \tag{11}$$

$$2O_2^{\bullet-} + 2H^+ \rightarrow H_2O_2 + O_2 \tag{12}$$

It is well established that a number of oxygen-containing functional groups, such as epoxides (C–O–C), hydroxy (OH), carboxylic (COOH) and carbonyl groups (C=O) are present on GO nanosheets [63]. However, other oxygen-containing groups, such as ketones and quinones have been detected. Aliyev et al. [64], using a combination of surface analytical techniques, clearly identified quinone groups on GO layers, and these may also contribute to the generation of H_2O_2 when GO/rGO is employed as the cathode material. Indeed, Nambi and co-workers [65] attributed some of the enhanced production of H_2O_2 and the degradation efficiency of E-Fenton to quinone functional groups on electrochemically exfoliated rGO.

Graphene oxide can be easily functionalised with ferrocene [66,67], and ferrocene-functionalised rGO has been deposited at graphite felt electrodes. For example, Nambi and co-workers [58] designed a cathode by fabricating ferrocene-functionalised rGO on graphite felt electrodes and studied its heterogeneous E-Fenton reaction for the degradation of ciprofloxacin at neutral pH conditions. The removal rate of ciprofloxacin was computed as 0.035 min^{-1} for the ferrocene-modified rGO electrode, significantly higher than the value of 0.004 min^{-1} obtained for the unmodified graphite felt and also higher than 0.010 min^{-1}, which was the rate constant observed with the reduced rGO-modified felt electrode. The authors concluded that the rGO and ferrocene participated in sequential steps, with rGO facilitating the production of H_2O_2, while the Fe^{2+} centre in ferrocene catalysed the decomposition of H_2O_2 to form $OH^{\bullet}$ and Fe^{3+}-centred ferricenium. The same group studied the ferrocene-functionalised rGO felt electrode as an E-Fenton catalyst using rotating disc voltammetry for the removal of ciprofloxacin and carbamazepine [59]. Using rotating disc voltammetry, which gives improved mass transfer, a continuous supply of reactive oxygen species was achieved without aeration of the solution, to give $OH^{\bullet}$ concentrations of 644 µM, 264 µM and 163 µM at pH values of 3.0, 7.0 and 9.0, respectively, facilitating the removal of contaminants over a wide pH range.

Using another approach, Mi et al. [4] combined rGO with WO_3 and Ce and deposited the hybrid composite, $rGO/Ce/WO_3$, on carbon felt. This composite was employed in E-Fenton to remove ciprofloxacin with complete degradation within 1 h, and a mineralisation degree of 98.55% within 8 h. It was suggested that the Ce^{3+} catalysed the decomposition of H_2O_2 to give the $OH^{\bullet}$ radical species (Equation (13)), while superoxide was generated from the reaction between Ce^{3+} and dissolved O_2 (Equation (14)) and between W^{5+} and O_2 (Equation (15)). The Ce^{4+} was recycled to Ce^{3+} through a simple reduction step (Equation (16)), while Fe^{2+} ions added to the E-Fenton cell also facilitated the reduction of Ce^{4+} (Equation (17)).

$$Ce^{3+} + H_2O_2 + H^+ \rightarrow Ce^{4+} + OH^{\bullet} + H_2O \tag{13}$$

$$Ce^{3+} + O_2 \rightarrow Ce^{4+} + O_2^{-\bullet} \tag{14}$$

$$W^{5+} + O_2 \rightarrow W^{6+} + O_2^{-\bullet} \tag{15}$$

$$Ce^{4+} + e^- \rightarrow Ce^{3+} \tag{16}$$

$$Ce^{4+} + Fe^{2+} \rightarrow Ce^{3+} + Fe^{3+} \tag{17}$$

Graphene/GO has also been combined with conducting polymers and used in E-Fenton. In a recent study, a simple electropolymerisation method was employed to deposit poly (3,4-ethylenedioxythiophene) (PEDOT), sodium polystyrene sulfonate (NaPSS) and GO on graphite felt electrodes [68]. This GO/PEDOT:NaPSS showed much greater rates of H_2O_2 production and greater efficiency in the degradation of methylene blue in heterogeneous E-Fenton compared with the GO-free PEDOT:NaPSS. This was attributed to a synergistic effect between PEDOT and GO, promoting higher electron transfer rates.

In addition to electrophoretic deposition and coating from graphene-containing slurries, there has been considerable interest in graphene-based inks that can be painted or printed onto substrates [69,70]. This approach has been employed to produce graphene-based cathodes in E-Fenton [6,71]. Conducting graphene-based inks have been formed in ethanol and water mixtures using Nafion as a binder and dispersant, and the resulting ink coated onto carbon cloth [71] and carbon fibres [6]. These ink-coated cathodes have been shown to give rise to a near doubling of the amount of H_2O_2 generated and a three-fold increase in the rate of phenol degradation. A graphene-based paste cathode has also been employed to degrade a pharmaceutical product [72]. The paste was formed by mixing rGO with graphite powder and paraffin, which served as the binding agent. Divyapriya et al. [73] used another approach, whereby a liquid crystal display glass was utilised as a supporting matrix for the deposition of a thin graphene-modified electrode without the need for binders or linkers. GO was drop casted on the glass and electrochemically reduced to rGO and then used for the oxidation and degradation of ciprofloxacin. The authors showed that the drop cast electrode exhibited good stability. Graphene oxide has also been deposited on stainless steel to create a membrane for the removal of paracetamol [74], and an electrode to oxidise and remove arsenic [5], while a PTFE membrane was modified with graphene and used as a catalytic membrane to both concentrate and oxidise an antibiotic [75].

In most of these approaches, the graphene-modified electrode is compared to the conducting carbon or graphite felt or carbon cloth substrate electrodes and the addition of rGO clearly enhances the rate of H_2O_2 generation to give a more efficient removal of the contaminants. Interestingly, it was shown by Wang et al. [57], who employed polyvinylidene difluoride to fabricate both carbon nanotube (CNT) modified felt and graphene-modified felt electrodes, that the graphene-modified carbon felt was superior in the degradation of an azo dye molecule, reaching degradation rates of 70.1%, compared to the lower rates of 55.3% evident with the CNT-modified electrode. While both modifications enhanced the azo dye degradation rate compared to the untreated carbon felt electrode, the graphene-modified electrode produced a higher quantity of H_2O_2. This suggests that the graphene-modified electrodes are not only more superior than the conducing graphite or carbon substrates used in their fabrication, but they may also be more efficient in producing H_2O_2 than carbon nanotubes. The impressive performance of the graphene-modified carbon/graphite felt electrodes is clearly illustrated in Table 1. For comparative purposes, some very recently optimised and high-performing graphite-based electrodes are included. Although the amount of H_2O_2 generated is expressed differently, sometimes as rates and in other cases as mass per volume, it is evident that a graphene-based system can be employed to give the more efficient generation of H_2O_2 with relatively high rate constants for the removal of a variety of contaminants.

2.2. Porous Graphene Electrodes

Porous graphene-based composites that can be employed as aerogels, foams and as gas diffusion electrodes are interesting and these new materials are beginning to emerge in E-Fenton.

Table 1. Summary of graphene-based felt electrodes in generating H_2O_2 or $OH^{\bullet}$ and in the removal of contaminants.

System	Pollutant	Experimental Conditions	k/min^{-1}	H_2O_2/$OH^{\bullet}$	Ref.
Ferrocene-rGO/graphite	Cipro-Floxacin	V = 150 mL, E = −1.5 V, A = 10 cm^2, t = 30 min, air sparging	0.035 (acidic) 0.222 (neutral)	$OH^{\bullet}$: 426 µM(acidic) 247 µM (neutral)	[58]
rGO-LCD	Cipro-Floxacin	V = 150 mL, E = −1.5 V, A = 10 cm^2, t = 30 min, 1 L min^{-1} air flow	0.019 (neutral) 0.034 (acidic)	H_2O_2: 45 mg L^{-1} (acidic) 20 mg L^{-1} (neutral)	[73]
rGO-paste	Cipro-Floxacin	V = 400 mL, t = 45 min, E = −0.62 V, 1.2 L min^{-1} O_2	0.0056 (acidic)	H_2O_2: 22 mg L^{-1} (acidic)	[72]
rGO ink/carbon	Phenol	V = 80 mL, t = 120 min, 0.2 L min^{-1} air flow, A = 6.3 cm^2, I = 1.25 A cm^{-2}	0.0157 (acidic)	H_2O_2: 2.81 mg L^{-1} cm^{-2}	[71]
rGO/graphite cloth	Orange II Methy blue Sulfadiazine Phenol	V = 100 mL, A = 5.0 cm^2, E = −0.9 V, t = 60 min	0.52 (acidic) 0.37 (acidic) 0.62 (acidic) 0.37 (acidic)	H_2O_2: 7.7 mg h^{-1} cm^{-2} (pH 7) 2.2 mg h^{-1} cm^{-2} (pH 5)	[40]
Flow-cell rGO	Sulfadiazine	Flow through system, 7 mL min^{-1}, I = 50 mA	-	H_2O_2: 4.4 mg h^{-1} cm^{-2} (pH 7)	[3]
rGO/C felt	Imatinib	V = 150 mL, A = 12 cm^2, air flow, I = 16.6 mA cm^{-2}, t = 8 h	0.22 (acidic)	-	[55]
rGO dip coated/C felt	Cipro-floxacin Carba-mazepine	V = 300 mL, disc electrode 80 mm diameter, E = −1.5 V, t = 180 min.	0.37 (acidic) 0.20 (neutral) 0.35 (acidic) 0.08 (neutral)	H_2O_2: 175 mg L^{-1} (pH 7) 81 mg L^{-1} (pH 3)	[59]
rGO/C felt	Reactive Black 5	V = 250 mL, A = 82 cm^2, E = −0.65 V, t = 180 min	-	H_2O_2: 0.26 mM	[57]
rGO C fibre Brush	Phenol	V = 250 mL, A = 46,665 cm^2, I = 1.25 mA, t = 180 min	0.06 (acidic)	H_2O_2: 4.23 mg L^{-1} cm^{-2}	[6]
Optimised graphite system				H_2O_2: 0.74 mg h^{-1} cm^{-2} 45 mg L^{-1}	[76] [77]

2.2.1. Graphene Aerogels

Aerogels are three-dimensional highly porous materials and graphene-based aerogels exhibit interesting properties of high thermal stability, surface area and electrical conductivity [78]. They are readily constructed by the self-assembly of rGO to form a three-dimensional macroporous architecture [79,80], through template-guided approaches, solvothermal sol–gel reactions, or patterning technologies [81]. A schematic illustrating the formation of an aerogel from graphene-containing dispersions using the hydrothermal route is provided in Figure 3. The synthesised aerogel can be further processed, formed into discs and used as cathodes. These emerging materials are finding applications in fuel cells [82], environmental remediation [81,83], batteries [84] and, more recently, as electrodes in E-Fenton [85–87]. In particular, the aggregation and restacking of the rGO layers are minimised by the three-dimensional structure inside the bulk aerogel, giving good stability, while the surface layers are available to the electrolyte ions and target pollutant molecules and can be tailored for high pollutant adsorption. In addition, these materials are promising in heterogeneous E-Fenton, where zero-valent iron nanocrystals/nanoparticles or Fe_3O_4 nanoparticles can be embedded and protected within the aerogel [85]. A number of studies has been reported where rGO aerogels have been assembled and successfully used in E-Fenton for the removal of various contaminants [85,87–89].

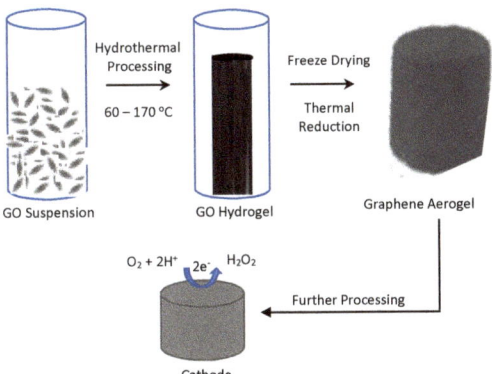

Figure 3. Schematic illustrating the formation of a graphene-based aerogel.

For example, Wen et al. [87] prepared three-dimensional macroporous rGO aerogels through the in situ assembly of rGO sheets. This electrode was employed as a cathode in E-Fenton to degrade the complex formed between ethylenediaminetetraacetic acid (EDTA) and Ni^{2+}, EDTA-Ni. Using surface analytical techniques and adsorption measurements, the authors confirmed that the three-dimensional structure possessed homogenous macropores with a high surface area of 280.15 m^2 g^{-1}. The cathode showed enhanced electrocatalytic activity, leading to the efficient generation of H_2O_2 and regeneration of Fe^{2+}. Likewise, Nazhif Mohd Nohan et al. [89] used a one-pot hydrothermal procedure to synthesise composites comprising CNTs and rGO aerogels. Studies revealed that the CNTs improved the surface area, pore volume and conductivity.

Graphene-based aerogels containing trapped Fe_2O_3 nanoparticles have also been successfully fabricated and employed in the heterogenous E-Fenton system for the degradation of Rhodamine B [88]. The addition of the Fe_2O_3 nanoparticles enhanced the degradation rate and this was attributed to the generation of H_2O_2 within the aerogel and the good diffusion and electrosorption of Rhodamine B within the aerogel to give high concentrations of the accumulated contaminant. The H_2O_2 was decomposed in the presence of Fe_2O_3 to give oxidising $OH^{\bullet}$ with little quenching, as high concentrations of the pollutant were available. Chemical oxygen demand elimination rates up to 82% were observed, while a degradation rate of 99% was reached after 30 min. Low iron leaching rates were observed even in acidic media and there was no significant loss in the catalytic activity over six cycles. In a separate study, a carbon aerogel with rGO, CNT and iron oxide nanoparticles, Fe_3O_4, was formed successfully using a sol–gel process and then employed in an E-Fenton cell as a cathode in the degradation of methyl blue [85]. The good removal efficiency, reaching 99% after a 60 min period, was attributed to the high adsorption ability of the graphene-based composite and the added strength introduced by the CNTs.

2.2.2. Three-Dimensional Graphene-Based Electrodes

In terms of three-dimensional graphene-based materials, there is some evidence that three-dimensional graphene-modified foams may be suitable in E-Fenton. In a comparison of graphene-based monolayer, multilayer and three-dimensional graphene-based foams as cathode materials in E-Fenton for wastewater treatment and the mineralisation of phenol, it was found that the three-dimensional foam exhibited the highest H_2O_2 electrogeneration yield, degradation and mineralisation rates [90]. The superiority of the three-dimensional graphene was attributed to its low interfacial charge transfer resistance, high surface area and porous structure. Interestingly, Roman et al. [28] have shown that a selective two-electron oxygen reduction reaction can be achieved with a selectivity of 94 ± 2% using three-dimensional out-of-plane graphene edge sites. This was achieved by tuning the synthesis conditions to control the size and density of the out-of-plane graphene flakes and edges to provide a three-dimensional fuzzy graphene. In addition to the high selectivity, the onset potential was measured as 0.79 V vs. reversible hydrogen electrode (RHE).

2.2.3. Three-Dimensional Graphene-Modified Electrodes

Gas diffusion electrodes (GDE) are being used increasingly in fuel cells [91], batteries [92,93] and in the generation of H_2O_2 [94,95]. They are attractive in E-Fenton, as the solubility of dissolved oxygen is very low in water and, by using these GDE electrodes, the oxygen in air could be employed instead. The GDE electrode consists of a gas diffusion layer and the solid catalyst in contact with the aqueous phase, consisting of solid, liquid and gaseous phases. Normally, the catalyst layer is porous and the aerogels described in Section 2.2.1 are finding applications as catalyst layers, facilitating interactions between the liquid and gas phase. A schematic diagram illustrating the difference between the oxygen reduction reaction at a conventional cathode and at a gas diffusion electrode is presented in Figure 4. As illustrated in this schematic, the air or gaseous phase is in contact with one side of the catalyst, enabling the diffusion of oxygen through the micropores of the diffusion layer to the catalyst phase, where it reacts with H^+ from the aqueous phase, while the conducting catalyst provides the electron, facilitating the formation of H_2O_2. Recently, this approach is finding applications in E-Fenton [96–98]. However, in many of these reports, pure oxygen is pumped to the surface of the gas diffusion cathode, rather than having the typical gas/air, solid and liquid phases, as illustrated in Figure 4.

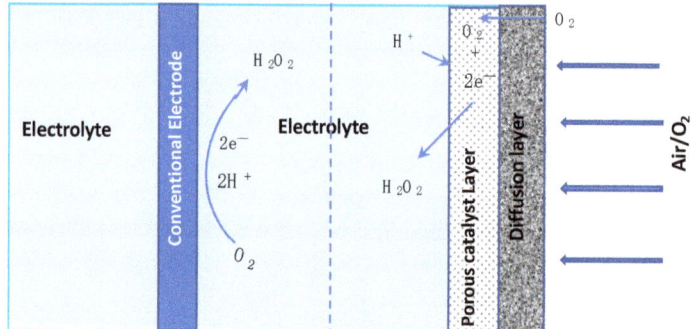

Figure 4. Schematic diagram of conventional solid-state electrode and porous gas diffusion electrode where the solid catalyst layer coexists with gas and liquid phases.

For example, a graphene–graphite diffusion electrode with high conductivity and electrocatalytic activity was formed for the continuous in situ generation of H_2O_2 through the oxygen reduction reaction, for the removal of Rhodamine B [56]. A removal rate of 98% was achieved after a 60 min period, which was higher than that obtained with a graphite-based gas diffusion cathode or graphite sheet cathode. The significant difference between the gas diffusion cathode and sheet cathode was attributed to the concentration of oxygen available for reaction, with the oxygen being supplied directly to the surface of the gas diffusion electrode, which, in turn, accelerated the generation rate of H_2O_2. The porous structure of the graphene-based diffusion electrode provided a reaction chamber for the efficient conversion of oxygen to H_2O_2. Liu et al. [98] used a graphene-based composite as a gas diffusion electrode for the removal of dimethyl phthalate from aqueous solution. Again, it was found that the H_2O_2 production was significantly improved. The apparent rate constant of dimethyl phthalate degradation was computed as 0.0322 min^{-1}. The graphene-based diffusion electrode showed excellent recovery and could be reused. However, the degradation rate decreased slightly with increasing use and this was attributed to the blocking of the microspores and channels by iron sludge as the E-Fenton reaction proceeded. In another study, a gas diffusion electrode, consisting of a cylindrical body with a built in air diffuser, was fabricated using carbon cloth treated with PTFE and coated with electrochemically exfoliated rGO for the removal of industrial electronic wastewater [99]. High

H_2O_2 concentrations of 495 mg L^{-1} were achieved to give a mineralisation rate of 80% over 80 min. Gas diffusion electrodes have also been fabricated by combining rGO with Fe_3O_4 [7] and by using boron-doped graphene-based aerogels [100] for the removal of Bisphenol A. The combination of available Fe_3O_4 particles adjacent to the electrogenerated H_2O_2 facilitated the efficient generation of OH• and the removal of Bisphenol A [7]. A comparison of the aerogels, the porous three dimensional graphene-based composites and gas diffusion electrodes is summarised in Table 2, where it is evident that these materials have high surface areas, with a very good generation of H_2O_2, and they have been employed in the removal of several pollutants.

Table 2. Summary of porosity and H_2O_2 generation rate for various porous graphene-based composites.

System	Surface Area/m^2 g^{-1}	Pore Diameter/nm	H_2O_2	Pollutant	Ref.
CNT/rGO	256.9	16.9	100 mg L^{-1}	Methylene blue	[89]
3D rGO	280.15	7.34	-	EDTA-Ni	[87]
3D rGO foam	-	(100–600) × 10^3	4.25 mg L^{-1} cm^{-3}	Phenol	[90]
rGO/GDC	132	-	28.19 mg h^{-1} cm^{-2}	Nalidixic acid	[101]
FeOOH aerogel	798–925	-	-	Sulfamethoxazole	[102]
rGO composite	459	3.9	85 mg L^{-1}	Phthalic acid esters	[103]

While many studies indicate the superiority of graphene/rGO in fabricating gas diffusion electrodes, there are examples where other carbon-based systems exhibit a somewhat higher catalytic activity for the generation of H_2O_2. In a comparative study, tert-butyl-anthraquinone (TBAQ) was used to modify four different carbon materials, carbon aerogel, CNT, carbon black and graphene-doped carbon black to fabricate gas diffusion electrodes for the production of H_2O_2 [104]. In this case, it was found that the highest H_2O_2 production and current efficiency were achieved with the CNT-gas diffusion electrode modified with 2% TBAQ, giving 2.15 mg h^{-1} cm^{-2} compared to 1.97 mg cm^{-2} h^{-1} of H_2O_2 for the corresponding graphene-based material. However, rates higher than 2.15 mg h^{-1} cm^{-2} can be seen in Table 2 for the graphene-based system. On the other hand, gas diffusion electrodes assembled using sulfur-doped carbon nanoparticles have been shown to exhibit superior electrocatalytic activity in an acidic medium for the oxygen reduction reaction [105].

2.3. Doping of Graphene-Based Materials

The doping of graphene/rGO, using heteroatoms such as N, B, P and Fe, has been widely studied for the development of catalysts for the oxygen reduction reaction [106,107]. Depending on the preparation conditions employed and the nature of the dopants selected, the mechanism can vary from a two-electron to a four-electron pathway. Much of the research focus for environmental applications is directed at N-doped graphene, as it has shown the more promising results, with many studies indicating that the presence of N enhances the production of H_2O_2 [108], while Fe and P co-doping favours the four-electron pathway [109]. Likewise, the addition of P and B to N-doped graphene composites reduces the production of H_2O_2 [106]. As N has a higher electronegativity than C, it attracts electrons, generating a partial positive charge on the C atoms, while B and P tend to donate electrons to the C. Both these partial positive and partial negative charges should promote the adsorption of oxygen. However, little is known about the oxygen reduction reaction at N-doped graphene composites or, indeed, the parameters that determine the selectivity of the oxygen reduction reaction at these materials. Several researchers have hypothesised that oxygen adsorbs through a side-on orientation on N-doped carbon [110,111]. However, it has also been shown from theoretical calculations that this side-on adsorption is unlikely and that end-on adsorption is more favourable [112,113]. The three main forms of N in N-doped graphene are pyridinic, pyrrolic and graphitic. Some authors claim that it is the graphitic N sites that facilitate oxygen adsorption and therefore the oxygen reduction reaction [112,114], while others have proposed that it is the pyridinic N that improves the chemisorption of oxygen [115].

It has also been shown that oxygen adsorption occurs on the carbon atoms at edges and these are far removed from the graphitic N atoms, making it difficult to explain how N doping facilitates the oxygen reduction reaction.

The four-electron transfer pathway is normally observed in alkaline solutions [107], while, in acidic environments, two-electron transfer is the more favoured reaction for carbon-based materials [116]. However, some N-doped or N-treated graphene-based electrodes have been found to catalyse the four-electron reduction in aqueous acid solutions [117,118]. Kurak and Anderson [119] have used a linear Gibbs energy relationship applied to N-doped graphene composites to predict the reversible potential for the formation of oxygen reduction reaction intermediates in acid. These authors concluded that there was no clear pathway for the direct four-electron reduction reaction, suggesting that transition metal impurities may be responsible for the observed direct four-electron pathway. Indeed, it has been suggested that the presence of Cu in N-doped graphene-based materials favours the two-electron pathway, but traces of Ti, Mo, Nb and Ru favour four-electron transfer [120]. Liu et al. [121] showed that by varying the N-doping temperature and microwave heating power, the number of electrons in the oxygen reduction reaction can be controlled, giving a two-electron reaction for the electrochemical degradation of organic contaminants, or the preferred four-electron transfer reaction for applications in metal air batteries. It has also been shown that the selectivity can be controlled by the degree of oxidation, with the more oxidised graphene facilitating a two-electron reduction pathway to give H_2O_2, but on reduction with $NaBH_4$, the same materials exhibit more selectivity for the four-electron pathway [122]. Kim et al. [123] suggested that the generation of H_2O_2 is connected with epoxy or ether groups in the N-doped rGO, while N-doped graphene composites, with abundant quaternary nitrogen species, show the selectivity of the two-electron reduction pathway; however, with pyridinic species, the four-electron pathway is preferred [124]. These studies highlight that, while the two-electron transfer reaction can be achieved with graphene or N-doped graphene-based materials, the selectivity of this reaction depends on how the graphene composite is formed and processed. Moreover, various methods have been employed in doping graphene-based materials with N and this may also influence the mechanism of the oxygen reduction reaction.

A number of N-doped graphene-based electrodes have been employed successfully in E-Fenton. Li and Zhang [39] employed N-doped graphene and N-doped graphene–graphite felt cathodes in the removal of phenacetin from wastewater, while Fe/N-doped graphene loaded with Fe/Fe_3C nanoparticles was used for the removal of phenol, methanol, acetone, dichloromethane and diethyl phthalate from real samples [125]. The authors concluded that the Fe/Fe_3C nanoparticles were encapsulated and protected, preventing the aggregation or leakage of the Fe, while the high-surface-area porous framework, with abundant channels and pores, provided access for the pollutants, while facilitating the formation of H_2O_2. A gas diffusion cathode was formed with N-doped graphene composites and CNTs for the removal of dimethyl phthalate [98]. The authors showed that the oxygen reduction reaction was facilitated with N-doping of the graphene composite, with the effective generation of H_2O_2 at a relatively low potential of 0.2 V vs. SCE. Yang et al. [126] employed N-doped graphene/graphite felt electrodes for the efficient generation of $OH^{\bullet}$, with a high yield of 6.2 mg h^{-1} cm^{-2} of H_2O_2. Su et al. [54] obtained an even higher yield of H_2O_2, reaching a value of 8.6 mg h^{-1} cm^{-2}, with a selectivity of 78% at neutral pH and a low energy consumption of 9.8 kW h kg^{-1}. In another recent study [127], N-doped graphene-based catalysts were found to accelerate the production of H_2O_2 and the generation of $OH^{\bullet}$. This N-doped graphene-modified graphite felt electrode was used in the mineralisation of 2,4-dichlorophenoxiacetic acid with an impressive rate of 88% at a pH of 7.0 after 480 min.

Graphene-based aerogels, where the graphene was doped with N and also co-doped with N and S were compared as cathodes in a heterogeneous E-Fenton cell [128]. The highest pollutant removal rate was observed with the N-doped graphene-based aerogels combined with a carbon–Fe_3O_4 catalyst, giving a removal rate of 71%, a mineralisation rate of 51% with good stability and low iron leaching of 0.33 mg L^{-1}. Interestingly, it was observed that the N/S-doped graphene facilitated the four-electron oxygen reduction reaction, while it was mainly the two-electron pathway that was observed with the N-doped graphene-based aerogel. While N-doped graphene appears to be superior in the promotion of the two-electron oxygen reduction reaction, Wu et al. [100] have shown that B-doped graphene-based materials may also have applications as cathodes in E-Fenton cells.

2.4. Graphene-Based Materials Combined with CNTs

Graphene/GO can be easily combined with a variety of other components such as metal oxides [129], metal nanoparticles [130] and metal organic frameworks [131]. While the addition of conducting metals or metal oxides can enhance the overall conductivity of the composite, metal-free carbon-based materials have considerable advantages in terms of cost and this has led to the development of graphene and CNT composites. CNTs can be combined with graphene-based materials to further enhance the electrocatalytic activity and conductivity, while acting as a structural support that prevents the graphene-modified sheets or flakes from restacking. These materials are finding applications in the fabrication of cathodes for E-Fenton and in the removal of contaminants. Bridged N-doped graphene and CNT composites, with microscopic three-dimensional structures, have been employed as gas diffusion electrodes and have been shown to enhance the oxygen reduction reaction, compared to graphene-based electrodes, CNT and graphite [98]. Other approaches include binder-free CNTs and graphene-based aerogel electrodes [89], iron oxide-containing CNTs, graphene-based aerogels [85] and graphene combined with CNTs on carbon felt electrodes [57]. In each of these cases, the addition of CNTs has been shown to be beneficial and to enable the more efficient removal of the contaminants.

2.5. Graphene-Based Materials Combined with Iron Oxides and Other Metal Oxides

As detailed earlier, the homogeneous E-Fenton reaction uses the Fe^{2+}/Fe^{3+} couple to catalytically decompose H_2O_2 into $OH^\bullet$, a potent oxidant (and possibly $OOH^\bullet$, which is a much weaker oxidising agent). While the Fe^{2+} and Fe^{3+} are soluble in acidic solutions, with any increase in pH, for example through the four-electron reduction at the cathode (Equation (4)), insoluble iron hydroxides begin to form, reducing the concentration of Fe^{2+} ions that are needed for the E-Fenton reaction. This is readily seen from the Pourbaix diagram presented in Figure 5, where the predicted stability phases of the soluble Fe^{2+} and Fe^{3+} are shown as a function of pH. The dashed lines show the pH dependence of the oxygen reduction and hydrogen ion reduction reactions.

Moreover, the reaction rate of Fe^{3+} to Fe^{2+} is approximately 6000 times slower than the oxidation of Fe^{2+} to Fe^{3+}, which inhibits efficient recycling between the Fe^{2+} and Fe^{3+} ions [1,132]. In addition, a large mass of iron-containing sludge is formed when the solution is neutralised. These issues can be largely resolved by using heterogenous E-Fenton [11,133]. In this case, solid iron-containing catalysts are immobilised onto a support, which helps to prevent the leaching of iron from the catalyst, minimising sludge formation, giving a wider working pH range and a more efficient conversion of Fe^{3+} to Fe^{2+} in the solid-state catalyst. As shown in Figure 5, at low pH values, leaching and dissolution are more significant, while this dissolution reaction becomes negligible as the pH is increased to neutral or slightly alkaline values. However, the principal aim in heterogeneous E-Fenton is to eliminate the dissolution of iron and the introduction of metal ions into the treated water. In Table 3, iron leaching rates are shown for different graphene-based materials together with the H_2O_2 or $OH^\bullet$ concentrations formed, where it is seen that relatively low leaching rates can be achieved.

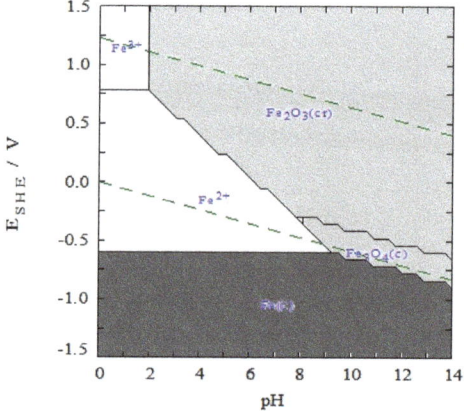

Figure 5. Pourbaix diagram for iron in water (dissolved iron concentration is 1.0×10^{-5} M at 298 K). Only Fe, Fe_3O_4, Fe_2O_3 are considered as the solid products, generated with the MEDUSA software based on the SOLGASWATER algorithm [134].

Table 3. Leaching rates of iron from iron-containing graphene-based composites.

System	Iron Leaching	H_2O_2	OH•	Ref.
Fe/Cu/FeO$_2$/rGO	2.0–3.1%	47.78 µM	-	[135]
Fe$_3$O$_4$/rGO	<1%, 0.02 mg L^{-1}	-	177.2 µM	[65]
Fe$_3$O$_4$/rGO	2.4%	-	-	[136]
Fe$_3$O$_4$/rGO	0.02 mg L^{-1}	-	-	[137]
Fe$_3$O$_4$/N-rGO, GDC	<9.5%, 0.009 mM	-	64 µM	[7]
Fe$_2$O$_3$/rGO aerogel	2.3 mg L^{-1}	4.3 mg L^{-1}	-	[88]
Fe$_3$O$_4$/CNT-rGO	<2 mg L^{-1}	40 mg L^{-1}	-	[85]
Fe$_3$O$_4$/N-rGO aerogel	0.33 mg L^{-1}	-	-	[128]

The oxides Fe_2O_3, Fe_3O_4 and FeOOH, and ferrocene, which all contain the Fe^{2+}/Fe^{3+} redox pair, have been employed [138–140], while graphene-containing composites have been combined with Fe_3O_4 [141,142], FeOOH [102], zero valent iron [143] and ferrocene [58]. In a recent study, Wang et al. [102] fabricated an γ-FeOOH graphene–polyacrylamide-carbonised aerogel (GPCA) for the degradation of sulfamethoxazole. The synthesised γ-FeOOH GPCA cathode had very good conductivity, a high surface area and very good dispersion of the iron component, which facilitated the regeneration of Fe^{2+}. This system was employed at a neutral pH to give a total organic carbon removal efficiency of 89%. Zero-valent iron has also been encapsulated within a three-dimensional graphene-based network to give a catalyst for the adsorption and degradation of sulfadiazine [143]. Iron oxide nanoparticles wrapped in graphene-based aerogel with α-Fe_2O_3 as the iron source have also been employed in heterogenous E-Fenton [88]. In this case, efficient removal of Rhodamine B was observed, with low iron leaching (<2.3 mg L^{-1}) in acidic solutions.

The iron oxide, Fe_3O_4, has been used more widely as the iron-containing catalyst [141,144]. For example, a quinone-functionalised graphene-based electrode modified with well-dispersed Fe_3O_4 nanoparticles was employed for the continuous electrogeneration of H_2O_2 and OH•. A degradation efficiency of 98% was observed for the removal of Bisphenol A within 90 min at neutral pH with less than 1% of iron leaching [65]. Akerdi et al. [141] have also employed well-dispersed Fe_3O_4 nanoparticles on GO and rGO to enhance the removal of two dyes, methylene blue and acid red. Shen et al. [145] have employed graphene–Fe_3O_4 hollow hybrid microspheres, while graphene oxide–Fe_3O_4 was employed as a heterogeneous catalyst for the E-Fenton degradation of two antibiotics, chloramphenicol and metronidazole [137]. Fe_3O_4 particles have also been utilised with N-doped graphene-based aerogels for the degradation of acetaminophen with a low iron leaching of 0.33 mg L^{-1} in heterogeneous E-Fenton [128]. These iron oxides have also been combined with graphene/CNT aerogel for the degradation of methyl blue [85]. In a more recent study, the efficient decomposition of Bisphenol A was achieved with a gas diffusion electrode and Fe_3O_4 particles at N-doped rGO as catalytic particle electrodes. The Fe_3O_4/N-rGO also served as the heterogeneous catalyst, resulting in the rapid regeneration of Fe^{2+} and high concentrations of $OH^{\bullet}$ oxidants [7].

There is also increasing interest in finding iron-free or, indeed, metal-free catalysts for heterogeneous E-Fenton. An iron-free rGO/MoS_2/$Ce_{0.75}Zr_{0.25}O_2$ composite was fabricated and used for the effective removal of ciprofloxacin. The decomposition of H_2O_2 to generate $OH^{\bullet}$ was facilitated by the redox pair Ce^{3+}/Ce^{4+} to give complete removal of ciprofloxacin within 5 h, with a mineralisation rate of 77% in 3 h [146]. HKUST (metal organic framework)-derived Cu nanoparticles were embedded within a three-dimensional rGO network to give the E-Fenton catalyst [147]. It was proposed that the zero-valent Cu was oxidised to Cu^+, which then catalysed the decomposition of the electrogenerated H_2O_2 to give $OH^{\bullet}$ and as a result the Cu^+ was converted to Cu^{2+}. The conducting catalyst layer, serving as the cathode, facilitated the reduction of Cu^{2+} to generate Cu^+, leading to the efficient recycling of the Cu^+/Cu^{2+} redox couple. Metal-free E-Fenton has also been recently proposed, as graphitic and pyridinic N sites on graphene-based materials appear to function as active sites for both the electrogeneration of H_2O_2 and the activation of $OH^{\bullet}$ radicals [54,126,148].

3. Conclusions and Future Perspectives

Graphene-based materials have attracted considerable interest, both from a fundamental viewpoint and in terms of their potential applications, and this family is one of the most studied, surpassing all other two-dimensional materials. It is no surprise that these materials are now finding applications in E-Fenton and, when combined with other carbon-based materials, they have the potential to give true iron- and metal-free E-Fenton catalysts. This would have a significant impact in terms of cost and environmental concerns. As shown earlier, rGO can be employed effectively in several ways, supported onto carbon or graphite electrodes or cloth, combined with CNTs, iron and other metal oxide catalysts, formed as aerogels and used in gas diffusion electrodes and doped with N and other elements, with N-doping appearing to be the best option in E-Fenton. It certainly appears that GO, with its very good conductivity when reduced to give rGO, high surface area, and good stability, will be used increasingly in future research in E-Fenton, leading to new and exciting developments. While the electrocatalytic generation of H_2O_2 is considered as an undesirable product in many research fields, such as batteries and fuel cells, tailoring the properties of the graphene-based composite with doping or by combining it with other materials to give a two-electron oxygen reduction reaction has significant potential not only in E-Fenton, but there are a number of other technologies, such as antimicrobial, medical, bleaching, gas scrubbing and refinery applications, that would benefit from the in situ generation of H_2O_2, eliminating the need for its storage.

In order for E-Fenton to emerge and be well integrated into real wastewater treatment facilities, new and more efficient Fenton catalysts are required. While heterogenous E-Fenton addresses many of the limitations of homogeneous E-Fenton, more efficient catalysts that are capable of generating high yields of H_2O_2, while catalysing the efficient conversion of H_2O_2 into $OH^{\bullet}$, reducing or eliminating the release of metal ions, such as Fe^{3+} or Fe^{2+}, into the water, are required. These catalysts will also need to exhibit high stability, enabling their use over multiple E-Fenton cycles, with low energy demand. This will require the development of new Fenton catalysts and graphene composites have a clear role to play in these developments, most likely in terms of composites formed using other new and emerging materials, including other two-dimensional materials, and by doping.

In terms of possible new materials that could, in the future, be combined with graphene-based materials, MXenes deserve a special mention. These are exciting two-dimensional materials [149,150] that could be combined with graphene-based materials and possibly exploited in E-Fenton. Indeed, it has been shown that the MXenes have a high density of oxygen adsorption sites [151]. The further development of three-dimensional graphene composites is another interesting possibility. The three-dimensional framework gives high porosity, high surface area and facilitates ion diffusion, with numerous active sites that can enhance the rate of electron transfer. This approach could be further refined with the appropriate heteroatom doping of graphene composites. While N-doped graphene has been employed successfully in E-Fenton, dual or multiple dopants may exert even more beneficial effects. Halogen doping, F, Cl, Br and I, has been used to tailor the catalytic activity of carbon and graphene-based composites [152]. The charge accumulation at the doped halogens create a strong dipole [153] and this may enhance the adsorption of oxygen, making halogen doping interesting for E-Fenton applications. The recent use of gas diffusion electrodes is also important, as these electrodes have the potential to deliver much higher amounts of oxygen to the Fenton catalyst. It is very clear that the application of graphene-based materials in E-Fenton is still in its infancy and new developments will be seen in the next decade.

However, the application of graphene-based materials, as composites, or combined with other materials, and/or with doping, in E-Fenton has a number of challenges that must be overcome before significant advances are made. While remarkable progress has been made in the synthesis of graphene flakes, GO and rGO, a cost-effective large-scale synthesis is needed before these applications become a reality. The cost effective production of graphene and its oxides is required, but control over reproducibility and quality is equally important. Wet graphite exfoliation methods, including chemical and electrochemical processes, are promising in terms of scalability and cost, but concerns still remain over the reproducibility of these approaches and the quality of the final graphene product. Nevertheless, there is evidence to show that the oxygen reduction reaction is promoted at graphene edges, rather than basal planes, making these wet graphite exfoliation methods possibly suitable for the scaling up and production of graphene-based cathodes that can be employed in E-Fenton.

It is also difficult to precisely control the doping levels of graphene-based materials. At another level, there are some concerns in terms of the environmental impact of GO and its potential adverse effects on aquatic ecosystems [154]. In particular, GO contains polar oxygen-containing groups, making it more soluble in water. These environmental concerns need to be addressed by fabricating highly stable graphene-based cathodes that prevent the leaching of GO flakes into the environment. Secondly, cathodes containing a graphene-based catalyst will require regeneration or some suitable disposal to prevent secondary pollution.

Nevertheless, with further advancements in the synthesis, scale-up and processing of graphene-based composites and electrodes, it is very likely that new Fenton catalysts and, indeed, metal-free Fenton catalysts can be fabricated, leading to innovations in advanced oxidation processes for the protection of water resources.

Author Contributions: The authors, T.Y. and C.B.B., contributed equally in producing this review. All authors have read and agreed to the published version of the manuscript.

Funding: This research was funded by the Irish Research Council.

Conflicts of Interest: The authors declare no conflict of interest.

References

1. Zhang, M.-H.; Dong, H.; Zhao, L.; Wang, D.-X.; Meng, D. A review on Fenton process for organic wastewater treatment based on optimization perspective. *Sci. Total Environ.* **2019**, *670*, 110–121. [CrossRef] [PubMed]
2. Zhao, X.; Liu, S.; Huang, Y. Removing organic contaminants by an electro-Fenton system constructed with graphene cathode. *Toxicol. Environ. Chem.* **2016**, *98*, 530–539. [CrossRef]
3. Ren, G.; Zhou, M.; Su, P.; Yang, W.; Lu, X.; Zhang, Y. Simultaneous sulfadiazines degradation and disinfection from municipal secondary effluent by a flow-through electro-Fenton process with graphene-modified cathode. *J. Hazard. Mater.* **2019**, *368*, 830–839. [CrossRef] [PubMed]

4. Mi, X.; Han, J.; Sun, Y.; Li, Y.; Hu, W.; Zhan, S. Enhanced catalytic degradation by using RGO-Ce/WO$_3$ nanosheets modified CF as electro-Fenton cathode: Influence factors, reaction mechanism and pathways. *J. Hazard. Mater.* **2019**, *367*, 365–374. [CrossRef] [PubMed]
5. Li, X.; Liu, F.; Zhang, W.; Lu, H.; Zhang, J. Electrocatalytical oxidation of arsenite by reduced graphene oxide via in-situ electrocatalytic generation of H$_2$O$_2$. *Environ. Pollut.* **2019**, *254*, 112958. [CrossRef]
6. Mousset, E.; Wang, Z.; Hammaker, J.; Lefebvre, O. Electrocatalytic phenol degradation by a novel nanostructured carbon fiber brush cathode coated with graphene ink. *Electrochim. Acta* **2017**, *258*, 607–617. [CrossRef]
7. Zhang, Y.; Chen, Z.; Wu, P.; Duan, Y.; Zhou, L.; Lai, Y.; Wang, F.; Li, S. Three-dimensional heterogeneous Electro-Fenton system with a novel catalytic particle electrode for Bisphenol A removal. *J. Hazard. Mater.* **2020**, *393*, 120448. [CrossRef]
8. Hammouda, S.B.; Fourcade, F.; Assadi, A.; Soutrel, I.; Adhoum, N.; Amrane, A.; Monser, L. Effective heterogeneous electro-Fenton process for the degradation of a malodorous compound, indole, using iron loaded alginate beads as a reusable catalyst. *Appl. Catal. B Environ.* **2016**, *82*, 47–58. [CrossRef]
9. Tang, Q.; Wang, D.; Yao, D.; Yang, C.; Sun, Y. Heterogeneous electro-Fenton oxidation of p-nitrophenol with a reusable fluffy clump steel wire. *Desalin. Water Treat.* **2016**, *57*, 15475–15485. [CrossRef]
10. Iglesias, O.; Meijide, J.; Bocos, E.; Sanromán, M.Á.; Pazos, M. New approaches on heterogeneous electro-Fenton treatment of winery wastewater. *Electrochim. Acta* **2015**, *169*, 134–141. [CrossRef]
11. Ganiyu, S.O.; Zhou, M.; Martínez-Huitle, C.A. Heterogeneous electro-Fenton and photoelectro-Fenton processes: A critical review of fundamental principles and application for water/wastewater treatment. *Appl. Catal. B Environ.* **2018**, *235*, 103–129. [CrossRef]
12. Waldt, C.T.; Ananthaneni, S.; Rankin, R.B. Towards quaternary alloy Au–Pd catalysts for direct synthesis of hydrogen peroxide. *Mater. Today Energy* **2020**, *16*, 100399. [CrossRef]
13. Kim, S.; Lee, D.-W.; Lee, K.-Y. Direct synthesis of hydrogen peroxide from hydrogen and oxygen over single-crystal cubic palladium on silica catalysts. *J. Mol. Catal. A Chem.* **2014**, *383*, 64–69. [CrossRef]
14. Wang, N.; Zheng, T.; Zhang, G.; Wang, P. A review on Fenton-like processes for organic wastewater treatment. *J. Environ. Chem. Eng.* **2016**, *4*, 762–787. [CrossRef]
15. Brillas, E.; Martínez-Huitle, C.A. Decontamination of wastewaters containing synthetic organic dyes by electrochemical methods. An updated review. *Appl. Catal. B Environ.* **2015**, *166*, 603–643. [CrossRef]
16. Babuponnusami, A.; Muthukumar, K. A review on Fenton and improvements to the Fenton process for wastewater treatment. *J. Environ. Chem. Eng.* **2014**, *2*, 557–572. [CrossRef]
17. Moreira, F.C.; Boaventura, R.A.R.; Brillas, E.; Vilar, V.J.P. Electrochemical advanced oxidation processes: A review on their application to synthetic and real wastewaters. *Appl. Catal. B Environ.* **2017**, *202*, 217–261. [CrossRef]
18. Nidheesh, P.V.; Zhou, M.; Oturan, M.A. An overview on the removal of synthetic dyes from water by electrochemical advanced oxidation processes. *Chemosphere* **2018**, *197*, 210–227. [CrossRef]
19. Le, T.X.H.; Bechelany, M.; Cretin, M. Carbon felt based-electrodes for energy and environmental applications: A review. *Carbon* **2017**, *122*, 564–591.
20. Nag, A.; Mitra, A.; Mukhopadhyay, S.C. Graphene and its sensor-based applications: A review. *Sens. Actuators A Phys.* **2018**, *270*, 177–194. [CrossRef]
21. Chang, Y.-M.; Lin, H.-W.; Li, L.-J.; Chen, H.-Y. Two-dimensional materials as anodes for sodium-ion batteries. *Mater. Today Adv.* **2020**, *6*, 100054. [CrossRef]
22. Jin, H.; Guo, C.; Liu, X.; Liu, J.; Vasileff, A.; Jiao, Y.; Zheng, Y.; Qiao, S.Z. Emerging two-dimensional nanomaterials for electrocatalysis. *Chem. Rev.* **2018**, *118*, 6337–6408. [CrossRef] [PubMed]
23. Geng, P.; Zheng, S.; Tang, H.; Zhu, R.; Zhang, L.; Cao, S.; Xue, H.; Pang, H. Transition metal sulfides based on graphene for electrochemical energy storage. *Adv. Energy Mater.* **2018**, *8*, 1703259. [CrossRef]
24. Li, D.; Lai, W.-Y.; Zhang, Y.-Z.; Huang, W. Printable transparent conductive films for flexible electronics. *Adv. Mater.* **2018**, *30*, 1704738. [CrossRef]
25. Divyapriya, G.; Nidheesh, P.V. Importance of Graphene in the Electro-Fenton Process. *ACS Omega* **2020**, *5*, 4725–4732. [CrossRef]
26. Li, J.; Li, X.; Chen, H.; Xiao, D.; Li, J.; Xu, D. Fe, N dual doped graphitic carbon derived from straw as efficient electrochemical catalysts for oxygen reduction reaction and Zn-air batteries. *J. Electroanal. Chem.* **2020**, *865*, 114133. [CrossRef]

27. Wang, N.; Ma, S.; Duan, J.; Zhai, X.; Guan, F.; Wang, X.; Hou, B. Electrocatalytic oxygen reduction to hydrogen peroxide by oxidized graphene aerogel supported cubic $MnCO_3$ for antibacteria in neutral media. *Electrochim. Acta* **2020**, *340*, 135880. [CrossRef]
28. San Roman, D.; Krishnamurthy, D.; Garg, R.; Hafiz, H.; Lamparski, M.; Nuhfer, N.T.; Meunier, V.; Viswanathan, V.; Cohen-Karni, T. Engineering three-dimensional (3d) out-of-plane graphene edge sites for highly selective two-electron oxygen reduction electrocatalysis. *ACS Catal.* **2020**, *10*, 1993–2008. [CrossRef]
29. Zhou, W.; Meng, X.; Gao, J.; Alshawabkeh, A.N. Hydrogen peroxide generation from O_2 electroreduction for environmental remediation: A state-of-the-art review. *Chemosphere* **2019**, *225*, 588–607. [CrossRef]
30. Rocha, R.S.; Valim, R.B.; Trevelin, L.C.; Steter, J.R.; Carneiro, J.F.; Forti, J.C.; Bertazzoli, R.; Lanza, M.R.V. Electrocatalysis of hydrogen peroxide generation using oxygen-fed gas diffusion electrodes made of carbon black modified with quinone compounds. *Electrocatalysis* **2020**. [CrossRef]
31. Kim, H.W.; Bukas, V.J.; Park, H.; Park, S.; Diederichsen, K.M.; Lim, J.; Cho, Y.H.; Kim, J.; Kim, W.; Han, T.H.; et al. Mechanisms of two-electron and four-electron electrochemical oxygen reduction reactions at nitrogen-doped reduced graphene oxide. *ACS Catal.* **2020**, *10*, 852–863. [CrossRef]
32. Sánchez-Sánchez, C.M.; Bard, A.J. Hydrogen peroxide production in the oxygen reduction reaction at different electrocatalysts as quantified by scanning electrochemical microscopy. *Anal. Chem.* **2009**, *81*, 8094–8100. [CrossRef] [PubMed]
33. Vasudevan, S.; Oturan, M.A. Electrochemistry: As cause and cure in water pollution-an overview. *Environ. Chem. Lett.* **2014**, *12*, 97–108. [CrossRef]
34. Sirés, I.; Brillas, E.; Oturan, M.A.; Rodrigo, M.A.; Panizza, M. Electrochemical advanced oxidation processes: Today and tomorrow. A review. *Environ. Sci. Pollut. Res.* **2014**, *21*, 8336–8367. [CrossRef]
35. Khataee, A.; Sajjadi, S.; Pouran, S.R.; Hasanzadeh, A.; Joo, S.W. A comparative study on electrogeneration of hydrogen peroxide through oxygen reduction over various plasma-treated graphite electrodes. *Electrochim. Acta* **2017**, *244*, 38–46. [CrossRef]
36. Randviir, E.P.; Banks, C.E. The oxygen reduction reaction at graphene modified electrodes. *Electroanalysis* **2014**, *26*, 76–83. [CrossRef]
37. Kim, H.W.; Ross, M.B.; Kornienko, N.; Zhang, L.; Guo, J.; Yang, P.; McCloskey, B.D. Efficient hydrogen peroxide generation using reduced graphene oxide-based oxygen reduction electrocatalysts. *Nat. Catal.* **2018**, *1*, 282–290. [CrossRef]
38. Ma, X.-X.; Su, Y.; He, X.-Q. Fe_9S_{10}-decorated N, S co-doped graphene as a new and efficient electrocatalyst for oxygen reduction and oxygen evolution reactions. *Catal. Sci. Technol.* **2017**, *7*, 1181–1192. [CrossRef]
39. Li, G.; Zhang, Y. Highly selective two-electron oxygen reduction to generate hydrogen peroxide using graphite felt modified with N-doped graphene in an electro-Fenton system. *New J. Chem.* **2019**, *43*, 12657–12667. [CrossRef]
40. Yang, W.; Zhou, M.; Cai, J.; Liang, L.; Ren, G.; Jiang, L. Ultrahigh yield of hydrogen peroxide on graphite felt cathode modified with electrochemically exfoliated graphene. *J. Mater. Chem. A* **2017**, *5*, 8070–8080. [CrossRef]
41. Zhang, C.J.; Nicolosi, V. Graphene and MXene-based transparent conductive electrodes and supercapacitors. *Energy Storage Mater.* **2019**, *16*, 102–125. [CrossRef]
42. Dreyer, D.R.; Park, S.; Bielawski, C.W.; Ruoff, R.S. The chemistry of graphene oxide. *Chem. Soc. Rev.* **2010**, *39*, 228–240. [CrossRef] [PubMed]
43. Eigler, S.; Hirsch, A. Chemistry with graphene and graphene oxide—Challenges for synthetic chemists. *Angew. Chem. Int. Ed.* **2014**, *53*, 7720–7738. [CrossRef] [PubMed]
44. Dideikin, A.T.; Vul', A.Y. Graphene oxide and derivatives: The place in graphene family. *Front. Phys.* **2019**, *6*, 149. [CrossRef]
45. Hummers, W.S.; Offeman, R.E. Preparation of graphitic oxide. *J. Am. Chem. Soc.* **1958**, *80*, 1339. [CrossRef]
46. Eda, G.; Chhowalla, M. Chemically derived graphene oxide: Towards large-area thin-film electronics and optoelectronics. *Adv. Mater.* **2010**, *22*, 2392–2415. [CrossRef]
47. Pei, S.; Cheng, H.-M. The reduction of graphene oxide. *Carbon N. Y.* **2012**, *50*, 3210–3228. [CrossRef]
48. Zhang, J.; Yang, H.; Shen, G.; Cheng, P.; Zhang, J.; Guo, S. Reduction of graphene oxide vial-ascorbic acid. *Chem. Commun.* **2010**, *46*, 1112–1114. [CrossRef]

49. Wang, Z.; Zhou, X.; Zhang, J.; Boey, F.; Zhang, H. Direct electrochemical reduction of single-layer graphene oxide and subsequent functionalization with glucose oxidase. *J. Phys. Chem. C* **2009**, *113*, 14071–14075. [CrossRef]
50. Guo, H.-L.; Wang, X.-F.; Qian, Q.-Y.; Wang, F.-B.; Xia, X.-H. A green approach to the synthesis of graphene nanosheets. *ACS Nano* **2009**, *3*, 2653–2659. [CrossRef]
51. Shao, Y.; Wang, J.; Engelhard, M.; Wang, C.; Lin, Y. Facile and controllable electrochemical reduction of graphene oxide and its applications. *J. Mater. Chem.* **2010**, *20*, 743–748. [CrossRef]
52. Le, T.X.H.; Bechelany, M.; Champavert, J.; Cretin, M. A highly active based graphene cathode for the electro-fenton reaction. *RSC Adv.* **2015**, *5*, 42536–42539. [CrossRef]
53. Gao, Y.; Zhu, W.; Wang, C.; Zhao, X.; Shu, M.; Zhang, J.; Bai, H. Enhancement of oxygen reduction on a newly fabricated cathode and its application in the electro-Fenton process. *Electrochim. Acta* **2020**, *330*, 135206. [CrossRef]
54. Su, P.; Zhou, M.; Lu, X.; Yang, W.; Ren, G.; Cai, J. Electrochemical catalytic mechanism of N-doped graphene for enhanced H_2O_2 yield and in-situ degradation of organic pollutant. *Appl. Catal. B Environ.* **2019**, *245*, 583–595. [CrossRef]
55. Yang, W.; Zhou, M.; Oturan, N.; Li, Y.; Oturan, M.A. Electrocatalytic destruction of pharmaceutical imatinib by electro-Fenton process with graphene-based cathode. *Electrochim. Acta* **2019**, *305*, 285–294. [CrossRef]
56. Zhang, Z.; Meng, H.; Wang, Y.; Shi, L.; Wang, X.; Chai, S. Fabrication of graphene@graphite-based gas diffusion electrode for improving H_2O_2 generation in Electro-Fenton process. *Electrochim. Acta* **2018**, *260*, 112–120. [CrossRef]
57. Wang, Y.-T.; Tu, C.-H.; Lin, Y.-S. Application of graphene and carbon nanotubes on carbon felt electrodes for the electro-fenton system. *Materials* **2019**, *12*, 1698. [CrossRef]
58. Divyapriya, G.; Nambi, I.; Senthilnathan, J. Ferrocene functionalized graphene-based electrode for the electro–Fenton oxidation of ciprofloxacin. *Chemosphere* **2018**, *209*, 113–123. [CrossRef]
59. Divyapriya, G.; Srinivasan, R.; Nambi, I.M.; Senthilnathan, J. Highly active and stable ferrocene functionalized graphene encapsulated carbon felt array—A novel rotating disc electrode for electro-Fenton oxidation of pharmaceutical compounds. *Electrochim. Acta* **2018**, *283*, 858–870. [CrossRef]
60. Xu, X.; Chen, J.; Zhang, G.; Song, Y.; Yang, F. Homogeneous electro-fenton oxidative degradation of reactive brilliant blue using a graphene doped gas-diffusion cathode. *Int. J. Electrochem. Sci.* **2014**, *9*, 569–579.
61. Luo, J.; Tung, V.C.; Koltonow, A.R.; Jang, H.D.; Huang, J. Graphene oxide based conductive glue as a binder for ultracapacitor electrodes. *J. Mater. Chem.* **2012**, *22*, 12993–12996. [CrossRef]
62. Le, T.X.H.; Bechelany, M.; Lacour, S.; Mehmet, N.O.; MarcCretin, A.O. High removal efficiency of dye pollutants by electron-Fenton process using a graphene-based cathode. *Carbon N. Y.* **2015**, *94*, 1003–1011. [CrossRef]
63. Liu, Z.; Rios-Carvajal, T.; Ceccato, M.; Hassenkam, T. Nanoscale chemical mapping of oxygen functional groups on graphene oxide using atomic force microscopy-coupled infrared spectroscopy. *J. Colloid Interface Sci.* **2019**, *556*, 458–465. [CrossRef]
64. Aliyev, E.; Filiz, V.; Khan, M.M.; Lee, Y.J.; Abetz, C.; Abetz, V. Structural characterization of graphene oxide: Surface functional groups and fractionated oxidative debris. *Nanomaterials* **2019**, *9*, 1180. [CrossRef]
65. Divyapriya, G.; Nambi, I.M.; Senthilnathan, J. An innate quinone functionalized electrochemically exfoliated graphene/Fe_3O_4 composite electrode for the continuous generation of reactive oxygen species. *Chem. Eng. J.* **2017**, *316*, 964–977. [CrossRef]
66. Guo, H.; Wang, Z.; Yang, W.; Li, J.; Jiang, D. A facile ratiometric electrochemical sensor for sensitive 4-acetamidophenol determination based on ferrocene-graphene oxide-Nafion modified electrode. *Anal. Methods* **2020**, *12*, 1353–1359. [CrossRef]
67. Wang, X.; Qi, Y.; Shen, Y.; Yuan, Y.; Zhang, L.; Zhang, C.; Sun, Y. A ratiometric electrochemical sensor for simultaneous detection of multiple heavy metal ions based on ferrocene-functionalized metal-organic framework. *Sens. Actuators B Chem.* **2020**, *310*, 127756. [CrossRef]
68. Liu, Y.; Li, K.; Xu, W.; Du, B.; Wei, Q.; Liu, B.; Wei, D. GO/PEDOT:NaPSS modified cathode as heterogeneous electro-Fenton pretreatment and subsequently aerobic granular sludge biological degradation for dye wastewater treatment. *Sci. Total Environ.* **2020**, *700*, 134536. [CrossRef]
69. Kamyshny, A.; Magdassi, S. Conductive nanomaterials for 2D and 3D printed flexible electronics. *Chem. Soc. Rev.* **2019**, *48*, 1712–1740. [CrossRef]

70. Huang, Q.; Zhu, Y. Printing conductive nanomaterials for flexible and stretchable electronics: A review of materials, processes, and applications. *Adv. Mater. Technol.* **2019**, *4*, 1800546. [CrossRef]
71. Mousset, E.; Ko, Z.T.; Syafiq, M.; Wang, Z.; Lefebvre, O. Electrocatalytic activity enhancement of a graphene ink-coated carbon cloth cathode for oxidative treatment. *Electrochim. Acta* **2016**, *222*, 1628–1641. [CrossRef]
72. Rani, V.; Das, R.K.; Golder, A.K. Fabrication of reduced graphene oxide-graphite paste electrode for H_2O_2 formation and its implication for ciprofloxacin degradation. *Surf. Interfaces* **2017**, *7*, 99–105. [CrossRef]
73. Divyapriya, G.; Thangadurai, P.; Nambi, I. Green approach to produce a graphene thin film on a conductive LCD matrix for the oxidative transformation of ciprofloxacin. *ACS Sustain. Chem. Eng.* **2018**, *6*, 3453–3462. [CrossRef]
74. Huong Le, T.X.; Dumee, L.F.; Lacour, S.; Rivallin, M.; Yi, Z.; Kong, L.; Bechelany, M.; Cretin, M. Hybrid graphene-decorated metal hollow fibre membrane reactors for efficient electro-Fenton—Filtration co-processes. *J. Memb. Sci.* **2019**, *587*, 117182. [CrossRef]
75. Jiang, W.-L.; Xia, X.; Han, J.-L.; Ding, Y.-C.; Haider, M.R.; Wang, A.-J. Graphene modified electro-Fenton catalytic membrane for in situ degradation of antibiotic florfenicol. *Environ. Sci. Technol.* **2018**, *52*, 9972–9982. [CrossRef]
76. Yu, F.; Tao, L.; Cao, T. High yield of hydrogen peroxide on modified graphite felt electrode with nitrogen-doped porous carbon carbonized by zeolitic imidazolate framework-8 (ZIF-8) nanocrystals. *Environ. Pollut.* **2019**, *255*, 113119. [CrossRef]
77. Zhou, W.; Rajic, L.; Meng, X.; Nazari, R.; Zhao, Y.; Wang, Y.; Gao, J.; Qin Alshawabkeh, N. Efficient H_2O_2 electrogeneration at graphite felt modified via electrode polarity reversal: Utilization for organic pollutants degradation. *Chem. Eng. J.* **2019**, *364*, 428–439. [CrossRef]
78. Cong, H.-P.; Ren, X.-C.; Wang, P.; Yu, S.-H. Macroscopic multifunctional graphene-based hydrogels and aerogels by a metal ion induced self-assembly process. *ACS Nano* **2012**, *6*, 2693–2703. [CrossRef]
79. Nardecchia, S.; Carriazo, D.; Ferrer, M.L.; Gutiérrez, M.C.; Del Monte, F. Three dimensional macroporous architectures and aerogels built of carbon nanotubes and/or graphene: Synthesis and applications. *Chem. Soc. Rev.* **2013**, *42*, 794–830. [CrossRef]
80. Li, C.; Shi, G. Three-dimensional graphene architectures. *Nanoscale* **2012**, *4*, 5549–5563. [CrossRef]
81. Wang, H.; Yan, X.; Zeng, G.; Wu, Y.; Liu, Y.; Jiang, Q.; Gu, S. Three-dimensional graphene-based materials: Synthesis and applications from energy storage and conversion to electrochemical sensor and environmental remediation. *Adv. Colloid Interface Sci.* **2015**, *221*, 41–59. [CrossRef]
82. Cavallo, C.; Agostini, M.; Genders, J.P.; Abdelhamid, M.E.; Matic, A. A free-standing reduced graphene oxide aerogel as supporting electrode in a fluorine-free Li_2S_8 catholyte Li-S battery. *J. Power Sources* **2019**, *416*, 111–117. [CrossRef]
83. Zhao, L.; Wang, Z.-B.; Li, J.-L.; Zhang, J.-J.; Sui, X.-L.; Zhang, L.-M. Hybrid of carbon-supported Pt nanoparticles and three-dimensional graphene aerogel as high stable electrocatalyst for methanol electrooxidation. *Electrochim. Acta* **2016**, *189*, 175–183. [CrossRef]
84. Jiang, T.; Bu, F.; Feng, X.; Shakir, I.; Hao, G.; Xu, Y. Porous Fe_2O_3 nanoframeworks encapsulated within three-dimensional graphene as high-performance flexible anode for lithium-ion battery. *ACS Nano* **2017**, *11*, 5140–5147. [CrossRef]
85. Chen, W.; Yang, X.; Huang, J.; Zhu, Y.; Zhou, Y.; Yao, Y.; Li, C. Iron oxide containing graphene/carbon nanotube-based carbon aerogel as an efficient E-Fenton cathode for the degradation of methyl blue. *Electrochim. Acta* **2016**, *200*, 75–83. [CrossRef]
86. Zhao, H.; Wang, Q.; Chen, Y.; Tian, Q.; Zhao, G. Efficient removal of dimethyl phthalate with activated iron-doped carbon aerogel through an integrated adsorption and electro-Fenton oxidation process. *Carbon N. Y.* **2017**, *124*, 111–122. [CrossRef]
87. Wen, S.; Niu, Z.; Zhang, Z.; Li, L.; Chen, Y. In-situ synthesis of 3D GA on titanium wire as a binder-free electrode for electro-Fenton removing of EDTA-Ni. *J. Hazard. Mater.* **2018**, *341*, 128–137. [CrossRef]
88. Cao, X.; Jiang, D.; Huang, M.; Pan, J.; Lin, J.; Chan, W. Iron oxide nanoparticles wrapped in graphene aerogel composite: Fabrication and application in electro-fenton at a wide pH. *Colloids Surf. A Physicochem. Eng. Asp.* **2020**, *587*, 124269. [CrossRef]
89. Nazhif Mohd Nohan, M.A.; Chia, C.H.; Hashimi, A.S.; Chin, S.X.; Khiew, P.S.; Azmi, A.; Laua, K.S.; Razalia, N.F. Highly stable binder free CNTs/rGO aerogel electrode for decolouration of methylene blue and palm oil mill effluent via electro-Fenton oxidation process. *RSC Adv.* **2019**, *9*, 16472–16478. [CrossRef]

90. Mousset, E.; Wang, Z.; Hammaker, J.; Lefebvre, O. Physico-chemical properties of pristine graphene and its performance as electrode material for electro-Fenton treatment of wastewater. *Electrochim. Acta* **2016**, *214*, 217–230. [CrossRef]
91. Bidault, F.; Brett, D.J.L.; Middleton, P.H.; Brandon, N.P. Review of gas diffusion cathodes for alkaline fuel cells. *J. Power Sources* **2009**, *187*, 39–48. [CrossRef]
92. Maja, M.; Orecchia, C.; Strano, M.; Tosco, P.; Vanni, M. Effect of structure of the electrical performance of gas diffusion electrodes for metal air batteries. *Electrochim. Acta* **2000**, *46*, 423–432. [CrossRef]
93. Tran, C.; Yang, X.-Q.; Qu, D. Investigation of the gas-diffusion-electrode used as lithium/air cathode in non-aqueous electrolyte and the importance of carbon material porosity. *J. Power Sources* **2010**, *195*, 2057–2063. [CrossRef]
94. Moreira, J.; Bocalon Lima, V.; Athie Goulart, L.; Lanza, M.R.V. Electrosynthesis of hydrogen peroxide using modified gas diffusion electrodes (MGDE) for environmental applications: Quinones and azo compounds employed as redox modifiers. *Appl. Catal. B Environ.* **2019**, *248*, 95–107. [CrossRef]
95. Tang, Q.; Wang, D.; Yao, D.M.; Yang, C.W.; Sun, Y.C. Highly efficient electro-generation of hydrogen peroxide using NCNT/NF/CNT air diffusion electrode for electro-Fenton degradation of p-nitrophenol. *Water Sci. Technol.* **2016**, *73*, 1652–1658. [CrossRef]
96. Yatagai, T.; Ohkawa, Y.; Kubo, D.; Kawase, Y. Hydroxyl radical generation in electro-Fenton process with a gas-diffusion electrode: Linkages with electro-chemical generation of hydrogen peroxide and iron redox cycle. *J. Environ. Sci. Heal. Part A Toxic Hazard. Subst. Environ. Eng.* **2017**, *52*, 74–83. [CrossRef]
97. Yu, X.; Zhou, M.; Ren, G.; Ma, L. A novel dual gas diffusion electrodes system for efficient hydrogen peroxide generation used in electro-Fenton. *Chem. Eng. J.* **2015**, *263*, 92–100. [CrossRef]
98. Liu, T.; Wang, K.; Song, S.; Brouzgou, A.; Tsiakaras, P.; Wang, Y. New electro-Fenton gas diffusion cathode based on nitrogen-doped graphene@carbon nanotube composite materials. *Electrochim. Acta* **2016**, *194*, 228–238. [CrossRef]
99. Garcia-Rodriguez, O.; Yang Lee, L.; Olvera-Vargasa, H.; Deng, F.; Wang, Z.; Lefebvrea, O. Mineralization of electronic wastewater by electro-Fenton with an enhanced graphene-based gas diffusion cathode. *Electrochim. Acta* **2018**, *276*, 12–20. [CrossRef]
100. Wu, P.; Zhang, Y.; Chen, Z.; Duan, Y.; Lai, Y.; Fang, Q.; Wang, F.; Li, S. Performance of boron-doped graphene aerogel modified gas diffusion electrode for in-situ metal-free electrochemical advanced oxidation of Bisphenol A. *Appl. Catal. B Environ.* **2019**, *255*, 117784. [CrossRef]
101. Zarei, M.; Beheshti Nahand, F.; Khataee, A.; Hasanzadeh, A. Removal of nalidixic acid from aqueous solutions using a cathode containing three-dimensional graphene. *J. Water Process Eng.* **2019**, *32*, 100978. [CrossRef]
102. Wang, Y.; Zhang, H.; Li, B.; Yu, M.; Zhao, R.; Xu, X.; Cai, L. γ-FeOOH graphene polyacrylamide carbonized aerogel as air-cathode in electro-Fenton process for enhanced degradation of sulfamethoxazole. *Chem. Eng. J.* **2019**, *359*, 914–923. [CrossRef]
103. Ren, W.; Tang, D.; Lu, X.; Sun, J.; Li, M.; Qiu, S.; Fan, D. Novel multilayer ACF@rGO@OMC cathode composite with enhanced activity for electro-Fenton degradation of phthalic acid esters. *Ind. Eng. Chem. Res.* **2016**, *55*, 11085–11096. [CrossRef]
104. Lu, X.; Zhou, M.; Li, Y.; Su, P.; Cai, J.; Pan, Y. Improving the yield of hydrogen peroxide on gas diffusion electrode modified with tert-butyl-anthraquinone on different carbon support. *Electrochim. Acta* **2019**, *320*, 134552. [CrossRef]
105. Kamaraj, R.; Vasudevan, S. Sulfur-doped carbon chain network as high performance electrocatalyst for electro-Fenton system. *ChemistrySelect* **2019**, *4*, 2428–2435. [CrossRef]
106. Choi, C.H.; Chung, M.W.; Kwon, H.C.; Park, S.H.; Woo, S.I. B, N- and P, N-doped graphene as highly active catalysts for oxygen reduction reactions in acidic media. *J. Mater. Chem. A* **2013**, *1*, 3694–3699. [CrossRef]
107. Dumont, J.H.; Martinez, U.; Artyushkova, K.; Purdy, G.M.; Dattelbaum, A.M.; Zelenay, P.; Mohite, A.; Atanassov, P.; Gupta, G. Nitrogen-doped graphene oxide electrocatalysts for the oxygen reduction reaction. *ACS Appl. Nano Mater.* **2019**, *2*, 1675–1682. [CrossRef]
108. Zheng, B.; Cai, X.-L.; Zhou, Y.; Xia, X.-H. Pure Pyridinic Nitrogen-Doped Single-Layer Graphene Catalyzes Two-Electron Transfer Process of Oxygen Reduction Reaction. *ChemElectroChem* **2016**, *3*, 2036–2042. [CrossRef]
109. He, F.; Li, K.; Xie, G.; Wang, Y.; Jiao, M.; Tang, H.; Wu, Z. Theoretical insights on the catalytic activity and mechanism for oxygen reduction reaction at Fe and P codoped graphene. *Phys. Chem. Chem. Phys.* **2016**, *18*, 12675–12681. [CrossRef]

110. Li, J.-C.; Hou, P.-X.; Liu, C. Heteroatom-doped carbon nanotube and graphene-based electrocatalysts for oxygen reduction reaction. *Small* **2017**, *13*, 1702002. [CrossRef]
111. Feng, L.; Yan, Y.; Chen, Y.; Wang, L. Nitrogen-doped carbon nanotubes as efficient and durable metal-free cathodic catalysts for oxygen reduction in microbial fuel cells. *Energy Environ. Sci.* **2011**, *4*, 1892–1899. [CrossRef]
112. Ikeda, T.; Boero, M.; Huang, S.-F.; Terakura, K.; Oshima, M.; Ozaki, J. Carbon alloy catalysts: Active sites for oxygen reduction reaction. *J. Phys. Chem. C* **2008**, *112*, 14706–14709. [CrossRef]
113. Sidik, R.A.; Anderson, A.B.; Subramanian, N.P.; Kumaraguru, S.P.; Popov, B.N. O_2 reduction on graphite and nitrogen-doped graphite: Experiment and theory. *J. Phys. Chem.* **2006**, *110*, 1787–1793. [CrossRef]
114. Vazquez-Arenas, J.; Galano, A.; Lee, D.U.; Higgins, D.; Guevara-Garcia, A.; Chen, Z. Theoretical and experimental studies of highly active graphene nanosheets to determine catalytic nitrogen sites responsible for the oxygen reduction reaction in alkaline media. *J. Mater. Chem. A* **2016**, *4*, 976–990. [CrossRef]
115. Sun, M.; Wu, X.; Xie, Z.; Deng, X.; Wen, J.; Huang, O.; Huang, B. Tailoring platelet carbon nanofibers for high-purity Pyridinic-N doping: A novel method for synthesizing oxygen reduction reaction catalysts. *Carbon, N. Y.* **2017**, *125*, 401–408. [CrossRef]
116. Noffke, B.W.; Li, Q.; Raghavachari, K.; Li, L.-S. A Model for the pH-dependent selectivity of the oxygen reduction reaction electrocatalyzed by N-doped graphitic carbon. *J. Am. Chem. Soc.* **2016**, *138*, 13923–13929. [CrossRef]
117. Kong, D.; Yuan, W.; Li, C.; Song, J.; Xie, A.; Shen, Y. Synergistic effect of Nitrogen-doped hierarchical porous carbon/graphene with enhanced catalytic performance for oxygen reduction reaction. *Appl. Surf. Sci.* **2017**, *393*, 144–150. [CrossRef]
118. Liu, Q.; Zhang, H.; Zhong, H.; Zhang, S.; Chen, S. N-doped graphene/carbon composite as non-precious metal electrocatalyst for oxygen reduction reaction. *Electrochim. Acta* **2012**, *81*, 313–320. [CrossRef]
119. Kurak, K.A.; Anderson, A.B. Nitrogen-treated graphite and oxygen electroreduction on pyridinic edge sites. *J. Phys. Chem. C* **2009**, *113*, 6730–6734. [CrossRef]
120. Bhatt, M.D.; Lee, G.; Lee, J.S. Density Functional Theory (DFT) Calculations for oxygen reduction reaction mechanisms on metal-, nitrogen- co-doped graphene ($M-N_2-G$ (M = Ti, Cu, Mo, Nb and Ru)) electrocatalysts. *Electrochim. Acta* **2017**, *228*, 619–627. [CrossRef]
121. Liu, C.L.; Hu, C.-C.; Wu, S.-H.; Wu, T.-H. Electron transfer number control of the oxygen reduction reaction on nitrogen-doped reduced-graphene oxides using experimental design strategies. *J. Electrochem. Soc.* **2013**, *160*, H547–H552. [CrossRef]
122. Favaro, M.; Carraro, F.; Cattelan, M.; Colazzo, L.; Durante, C.; Sambi, M.; Gennaro, A.; Agnoli, S.; Granozzi, G. Multiple doping of graphene oxide foams and quantum dots: New switchable systems for oxygen reduction and water remediation. *J. Mater. Chem. A* **2015**, *3*, 14334–14347. [CrossRef]
123. Kim, H.W.; Park, H.; Roh, J.S.; Shin, J.E.; Lee, T.H.; Zhang, L.; Cho, Y.H.; Yoon, H.W.; Bukas, V.J.; Guo, J.; et al. Carbon Defect Characterization of Nitrogen-Doped Reduced Graphene Oxide Electrocatalysts for the Two-Electron Oxygen Reduction Reaction. *Chem. Mater.* **2019**, *31*, 3967–3973. [CrossRef]
124. Yasuda, S.; Yu, L.; Kim, J.; Murakoshi, K. Selective nitrogen doping in graphene for oxygen reduction reactions. *Chem. Commun.* **2013**, *49*, 9627–9629. [CrossRef] [PubMed]
125. Huang, X.; Niu, Y.; Hu, W. Fe/Fe_3C nanoparticles loaded on Fe/N-doped graphene as an efficient heterogeneous Fenton catalyst for degradation of organic pollutants. *Coll. Surf. A Physicochem. Eng. Asp.* **2017**, *518*, 145–150. [CrossRef]
126. Yang, W.; Zhou, M.; Liang, L. Highly efficient in-situ metal-free electrochemical advanced oxidation process using graphite felt modified with N-doped graphene. *Chem. Eng. J.* **2018**, *338*, 700–708. [CrossRef]
127. Yang, W.; Zhou, M.; Oturan, N.; Li, Y.; Su, P.; Oturan, M.A. Enhanced activation of hydrogen peroxide using nitrogen doped graphene for effective removal of herbicide 2,4-D from water by iron-free electrochemical advanced oxidation. *Electrochim. Acta* **2019**, *297*, 582–592. [CrossRef]
128. Fernández-Sáez, N.; Villela-Martinez, D.E.; Carrasco-Marín, F.; Pérez-Cadenas, A.F.; Pastrana-Martínez, L.M. Heteroatom-doped graphene aerogels and carbon-magnetite catalysts for the heterogeneous electro-Fenton degradation of acetaminophen in aqueous solution. *J. Catal.* **2019**, *378*, 68–79. [CrossRef]
129. Tong, Y.; Chen, P.; Zhou, T.; Xu, K.; Chu, W.; Wu, C.; Xie, Y. A bifunctional hybrid electrocatalyst for oxygen reduction and evolution: Cobalt oxide nanoparticles strongly coupled to B, N-decorated graphene. *Angew. Chemie Int. Ed.* **2017**, *56*, 7121–7125. [CrossRef]

130. Bagheri, H.; Hajian, A.; Rezaei, M.; Shirzadmehr, A. Composite of Cu metal nanoparticles-multiwall carbon nanotubes-reduced graphene oxide as a novel and high performance platform of the electrochemical sensor for simultaneous determination of nitrite and nitrate. *J. Hazard. Mater.* **2017**, *324*, 762–772. [CrossRef]
131. Xu, Y.; Tu, W.; Zhang, B.; Yin, S.; Huang, Y.; Kraft, M.; Xu, R. Nickel nanoparticles encapsulated in few-layer nitrogen-doped graphene derived from metal–organic frameworks as efficient bifunctional electrocatalysts for overall water splitting. *Adv. Mater.* **2017**, *29*, 1605957. [CrossRef]
132. Martínez-Huitle, C.A.; Rodrigo, M.A.; Sirés, I.; Scialdone, O. Single and coupled electrochemical processes and reactors for the abatement of organic water pollutants: A critical review. *Chem. Rev.* **2015**, *115*, 13362–13407. [CrossRef] [PubMed]
133. Poza-Nogueiras, V.; Rosales, E.; Pazos, M.; Sanromán, M. Current advances and trends in electro-Fenton process using heterogeneous catalysts—A review. *Chemosphere* **2018**, *201*, 399–416. [CrossRef] [PubMed]
134. Eriksson, G. An algorithm for the computation of aqueous multi-component, multiphase equilibria. *Anal. Chim. Acta* **1979**, *112*, 375–383. [CrossRef]
135. Nazari, P.; Rahman Setayesh, S. Efficient Fe/CuFeO$_2$rGO nanocomposite catalyst for electro-Fenton degradation of organic pollutant: Preparation, characterization and optimization. *Appl. Organomet. Chem.* **2019**, *33*, e5138. [CrossRef]
136. Nazari, P.; Setayesh, S.R. Effective degradation of Reactive Red 195 via heterogeneous electro-Fenton treatment: Theoretical study and optimization. *Int. J. Environ. Sci. Technol.* **2019**, *16*, 6329–6346. [CrossRef]
137. Görmez, F.; Görmez, Ö.; Gözmen, B.; Kalderis, D. Degradation of chloramphenicol and metronidazole by electro-Fenton process using graphene oxide-Fe$_3$O$_4$ as heterogeneous catalyst. *J. Environ. Chem. Eng.* **2019**, *7*, 102990. [CrossRef]
138. Zhang, G.; Wang, S.; Yang, F. Efficient adsorption and combined heterogeneous/homogeneous fenton oxidation of Amaranth using supported nano-FeOOH as cathodic catalysts. *J. Phys. Chem. C* **2012**, *116*, 3623–3634. [CrossRef]
139. Özcan, A.; Atılır Özcan, A.; Demirci, Y.; Şener, E. Preparation of Fe$_2$O$_3$ modified kaolin and application in heterogeneous electro-catalytic oxidation of enoxacin. *Appl. Catal. B Environ.* **2017**, *200*, 361–371. [CrossRef]
140. Sun, M.; Ru, X.-R.; Zhai, L.-F. In-situ fabrication of supported iron oxides from synthetic acid mine drainage: High catalytic activities and good stabilities towards electro-Fenton reaction. *Appl. Catal. B Environ.* **2015**, *165*, 103–110. [CrossRef]
141. Akerdi, A.G.; Es'Haghzade, Z.; Bahrami, S.H.; Arami, M. Comparative study of GO and reduced GO coated graphite electrodes for decolorization of acidic and basic dyes from aqueous solutions through heterogeneous electro-Fenton process. *J. Environ. Chem. Eng.* **2017**, *5*, 2313–2324. [CrossRef]
142. Wu, Z.-S.; Yang, S.; Sun, Y.; Parvez, K.; Feng, X.; Mullen, K. 3D nitrogen-doped graphene aerogel-supported Fe$_3$O$_4$ nanoparticles as efficient electrocatalysts for the oxygen reduction reaction. *J. Am. Chem. Soc.* **2012**, *134*, 9082–9085. [CrossRef] [PubMed]
143. Yang, Y.; Xu, L.; Li, W.; Fan, W.; Song, S.; Yang, J. Adsorption and degradation of sulfadiazine over nanoscale zero-valent iron encapsulated in three-dimensional graphene network through oxygen-driven heterogeneous Fenton-like reactions. *Appl. Catal. B Environ.* **2019**, *259*, 118057. [CrossRef]
144. Ghanbarlou, H.; Nasernejad, B.; Fini, M.N.; Simonsen, M.E.; Muff, J. Synthesis of an iron-graphene based particle electrode for pesticide removal in three-dimensional heterogeneous electro-Fenton water treatment system. *Chem. Eng. J.* **2020**, *395*, 125025. [CrossRef]
145. Shen, J.; Li, Y.; Zhu, Y.; Hu, Y.; Li, C. Aerosol synthesis of Graphene-Fe$_3$O$_4$ hollow hybrid microspheres for heterogeneous Fenton and electro-Fenton reaction. *J. Environ. Chem. Eng.* **2016**, *4*, 2469–2476. [CrossRef]
146. Mi, X.; Yang, M.; Xie, L.; Li, Y.; Sun, Y.; Zhan, S. RGO/MoS$_2$Ce$_{0.75}$Zr$_{0.25}$O$_2$ electro-Fenton cathode with higher matching and complementarity for efficient degradation of ciprofloxacin. *Catal. Today* **2020**, *339*, 371–378. [CrossRef]
147. Yang, Y.; Liu, Y.; Fang, X.; Miao, W.; Chen, X.; Sun, J.; Ni, B.J.; Mao, S. Heterogeneous Electro-Fenton catalysis with HKUST-1-derived Cu@C decorated in 3D graphene network. *Chemosphere* **2020**, *243*, 125423. [CrossRef]
148. Haider, M.R.; Jian, W.-L.; Han, J.-L.; Sharif, H.M.A.; Ding, Y.-C.; Chen, H.-Y.; Wang, A.-J. In-situ electrode fabrication from polyaniline derived N-doped carbon nanofibers for metal-free electro-Fenton degradation of organic contaminants. *Appl. Catal. B Environ.* **2019**, *256*, 117774. [CrossRef]
149. Yu, T.; Breslin, C.B. Review—Two-dimensional titanium carbide MXenes and their emerging applications as electrochemical sensors. *J. Electrochem. Soc.* **2020**, *167*, 037514. [CrossRef]

150. Zhou, S.; Yang, X.; Pei, W.; Liu, N.; Zhao, J. Heterostructures of MXenes and N-doped graphene as highly active bifunctional electrocatalysts. *Nanoscale* **2018**, *10*, 10876–10883. [CrossRef]
151. Zhang, Z.; Li, H.; Zou, G.; Fernandez, C.; Liu, B.; Zhang, Q.; Hu, J.; Peng, Q. Self-Reduction Synthesis of New MXene/Ag Composites with Unexpected Electrocatalytic Activity. *ACS Sustain. Chem. Eng.* **2016**, *4*, 6763–6771. [CrossRef]
152. Liu, H.; Tang, Y.; Zhao, W.; Ding, W.; Xu, J.; Liang, C.; Zhang, Z.; Lin, T.; Huang, F. Facile synthesis of nitrogen and halogen dual-doped porous graphene as an advanced performance anode for lithium-ion batteries. *Adv. Mater. Interfaces* **2018**, *5*, 1701261. [CrossRef]
153. Olanrele, S.O.; Lian, Z.; Si, C.; Chen, S.; Li, B. Tuning of interactions between cathode and lithium polysulfide in Li-S battery by rational halogenation. *J. Energy Chem.* **2020**, *49*, 147–152. [CrossRef]
154. Li, M.; Zhu, J.; Wang, M.; Fang, H.; Zhu, G.; Wang, Q. Exposure to graphene oxide at environmental concentrations induces thyroid endocrine disruption and lipid metabolic disturbance in Xenopus laevis. *Chemosphere* **2019**, *236*, 124834. [CrossRef]

© 2020 by the authors. Licensee MDPI, Basel, Switzerland. This article is an open access article distributed under the terms and conditions of the Creative Commons Attribution (CC BY) license (http://creativecommons.org/licenses/by/4.0/).

Review

Graphene-Based Materials Immobilized within Chitosan: Applications as Adsorbents for the Removal of Aquatic Pollutants

Daniele C. da Silva Alves [1,2], Bronach Healy [1], Tian Yu [1] and Carmel B. Breslin [1,*]

[1] Department of Chemistry, Maynooth University, Maynooth, W23 F2H6 Kildare, Ireland; Daniele.CostaDaSilvaAlves.2021@MUMAIL.IE (D.C.d.S.A.); bronach.healy.2017@mumail.ie (B.H.); Tian.Yu.2020@mumail.ie (T.Y.)

[2] School of Chemistry and Food, Federal University of Rio Grande, Rio Grande 96.203-900, Brazil

* Correspondence: Carmel.Breslin@mu.ie

Abstract: Graphene and its derivatives, especially graphene oxide (GO), are attracting considerable interest in the fabrication of new adsorbents that have the potential to remove various pollutants that have escaped into the aquatic environment. Herein, the development of GO/chitosan (GO/CS) composites as adsorbent materials is described and reviewed. This combination is interesting as the addition of graphene to chitosan enhances its mechanical properties, while the chitosan hydrogel serves as an immobilization matrix for graphene. Following a brief description of both graphene and chitosan as independent adsorbent materials, the emerging GO/CS composites are introduced. The additional materials that have been added to the GO/CS composites, including magnetic iron oxides, chelating agents, cyclodextrins, additional adsorbents and polymeric blends, are then described and discussed. The performance of these materials in the removal of heavy metal ions, dyes and other organic molecules are discussed followed by the introduction of strategies employed in the regeneration of the GO/CS adsorbents. It is clear that, while some challenges exist, including cost, regeneration and selectivity in the adsorption process, the GO/CS composites are emerging as promising adsorbent materials.

Keywords: graphene oxide; chitosan; adsorbent; adsorption; environmental contaminants; magnetic adsorbents; 3D graphene; cyclodextrins; heavy metal ions; dyes

Citation: Alves, D.C.d.S.; Healy, B.; Yu, T.; Breslin, C.B. Graphene-Based Materials Immobilized within Chitosan: Applications as Adsorbents for the Removal of Aquatic Pollutants. *Materials* 2021, *14*, 3655. https://doi.org/10.3390/ma14133655

Academic Editors: Federico Cesano and Domenica Scarano

Received: 15 May 2021
Accepted: 26 June 2021
Published: 30 June 2021

Publisher's Note: MDPI stays neutral with regard to jurisdictional claims in published maps and institutional affiliations.

Copyright: © 2021 by the authors. Licensee MDPI, Basel, Switzerland. This article is an open access article distributed under the terms and conditions of the Creative Commons Attribution (CC BY) license (https://creativecommons.org/licenses/by/4.0/).

1. Introduction

The rapid development of industry, agriculture and urbanization, coupled with multiple human activities that can negatively impact the environment, has led to environmental pollution, and especially the contamination of water bodies [1]. Harmful pollutants in water, such as heavy metal ions, organic pollutants and chemical dyes, represent sources of toxicity and create the potential for bio-accumulation and contamination of the aquatic food chain [2]. The continuous developments in novel techniques and routes that are capable of providing clean and safe water have become a significant interest for scientists [3]. Various technologies, such as electrochemical precipitation [4], ion exchange [5], reverse osmosis [6] and photocatalytic degradation [7], have been explored in the removal of pollutants from aqueous solutions. However, these methods have certain limitations. For example, the additional reagents employed in chemical reduction/oxidation may cause secondary pollution [8], while the electrochemical treatment has additional operating costs and the precipitation of sludge is difficult to avoid and requires careful management [9]. Thus, adsorption is considered to be one of the most promising treatment technologies because of its cost-efficiency, ease of operation, simplicity, flexibility and the absence of secondary pollution [10,11].

Graphene, a two dimensional (2D) carbonaceous material with a high specific surface area and very good stability, is emerging as a candidate in the fabrication of adsorbent materials [12]. It has been used as an adsorbent to remove various pollutants from water [13,14] and it also has very good adsorption capacity for gaseous molecules [15,16]. Likewise, graphene oxide (GO) has drawn much attention as it possesses a high surface area, a π-electron system and abundant oxygen-containing functional groups. In addition, studies have shown that the performance of GO as an adsorbent material can be improved by functionalization of GO with a number of reagents [17,18]. Nevertheless, for regeneration and reuse, the collection of GO-based materials from water is an issue, and new simple and efficient removal methods are needed before graphene or GO can be employed in practical applications of water treatment [19]. Moreover, there are strong π–π interactions between GO sheets which result in aggregation, lowering of the surface area, poor dispersion in aqueous media and reduced adsorption efficiency, thus limiting its further use in wastewater treatment [20]. Therefore, new strategies aimed at minimizing aggregation, leaching and recovery of the graphene and/or GO sheets are required, and their functionalization and combination with, or immobilization within, other materials may provide possible solutions. Chitosan is one possible companion material.

Chitosan (CS), a well-known biopolymer, is a promising environmentally-friendly adsorbent due to its biodegradability, non-toxicity and physicochemical properties [21]. It can be used as an immobilization matrix for graphene since it has good receptivity to changes in its structure, while its functional groups, –OH and –NH_2, can not only act as active adsorption sites, but can also participate in electrostatic interactions and hydrogen bonding with the functional groups on GO, anchoring the GO within the chitosan matrix. On the other hand, this biopolymer has poor thermal stability and mechanical properties, but these properties can be improved and enhanced when chitosan is impregnated with graphene. The combination of these two materials is clearly beneficial with an improvement in the mechanical, thermal and chemical stability of chitosan. In return, the biopolymer acts as a stabilizer for the GO sheets minimizing their aggregation [22].

Given the scientific interest in graphene and GO and the increasing attention that these materials are receiving as adsorbents for the elimination of various contaminants from aquatic environments, a number of review articles have already been published describing the applications of GO as an adsorbent material [23–27]. Moreover, the performances of magnetic GO [28] and 3D graphene-based adsorbents [29,30] have been reviewed recently. Here, we concentrate on the applications of GO/CS composites as adsorbents for the removal of pollutants from aqueous environments. Initially, we focus on the adsorption properties of GO as many of the approaches used to improve its potential as an adsorbent can be applied to the GO/CS system. This is followed by a short introduction to chitosan, its properties and applications as an immobilizing matrix for graphene. Next, we review the GO/CS adsorbents and the additional support materials utilized with GO/CS, providing a comprehensive description of the progress being made in combining these two complementary materials and their performances as adsorbents for the removal of aquatic pollutants.

2. Graphene as an Adsorbent Material

In recent years, graphene has attracted considerable attention [31], and it has been employed in numerous applications, ranging from energy storage [32], sensors [33] and electro-Fenton [9] to microwave absorbers [34]. It is also finding applications as an adsorbent material for environmental applications and it is a very welcomed emerging material in this sector as the quality of water continues to decline with increasing amounts of pollutants escaping into the aquatic environment [35–37].

2.1. Adsorption at Graphene Oxide

Graphene can be synthesized using techniques such as chemical vapour deposition [38] and epitaxial growth [39] to give pristine graphene, but its derivative, graphene oxide (GO), which is decorated with oxygen-containing functional groups, is readily formed using the well-known modified Hummers method [40–42]. This process can be used to produce GO on a large scale and involves the oxidation of bulk graphite, a cost effective and abundant material. The resulting GO can then be exfoliated to give GO sheets. Single GO sheets can be generated but the exfoliated GO normally consists of more than one sheet and may exist as a few or multiple sheets. It is this oxidized form of graphene, GO, decorated with hydroxyl, carbonyl, carboxyl, phenol, epoxy, lactone and quinone groups [43], that is attracting considerable attention in the removal of contaminants from water. As an adsorbent material, GO has a number of attractive properties, including a high theoretical specific surface area [44], and the potential for high adsorption capacity. This combined with its very good stability, good thermal and mechanical properties and facile functionalization makes GO an especially interesting material for the removal of contaminants from water [45].

The impressive adsorption capacity of GO has been explained in terms of electrostatic, ion exchange, π–π and hydrophobic interactions [17,46,47]. At near neutral pH, GO adopts an overall negative charge, as the acidic groups, such as –COOH, are dissociated. This facilitates the adsorption of positively charged species, such as heavy metal cations, cationic dyes and other cationic molecules through electrostatic interactions [48,49]. Indeed, it has been reported that GO remains negatively charged throughout a wide pH range, typically between 2–11 [50]. However, as the pH increases, the GO becomes more negatively charged, as the equilibrium shifts in favour of the dissociated carboxylate anion, resulting in a higher removal efficiency for cationic dyes. For example, the cationic dye molecule MB (methylene blue) exhibits a maximum adsorption capacity at a pH of 10.0 with GO powders dispersed in the MB-containing solution [50]. Ion exchange has also been proposed as the adsorption mechanism for the removal of heavy metal cations at low pH, where the metal cations are exchanged with the protons in the COOH and OH functional groups [51].

The π–π interactions between aromatic ring structures and the GO sheets can also facilitate the adsorption of pollutants with aromatic ring segments. Nevertheless, the electrostatic interactions tend to be stronger, and this is seen when the aromatic pollutants also contain a cationic group [46]. Hydrophobic effects are seen when the GO polar groups are removed and this can be achieved by reducing the GO sheets to give rGO with a low density of polar functional groups and very good hydrophobicity. In this case, hydrophobic interactions play a role in the adsorption process when the pollutant molecule possesses suitable hydrophobic groups. This is especially relevant in the remediation of oil spills in water, and various graphene-based adsorbents with good hydrophobic properties have been employed to remove oils from water [44,52].

Nevertheless, as an adsorbent material, GO has several issues when it is used without any further modifications or not combined with other additives. For example, the GO sheets tend to aggregate and this reduces the adsorption efficiently. Moreover, the number of oxygen-containing groups on GO is relatively low, and while the density of these groups will depend on the degree of oxidation of the GO sheets, there is always an insufficient number of these groups to bind with cationic pollutants. Besides, these oxygen containing functional groups, such as hydroxyl, carboxyl, carbonyl, epoxy and quinone groups, are less effective in binding pollutant molecules compared with nitrogen-based functional groups, such as amines. Therefore, it is no surprise that GO has been functionalized with various nitrogen-containing groups and these have included amino acids [53–55], ethylenediamine [56], thiourea [57] and a variety of amino-containing reagents [58,59]. The adsorption capacity of these nitrogen-containing groups is pH dependent, as protonated amine groups, NH_3^+, are generated when the pH is decreased. However, at near neutral pH values, these groups become deprotonated, facilitating the binding of heavy metal cations through chelation.

Selectivity in the adsorption process is especially significant in terms of water treatment, where the water may contain cations, such as Na^+ and Mg^{2+}, and anionic species, such as Cl^- and SO_4^{2-}. These have the potential to compete with the adsorption of the targeted contaminants. Different strategies, mainly focussed on molecular recognition, are emerging as possible solutions to the lack of the selectivity associated with GO. One of the more well developed approaches involves the use of cyclodextrins. Recently, there has been an increasing number of publications describing the incorporation of cyclodextrins at GO-based adsorbents [60–62]. Cyclodextrins (CDs) are macrocyclic oligosaccharides with a distinctive truncated cone structure that possess a cavity that can include molecules [63,64]. The cavity size differs to give the well-known α, β and γ-CDs. These CDs can incorporate a large variety of guest molecules (host-guest inclusion complexation) making them interesting in drug delivery [65], the development of sensors, [66,67] and as adsorbent materials [68]. In addition, the hydroxyl groups on the CD (7 primary and 14 secondary –OH groups for β-CD) are known to form stable complexes with metal ions. Examples of some β-CD modified GO-based materials and their performances in the adsorption of several contaminants are summarized in Table 1. These CD modified GO-based adsorbents can be formed using a simple self-assembly method, where the CDs, or functionalized CDs, are physically mixed with GO [69]. Self-assembly is favoured as hydrogen bonding occurs between the –OH groups on the CDs and the oxygen-containing groups on GO [66,70]. Alternatively, crosslinking can be used to link the CDs through covalent attachment to the GO [71].

Table 1. Summary of some CD-modified GO adsorbents and their adsorption capacity.

Adsorbent	Adsorbate	Adsorption q_m (mg g^{-1})	Ref.
β-CD/GO	Bisphenol A	373.4	[72]
β-CD/GO	Methyl blue Methyl orange Basic fuchsin	580.4 328.2 425.8	[73]
β-CD/GO	Cd(II)	196.0	[74]
β-CD/GO	p-Nitrophenol	117.28	[75]
β-CD/poly(acrylic acid)/GO	Methylene blue Safranine T	247.99 175.49	[76]
β-CD/poly (L-glutamic acid) magnetic/GO	17β-estradiol	298.9	[77]

Another emerging option is to use molecular imprinted polymers (MIPs) as the recognition element. MIPs are synthetic polymers tailored to recognise and bind a specific target [78]. This is achieved by polymerization of the monomers in the presence of a template molecule, which is structurally related to the target adsorbent. Once this template is removed from the polymer, binding sites complementary to the target adsorbent are generated to give very impressive selectivity. Besides, the formation of MIPs at GO has the potential to give large surface areas and high adsorption. For example, Cheng et al. [79] formed GO/MIP using GO as the support, bis(2-ethylhexyl) phthalate as the template molecule, methacrylic acid as the functional monomer and ethylene dimethacrylate as the cross-linking agent. The GO/MIP was then employed in the selective extraction of bis(2-ethylhexyl) phthalate. Similarly, Cheng et al. [79] used GO/MIP for the selective adsorption of naphthalene-derived plant growth regulators in apples.

These additional functional groups, CDs and MIP can enhance the removal of several contaminants; however, the GO sheets/nanosheets are nevertheless difficult to remove following the adsorption step. The GO sheets are often used as powders, requiring techniques that are not always suitable in real wastewater treatment, such as centrifugation and filtration, for their removal. In addition, they can leach into the aquatic environment and cause secondary pollution [80]. This leaching can also occur during the adsorption process, making it difficult to control. Even though the toxic effects of GO are poorly understood, there is substantial evidence to show that GO has the ability to penetrate through the cell membranes of aquatic organisms [81,82]. Moreover, GO has the capacity to carry polycyclic aromatic hydrocarbons to aquatic organisms, causing significant toxicity [83]. In order to minimize the leaching of GO, the GO sheets must be immobilized within a support that will facilitate their removal from the aquatic environment, while at the same time maintaining their attractive adsorption potential.

2.2. Recovery of the GO Adsorbent

One approach that is attracting attention is the use of 3D graphene materials [84–86]. The development of these 3D hierarchical architectures is challenging as it relies on maintaining the properties of the individual GO nanosheets. These 3D graphene structures include sponges, aerogels and foams, and provided this 3D structure is maintained and does not collapse, these materials are more easily recovered from the solution phase. Moreover, the high surface areas and porous structures give rise to the efficient transport and trapping of the pollutants, while the 3D structures can be further functionalized using covalent and non-covalent methodologies [87,88]. The various methods used in the formation of porous 3D graphene structures can be found in a recent review by Lin et al. [30], highlighting the increasing interest in 3D graphene-based materials. In Figure 1, a schematic illustration of a 3D GO-based magnetic polymeric aerogel is shown, where the 3D network was formed using freeze-drying and contains magnetic Fe_3O_4 nanoparticles, polyvinyl alcohol (PVA), cellulose and GO sheets.

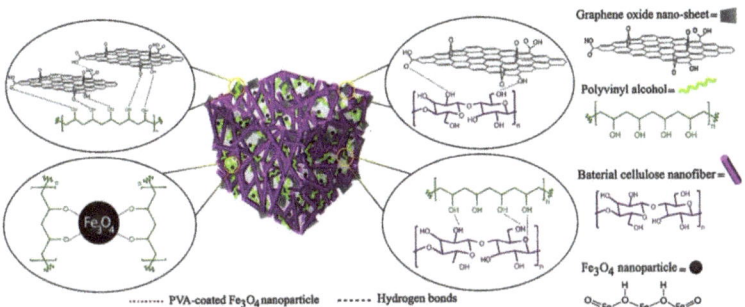

Figure 1. Schematic illustration of a 3D magnetic GO-polymer aerogel. Reproduced with permission from Arabkhani and Asfaram [86], J. Hazard. Mater.; published by Elsevier, 2020.

Another strategy involves the use of magnetic graphene-based adsorbents and these materials are attracting considerable interest in the adsorption of pollutants from water [89–91]. The magnetic properties are normally introduced using Fe_3O_4, a ferromagnetic black iron oxide, which possesses very good compatibility, low toxicity, very good magnetic properties and which can be generated as rods, wires, spheres and nanoparticles [92]. The addition of the magnetic particles not only facilitates the separation of the adsorbent from the aquatic environment through a simple magnetic process [93], but also can form between the GO sheets and reduce the inevitable aggregation of the sheets.

2.3. Immobilization of GO within Biopolymers

Clearly, GO can be modified with various additives aimed at improving adsorption, the selectivity of the adsorption process and the recovery and removal of the adsorbent following the adsorption step. However, the immobilization of GO using support materials that can provide a stable matrix to securely anchor the GO, while maintaining its impressive high surface area, mechanical strength and adsorption capacity is still a challenge. Biopolymeric/GO-based adsorbents are now attracting increasing attention as they are environmentally acceptable and can be easily synthesized. Besides, the biopolymer backbone possesses interesting functional groups that have the ability to bind environmental contaminants. Several biopolymers have been combined with GO and employed in the adsorption of environmental contaminants, including cellulose [94], alginate [95], gum [96] and lignin [97]. However, within the family of biopolymers, chitosan is the leading candidate and is attracting increasing interest as a support matrix for GO, rGO and graphene. Using Scopus and the key words 'graphene' and 'chitosan', a total of 2226 publications were found, with 75 publications in 2012 and 413 in 2020. These chitosan/GO composites are gaining attention in a number of applications and especially as adsorbent materials. They can be used as powders and dispersed in the solution phase, but also employed as solid adsorbents, enabling their removal and recovery from the treated water, while limiting the aggregation of the GO sheets, arising from intermolecular forces. Their impressive adsorption potential is described following a short introduction to chitosan and its properties related to its adsorption potential.

3. Chitosan as an Immobilization Matrix

Chitosan, a readily available eco-friendly and non-toxic polysaccharide, is fabricated from chitin, and consists of β–(1–4)–D–glucosamine. It has very good chemical stability combined with chelating properties [98]. Indeed, chitosan has been employed as an effective adsorbent for the removal of dyes [99], phenol [100], heavy metal ions [101], antibiotics [102] and pesticides [103] from water. It is an effective adsorbent as it has a high surface area and possesses a large density of hydroxyl (–OH) and primary amine (–NH_2) groups, as illustrated in Figure 2. These functional groups can serve as active adsorption sites which facilitate the adsorption and removal of both positively and negatively charged molecules through electrostatic interactions [104,105]. In particular, amine groups are strongly attracted to metal ions through ion–dipole interactions [101], while the protonation of these amine groups (–NH_3^+) facilitate electrostatic attraction of anionic compounds, including halides [106] and anionic dyes [107].

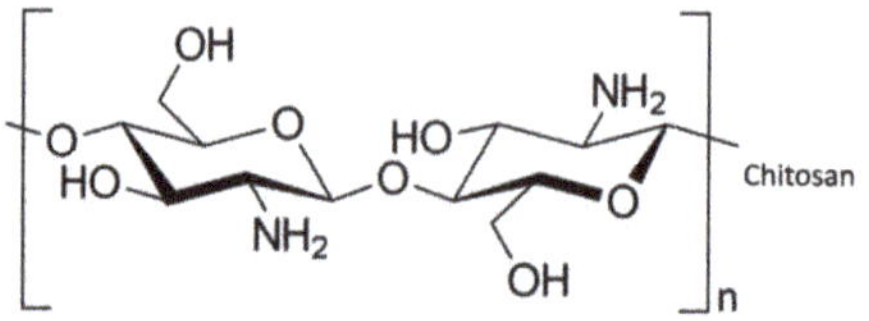

Figure 2. Chemical structure of chitosan.

Chitosan is an interesting immobilization matrix for graphene as it can be modified chemically through cross-linking, complexation and grafting, while different functional groups can also be introduced [105,108]. Furthermore, parameters such as the molecular weight (MW), deacetylation degree (DD), solubility, crystallinity, particle size and surface area can all be optimized to enhance the adsorption capacity [109]. The MW of chitosan depends on the deacetylation process, the source and the preparation procedures employed in its formation [110]. It can affect many of the physicochemical properties of chitosan, including its crystallinity, solubility, viscosity, tensile strength and elasticity, and impact on the applications of chitosan as an adsorbent material [111,112]. The DD, which is related to the acetyl content in chitosan, can be altered by varying the alkaline treatment step in the chitin deacetylation process [113]. An increase in DD, usually achieved through a repeated or prolonged alkaline treatment, gives rise to an increase in the density of free amino groups. Consequently, the polycationic nature is increased. This higher charge density along the chitosan chain alters the chain flexibility [114]. Indeed, it has been demonstrated that chitosan chains with higher DD have a more regular packing of the polymer chains, which promotes crystallinity in chitosan [115]. Although the stiffness and tensile strength of chitosan are improved on increasing the crystallinity, a reduction in elongation and an increase in the brittleness can also occur. Since the cationic properties of chitosan are connected with DD, it has been shown in several reports that DD affects the adsorption properties of chitosan [116]. For example, Gonçalves et al. [117], in studying the adsorption of dyes at chitosan powders with different DD levels, observed an increase in the adsorption capacity of the dyes on increasing the DD from 75% to 95%. Likewise, Piccin et al. [118] observed a higher adsorption capacity as the DD was increased from 42 to 84%. Furthermore, it has been shown that chitosan composites with higher DD are more stable and reusable, even after 15 adsorption-regeneration steps [113]. Nevertheless, the highly hydrophilic character associated with a high DD may have some limitations [119]. Indeed, Iamsamai et al. [120] showed that better dispersion of multiwalled carbon nanotubes (MWCNTs) was achieved with lower DD levels (61% DD). Therefore, DD can play a key role when chitosan is combined with other materials and employed as an adsorbent [121].

The solubility of chitosan depends on a number of factors, including the density and distribution of amino and N-acetyl groups on the polymeric chitosan chains and the ionic strength of the solution phase [122]. Hence, solubility is linked with both the DD and MW. Under acidic conditions, the amine groups become protonated and the chitosan becomes more soluble. As more amino groups become protonated, achieved with higher DD, stronger electrostatic repulsion occurs between the neighbouring chains and this results in dissolution of the polymer [109]. On the other hand, as the pH of the solution is increased to a value in the vicinity of 6.0, precipitation of the solubilized chitosan occurs as the amine groups become less protonated [123]. On increasing the MW, higher levels of inter- and intra-molecular hydrogen bonding occurs within the chains, and this results in chain entanglement and a concomitant reduction in solubility [124]. Indeed, it has been shown that the solubility of chitosan depends on the pH and ionic strength [125], temperature, time of deacetylation, alkali concentration, previous treatments applied to the isolation of chitin and particle size [126]. Therefore, the solubility is important as it imparts the chitosan with excellent gel-forming properties and these are important in the formation of the GO/CS hydrogel composites.

The porosity and surface area of chitosan can also play a central role in the adsorption process and can have a significant effect on the adsorption capacity [127]. It is well known that the particle size of chitosan is an important characteristic that is related to the porosity, pore size and pore volume [128] and these are fundamental for adsorption applications [129]. For example, a low uptake of pollutants was observed with larger particle sizes and this was explained in terms of a lower surface area [130], while an increase in the surface area results in the formation of new active sites, which allows more binding of contaminants, and consequently an increase in the overall performance of the adsorbent is seen [131]. Nevertheless, it is challenging to obtain a highly porous chitosan material, with good mechanical stability combined with the possibility of regeneration and reuse [132].

The relatively poor mechanical stability of chitosan hydrogels is limiting its applications and the addition of reinforcing fillers is a possible strategy to enhance its mechanical properties [133]. GO has been used to form GO/CS composites. This is an interesting combination as the GO can self-assemble with the chitosan chains especially in acidic solutions where the amine groups are protonated. This provides good stabilization and this combination is also very well suited to freeze-drying [134]. In addition, GO can improve the adsorption capacity due to its inherent ability to adsorb certain classes of water pollutants [135]. This GO/CS combination is interesting as the GO can enhance the physiochemical properties of chitosan, while the chitosan can immobilize the GO sheets, minimizing aggregation and minimizing the leaching of GO. In the following section, these GO/CS composites are described, with a focus on their synthesis, preparation and modification methods together with some of their properties and their ability to form adsorbents for the removal of water contaminants.

4. Graphene/Chitosan Adsorbents

As detailed in Section 3, chitosan has a number of unique properties that make it an interesting candidate as an adsorbent material, and also as an immobilization matrix for graphene, GO, rGO and 3D porous graphene monoliths [13,136]. Impregnation of graphene into chitosan is a good way to achieve both mutual stabilization and an enhancement in the adsorption capacity [137].

4.1. Formation of Graphene/Chitosan Adsorbents

Graphene and chitosan composites can be formed using a number of strategies, and the main elements involved in these processes are summarized in Figure 3a. In many cases, these different elements are combined to give the final GO/CS composite. Generally, the first step is dissolution of chitosan in an aqueous solution of acetic acid (1–3% v/v), which is one of the most frequently employed solvents for the solubilization of chitosan, to give a yellowish coloured homogeneous solution [138]. Then, GO, or functionalized GO, is added with sonication to form a homogeneous mixture [139]. This mixture can be used to give GO/CS beads [140] or further processed using vacuum-assisted self-assembled filtration (VASA) [141], solvothermal and hydrothermal reactions [142], freeze drying [134] or combinations of these. Some of the advantages and disadvantages of these synthetic processes are summarised in Table 2. Micrographs obtained using scanning electron microscopy (SEM) are presented in Figure 4 and these highlight the surface morphologies of the synthesised GO/CS-based adsorbents. The micrographs presented in Figure 4a,b represent the morphology of GO/CS combined with lignosulfonate (LS) and illustrate the interconnected three dimensional porous network of the fabricated GO/CS/LS. As shown in the inset of Figure 4b, the porous GO/CS/LS is sufficiently light to be supported on a leaf. In Figure 4c–e, GO/CS beads are shown, displaying a spherical shape with little or no defects, while higher magnifications of the surface and cross sections show the typical interconnected network.

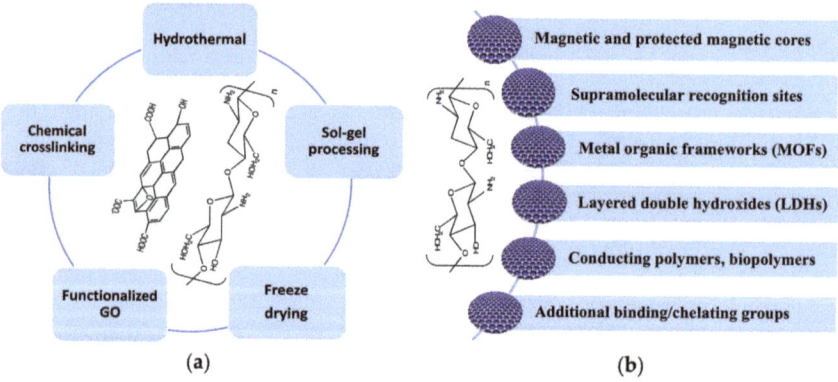

Figure 3. Summary of the (**a**) main processes used to fabricate GO/CS and (**b**) other additives combined with GO/CS.

Table 2. Summary of the advantages and disadvantages of the synthetic processes.

Synthetic Step	Advantages/Disadvantages
Sol gel	Simple, other reagents can be easily added/mixing of GO and chitosan gives rise to an increase in solution viscosity, which can give rise to inhomogeneity in the final GO/CS hydrogel.
Hydrothermal	No need for crosslinking agents/some cost considerations with the relatively high temperatures in the vicinity of 120 °C.
Crosslinking Agents	Increase in mechanical properties/reduces the number of chelating sites that are required to bind and trap the pollutants and can be toxic.
Functionalised GO	Modifiable oxygenated functional groups of GO are ideal for functionalised, rich chemistry, can be used to crosslink single graphene sheets/synthesis can be time consuming.
Freeze drying	Scaffolds with defined pore size, highly suited to enhanced adsorption/freeze drying can be slow.

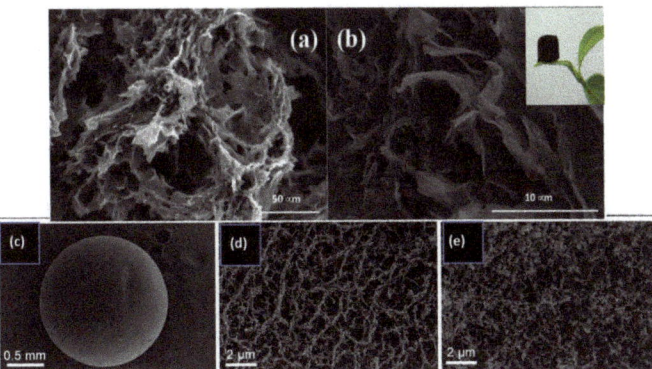

Figure 4. SEM micrographs of GO/CS/LS at magnifications of (**a**): ×600, (**b**): ×5000 and supported on a leaf. Reproduced with permission from Yan et al. [143], Int. J. Biol. Macromol.; published by Elsevier 2019. GO/CS (**c**) beads/spheres, (**d**) surface and (**e**) cross-sectional structures. Reproduced with permission from Wu et al. [140], Mater. Sci Eng. C; published by Elsevier 2020.

The formation of these GO/CS hydrogel composites is aided by electrostatic interactions, hydrogen bonding and covalent interactions between chitosan and GO [135]. At near neutral pH, the amino groups of chitosan are protonated and these attract the negatively charged groups, such as –COO⁻, on GO, to give a stable hydrogel composite [144]. These electrostatic interactions combined with hydrogen bonding facilitate the formation of the GO/CS hydrogel which provides stable composites with excellent mechanical and thermal properties [145]. Furthermore, the introduction of chitosan to GO nanosheets provides an increase in the surface area, pore size and total pore volume, which facilitate efficient adsorption onto the GO/CS hydrogel surface [146]. Thus, the effective intermolecular interactions between GO and chitosan play an important role in the specific structural formation of the hydrogel nanocomposites and these are illustrated in Figure 5.

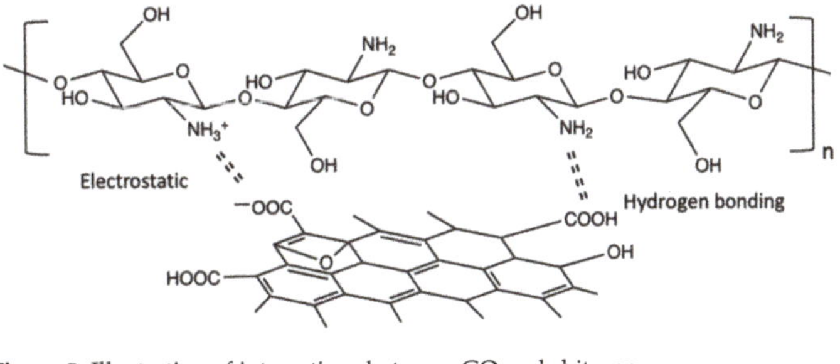

Figure 5. Illustration of interactions between GO and chitosan.

Crosslinking agents have also been added to further enhance the mechanical properties of GO/CS. Reagents such as trisodium citrate, sodium tripolyphosphate [147], glutaraldehyde [104,148], genipin [149,150], borax [151] and N-(3-dimethylaminopropyl)-N-ethylcarbodiimide hydrochloride [152] have been successfully employed. Some of the more commonly used crosslinking reagents are summarised in Table 3, where it is seen that they are normally employed at low concentrations, and both temperature and time can be varied to achieve the crosslinking step. The GO can also be functionalized to facilitate cross-linking with chitosan and one of the more well-known reactions employed is amidation, where the functional groups on GO are activated to give acetyl chloride (–COCl) functionalized GO, which can then be connected with a nitrogen containing group through the formation of –CONH– linkages [152].

These cross-linking agents can play different roles and influence the number of functional groups available to bind with pollutants. For example, borax undergoes hydrolysis to yield the tetrahydroxyborate ions, $B(OH)_4^-$, which react with the hydroxyl groups in chitosan and GO to produce orthoborate chemical bonds. On the other hand, N-(3-dimethylaminopropyl)-N-ethylcarbodiimide hydrochloride, which is water soluble and biocompatible, can be used as a synthetic grafting agent [152], coupling the carboxyl (GO) and amino groups (CS) to give amide bonds. Moreover, the timing of this step can influence the performance of the GO/CS adsorbent. For example, Salzano de Luna et al. [153] demonstrated that freeze dried GO/CS composites were more effective in the adsorption of dyes when the cross-linking was carried out following the freeze drying step. This was attributed to a higher degree of pore interconnectivity in the GO/CS composites crosslinked after freeze-drying. However, the compressive modulus was reduced. These studies highlight the significant contribution of cross-linking, not only in terms of mechanical properties, but also in terms of pore interconnectivity and the provision of chelating binding sites for the uptake and removal of pollutants.

Table 3. Summary of some crosslinking agents employed in the fabrication of GO/CS adsorbents.

Crosslinking Agent	Crosslinking Conditions	Advantages/ Disadvantage	Ref.
Gluteraldehyde	2% solution (wt %), 1 h at 60 °C	Inexpensive but exhibits toxicity	[148]
Gluteraldehyde	1% solution (wt %), 6 h at 25 °C		[141]
Gluteraldehyde	2% solution (wt %), 8 h at 30 °C		[154]
Glycidoxypropyltri-methoxysilane (KH-560)	0.22 g KH-560 with 0.12 g GO, 1.5 g CS at 50 °C	Commonly used coupling agent, some toxicity	[155]
Genipin	1% solution (wt %) added dropwise, 1 h at 25 °C	Negligible toxicity	[150]
Borax	10% solution (wt%) 1 h at 25 °C	Toxic	[151]

As illustrated in Figure 3b, additional components are added to further enhance the performance of the GO/CS-based adsorbents. These vary from magnetic particles to blending with other polymers. In most cases, GO is employed; however, the more conducting rGO has also been used and decorated with uncapped metal nanoparticles and then combined with chitosan [156]. As the metal nanoparticles are formed through reduction of the appropriate salt in solution, the rGO is oxidized to GO to give a simple methodology for the decoration of GO with uncapped nanoparticles. Nevertheless, it is generally accepted that the GO/CS-based adsorbents have a higher adsorption capacity due to a combination of π–π stacking, electrostatic interactions and hydrogen bonding with the water contaminants [12,50]. Indeed, Guo et al. [18] found that while both GO/CS and rGO/CS could adsorb dyes, the GO/CS appeared as the more efficient adsorbent.

4.2. Magnetic Chitosan/GO

Magnetic materials have emerged as exciting new materials in several applications [157] and it is no surprise that they have been employed to give magnetic GO/CS (MGO/CS) adsorbents [158,159]. As previously mentioned in Section 2.2, magnetism can be employed to achieve removal of the adsorbent from the aquatic environment [93]. Depending on the synthetic conditions utilized to fabricate the GO/CS adsorbents, it may be difficult to completely remove these adsorbents using techniques such as sedimentation or filtration. The introduction of magnetism provides a simple but effective solution and this has become the focus of a number of recent investigations.

The MGO/CS composites can be formed using ex-situ methods [160], where the Fe_3O_4 nanoparticles are chemically synthesized and then combined with the GO/CS hydrogel [161]. Alternatively, in-situ methods [162,163] can be used. For example, Singh et al. [159] used the amide functional groups (formed between the epoxy (GO) and amine (CS) groups) for the in-situ reduction of iron ions to iron oxides. The synthesized Fe_3O_4 nanoparticles have been observed to immobilize on the GO sheets [158,164], but agglomeration is normally seen [92,165,166]. However, Jiang et al. [167] have shown that the severe aggregation, arising from the magnetic properties of the Fe_3O_4 particles, can be minimized when the particles are coated with silica. This silica layer not only inhibits aggregation, but also protects the magnetic cores and these silica coated Fe_3O_4 particles have been combined with GO/CS and employed in the removal of alkaloids [168], as illustrated in Figure 6. The MGO/CS can be formed as beads [169,170] and as various powders and nanocomposites [171,172]. In addition to providing magnetic separation the added Fe_3O_4 nanoparticles, provided they are well dispersed and not agglomerated, can give rise to an improvement in the surface area of the MGO/CS adsorbents [173]. Indeed, these MGO/CS adsorbents have been described as mesoporous materials with surface areas ranging from 37.28 m^2 g^{-1} [174], 74.345 m^2 g^{-1} [175], 388.3 m^2 g^{-1} [176], 392.5 m^2 g^{-1} [173] to 402.1 m^2 g^{-1} [177] and pore volumes varying from 0.084 cm^3 g^{-1} [174], 0.3852 cm^3 g^{-1} [173] to 0.4152 cm^3 g^{-1} [177].

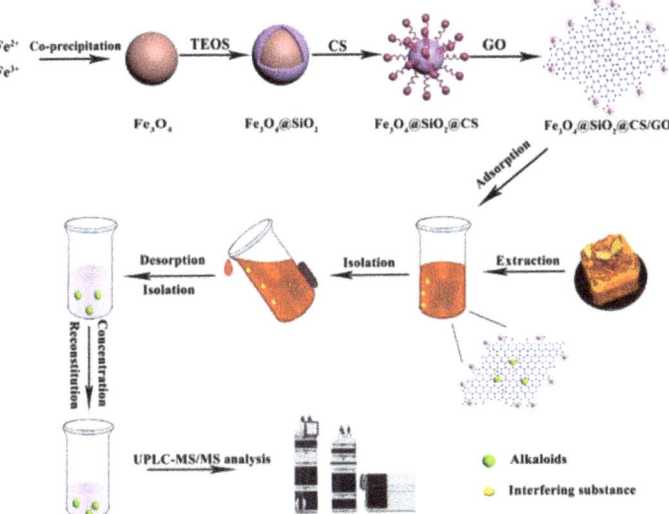

Figure 6. Schematic for the procedure for synthesis of Fe_3O_4/SiO$_2$/CS/GO and its application in the removal of alkaloids. Reproduced with permission from Tang et al. [168], Int. J. Biol. Macromol.; published by Elsevier 2020.

The performance of various MGO/CS adsorbents is illustrated in Table 3, where the adsorption capacity, q_m, is expressed in terms of the ratio of the mass of adsorbate removed to the mass of adsorbent used. These experimental adsorption data are normally fitted to two main adsorption models, the Langmuir [178,179] and Freundlich [180] isotherms, with the Langmuir isotherm being more frequently utilized. In the Langmuir isotherm, all the adsorption sites are considered equivalent with no interactions between adjacent sites. This gives rise to monolayer adsorption. On the other hand, the Freundlich isotherm is based on multilayer adsorption. Although the GO/CS-based adsorbents have a number of different and distinct adsorption sites, the experimental data fit very well with the monolayer adsorption process inherent in the Langmuir model.

As illustrated in Table 4, the MGO/CS adsorbents are often further modified with cyclodextrins. For example, Li et al. [181] have described the removal of Cr(VI) under acidic conditions to the attraction of the negatively charged chromate anions to the protonated chitosan, reduction of Cr(VI) to Cr(III) at the GO sheets, and the binding of anionic Cr(VI) and cationic Cr(III) at the cyclodextrins. The hydrophobic and inclusion complexation characteristics of the cyclodextrin may also be relevant in the removal of dye molecules and these properties have been exploited to give good removal of methylene blue [177] and hydroquinone [182] at the CD modified MGO/CS. However, it is difficult to make a direct comparison between the adsorption capacity of the adsorbent materials presented in Table 3. Not only are the properties of chitosan, which is very dependent on its production and isolation from chitin, not identical, but the conditions used in the adsorption experiments are different. Interestingly, on comparing the adsorption capacity of methylene blue, which is frequently used as a model pollutant, it is clearly evident that there is considerable variations in its adsorption, with values ranging from 43.34 to 2478 mg g^{-1}. These results highlight the importance of well-dispersed and interconnected GO sheets throughout the chitosan to give a porous matrix, while the characteristics of the chitosan employed (for example DD and MW) and the nature of functionalized groups are also very important. Indeed, the highest adsorption capacity is evident with methacrylic acid-functionalized chitosan, which is cross-linked with N,N-methylenebis(acrylamide), to give a solid matrix that facilitates the stable dispersion of both GO sheets and Fe_3O_4 nanoparticles.

Table 4. Summary of some magnetic GO/CS adsorbents. The GO/CS is employed as a solid adsorbent in contact with the targeted pollutant dissolved in the solution phase.

Adsorbent	Adsorbate	Adsorption Conditions	Adsorption q_m (mg g^{-1})	Ref.
MGO/CS	Rifampicin	Batch mode, 55 °C, 200 rpm, 10 mg adsorbent, 20 mg/L adsorbate	102.11	[166]
MGO/CS	As(III)	Batch, 25 °C, 250 rpm, 10 mg/L adsorbate, V = 100 mL	45	[178]
MGO/CS	Methylene blue Eriochrome black T	Batch, 26 °C, 130 rpm, 50 mg adsorbent, 100 mg/L adsorbate, V = 50 mL	289 292	[183]
MGO/CS	Cr(VI)	Batch, 21 °C, 500 mg adsorbent, 40 mg/L adsorbate, V = 10 mL	100.51	[184]
GO/CS/$ZnFe_2O_4$	Basic fuchsin	Batch, 25 °C, 50 mg adsorbent, 50 mg/L adsorbate, V = 25 mL	335.57	[185]
Methacrylic acid functionalized-MGO/CS	Methylene blue	Batch, 25 °C, 120 rpm, 10 mg adsorbent, 100 mg/L adsorbate, V = 20 mL	2478	[162]
MGO/CS/ Lignosulfonate	Methylene blue	Batch, 30 °C, 160 rpm, 10 mg adsorbent, V = 20 mL	253.53	[186]
MGO/CS/ Ethylenediamine	Cu(II)	Batch, 25 °C, 300 rpm, 10 mg adsorbent, 100 mg/L adsorbate, V = 50 mL	217.4	[187]
MGO/CS/SiO_2	Dopamine Clenbuterol Orciprenaline Methylene blue Crystal violet	Batch, 20 °C, 180 rpm, 10 mg adsorbent, V = 100 mL	127.34 109.56 150.21 300.42 347.35	[171]
MGO/CS/SiO_2	methyl violet	Batch, 52 °C, 150 rpm, 10 mg adsorbent, 10 mg/L adsorbate, V = 5 mL	243.8	[188]
MGO/CS/SiO_2/ ionic liquid	Morphine Codeine Ephedrine Amphetamine Benzoylecgonine	Batch, 25 °C, 150 rpm, 15 mg adsorbent, 10 mg/L adsorbate, V = 5 mL	7.2 8.4 9.2 5.8 11.2	[189]

Table 4. Cont.

Adsorbent	Adsorbate	Adsorption Conditions	Adsorption q_m (mg g^{-1})	Ref.
3D-MGO/CS	Disperse blue 367	Batch, 25 °C, 150 rpm, 150 mg adsorbent, 60 mg/L adsorbate	298.27	[190]
β-CD-MGO/CS	Bisphenol A Bisphenol F	Batch, 30 °C, 200 rpm, 20 mg adsorbent, 20 mg/L adsorbate, V = 50 mL	326.8 328.3	[191]
β-CD–MGO/CS	Hydroquinone	Batch mode, 180 rpm, 100 mg adsorbent, V = 100 mL	148	[182]
β-CD–MGO/CS	Cr(VI)	Batch, 180 rpm, 100 mg adsorbent, 50 mg/L adsorbate, V = 100 mL	67.66	[181]
β-CD–MGO/CS	Cr(VI)	Batch, 150 rpm, 100 mg adsorbent, 100 mg/L adsorbate, V = 100 mL	120	[192]
β-CD–MGO/CS	Malachite green	Batch, 25 °C, 150 rpm, 5 mg adsorbent, V = 20 mL	740.74	[60]
β-CD–MGO/CS	p-Phenylene-diamine	Batch, 45 °C, 5 mg adsorbent, 100 mg/L adsorbate, V = 20 mL	1102.58	[193]
β-CD–MGO/CS	Methylene blue	Batch, 25 °C, 180 rpm, 10 mg adsorbent, V = 25 mL	43.34	[177]

Surface ion imprinted MGO/CS adsorbents have also been fabricated with a view to providing enhanced selective uptake and removal of certain heavy metal ions. Wang et al. [194] used this strategy to synthesize Pb-MGO/CS for the selective removal of Pb(II). While the MGO/CS showed no selectivity for the adsorption of Pb(II), the imprinted adsorbent exhibited specific recognition for Pb(II) in a mixed metal ion solution. This was attributed to the cavities created on removal of the Pb template, with the appropriate size, shape and coordination geometry to capture Pb(II). Ion imprinting has also been employed with the MGO/CS adsorbents in the selective removal of Cu(II) [195].

4.3. Chitosan/GO with 3D Architectures

As detailed in Section 2.2, 3D GO composites possess a unique 3D porous structure decorated with surface functional groups that have the ability to bind and remove pollutants from water. The 3D network consists of interconnected, curved, wrinkled and distorted graphene sheets to provide a porous 3D structure with high specific surface area [196]. Depending on the shape of the material, sponges, foams, or porous graphene films can be obtained [197]. Although these 3D structures are more easily recovered from the solution phase following the adsorption step compared to GO powders, they can exhibit rather poor stability in water. Typically additional supports are added, but these must be selected and chosen so that the distorted GO sheets are available and free to act as adsorption sites. In this regard, chitosan is particularly suitable as it can be processed by freeze-drying, to give aerogels [21].

These 3D GO/CS porous networks are normally formed by mixing/sonicating all the components, including GO or rGO, chitosan and any crosslinking agents or other additives, followed by some thermal processing and a final freeze-drying step [198–200]. Alternatively, a 3D scaffold can be employed as a template. For example, a polylactic acid (PLA) 3D scaffold was printed and then immersed in a GO/CS mixture, followed by freeze-drying to give a 3D sponge [201], while aerogel microspheres were prepared by combining electrospraying and freeze-casting [202]. Likewise, Kovtun et al. [203] used aerogels comprising 3D chitosan-gelatin. These were then modified with GO by either embedding the GO sheets within the aerogel or by coating the surface of the aerogel with GO sheets. These two methods were compared in the adsorption of ofloxacin and ciprofloxacin (fluoroquinolonic antibiotics) and Pb(II). It was concluded that the adsorption of Pb(II) was fast at the GO surface coated aerogels, where the adsorption sites were easily and rapidly accessed. On the other hand, diffusion of Pb(II) to the embedded GO required a longer contact time for adsorption.

These 3D GO/CS adsorbents have shown relatively good stability and reuse in the removal of reactive black 5 dye [134]. Likewise, good recyclability and stability was achieved with GO/CS sponges, with a regeneration efficiency in excess of 80% over five cycles in the removal of heavy metal ions [204]. Other contaminants that have been removed successfully using 3D GO/CS composites include dyes [146,153], Cu(II) [205], Cr(VI) [206], tetracycline [207] and 4-nonylphenol [208]. The incorporation of β-cyclodextrins into this 3D network can be employed to further enhance adsorption. For example, 3D-GO/CS/β-CD was employed as an effective adsorbent for MB yielding an ultrahigh adsorption capacity of 1134 mg g^{-1} [198]. Other additives include montmorillonite, a well-known adsorbent [199], kaolin, a filler employed to increase the mechanical strength of the adsorbent [209], and magnetic nanoparticles [200]. One of the more significant advantages of using these 3D GO/CS networks is the reuse and recyclability of the adsorbent without the need for a complex and time-consuming filtration process [198]. Moreover, they exhibit large surface areas, while the interconnected pores enable diffusion of the adsorbate throughout the 3D network leading to high adsorption capacity.

4.4. GO/Chitosan and Additional Chelating Agents

The GO/CS adsorbents have a number of binding sites for adsorption, but some of these are consumed during the crosslinking steps and this can involve the removal of essential $-$OH and $-$NH$_3^+$ groups, as illustrated earlier. However, the density of appropriate functional groups can be enhanced through the addition of chelating agents and this strategy has been employed with the aim of increasing the adsorption capacity. Some of the chelating agents employed with the GO/CS adsorbents and their performances in the adsorption of various contaminants are provided and summarized in Table 5. EDTA, an amino carboxylic acid with four carboxylic acid groups, is one of the more popular chelating agents as it can scavenge or chelate various cationic species.

Another emerging chelating additive is polydopamine (PDA). This is easily generated from the oxidation and self-polymerization of dopamine in slightly alkaline solutions [210]. Polydopamine provides both amine and catechol groups and can also participate in hydrogen bonding and π–π stacking interactions, making it a very good chelating agent that can be easily combined with GO/CS. A schematic illustration of these interactions and the assembly of GO/CS/DPA is shown in Figure 7, where the added PDA is sandwiched between the GO sheets, serving to minimise aggregation [211]. Moreover, the PDA with abundant catechol and amino groups can serve as an active surface for functionalization. For example, Cao et al. [212] used a Michael addition reaction with the thiol group of 1H,1H,2H,2H-perfluorodecanethiol (PFDT) to create perfluorinated rGO/CS-PDA with superhydrophobic properties. This functionalized adsorbent was then utilized as a cost-effective and environmentally-acceptable approach for the separation of oil/water mixtures.

Lignosulfonate (LS) is a further example of a chelating agent with abundant sulfonic ($-$SO$_3^-$) and hydroxyl ($-$OH) groups and with a strong affinity for the binding of metal ions and charged molecules. This polyelectrolyte has been combined with GO/CS and employed in the adsorption of the cationic methylene blue [186]. As shown in Table 5, these additional chelating agents, PDA, EDTA and LS, are effective in the adsorption of a variety of heavy metal ions. In particular, very good removal of Pb(II) is evident, while the presence of these additional chelating agents also facilitate the removal of tri-valent cations.

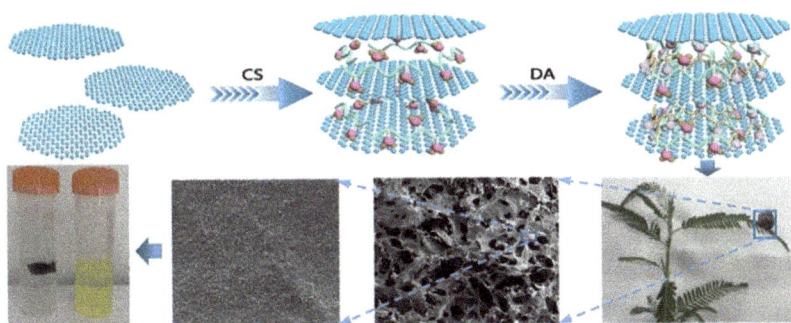

Figure 7. Schematic illustration of the preparation of graphene oxide/chitosan/polydopamine (GO/CS/PDA). Reproduced with permission from Li et al. [211], Int. J. Biol. Macromol; published by Elsevier, 2020.

Table 5. Chelating agents combined with GO/CS-based adsorbents, where the solid adsorbents are added to the pollutant-containing solution.

Chelating Agent	Adsorbent/Adsorption Experiment	Adsorbate	Adsorption q_m (mg g^{-1})	Ref.
Ethylenediaminetetraacetic acid (EDTA)	MGO/CS/EDTA Batch, 25 °C, 180 rpm, 0.33 g/L adsorbate, V = 30 mL	Pb(II) Cu(II) As(III)	206.5 207.3 42.7	[179]
EDTA	GO/CS/EDTA Batch, 25 °C, 160 rpm, 20 mg adsorbent, V = 50 mL	Cr(VI)	86.2	[180]
EDTA	MGO/CS/EDTA Batch, 33 °C, 0.14 g/L adsorbent, 114 mg/L adsorbate	Rhodamine B	1085.3	[213]
EDTA	MGO/CS/EDTA Batch, 20 mg adsorbent, V = 15 mL	Pb(II)	970	[214]
EDTA	MGO/CS/EDTA Batch, 49.2 °C, 40 Hz sonication, 9.5 mg adsorbent, V = 50 mL	Pb(II)	666.6	[161]
Polydopamine (PDA)	GO/CS/Polyvinyl alcohol (PVA)/PDA Batch, 40 °C, 150 rpm, 50 mg adsorbate	Cu(II) Pb(II) Cd(II)	210.9 236.2 214.9	[215]
PDA	GO/CS/PDA aerogel Batch, 25 °C, 120 rpm, 0.3 g/L adsorbent	U(VI)	415.9	[216]
PDA	GO/CS/PDA Batch, 25 °C, 150 rpm, 15 mg adsorbent, V = 20 mL	Cr(VI)	312.0	[211]
PDA	MWCNT/PDA/GO/CS Batch, 25 °C, 10 mg adsorbent, V = 10 mL	Gd(III)	150.8	[217]
PDA	GO/CS/PDA Batch, 30 °C, adsorbate 500 mg/L, V = 100 mL	Cu(II) Ni(II) Pb(II)	170.3 186.8 312.8	[218]
Lignosulfonate (LS)	MGO/LS/CS Batch, 30 °C, 160 rpm, 10 mg adsorbent, V = 20 mL	Methylene blue	50	[186]
LS	GO/LS/CS Batch, 30 °C, 130 rpm, 0.2 g/L adsorbent, V = 25 mL	Methylene blue	1023.9	[143]

4.5. GO/Chitosan Combined with Other Adsorbent Materials

The GO/CS hydrogel provides an attractive matrix for encapsulating other adsorbent materials and especially powdered materials that are difficult to separate and recover after the adsorption process, limiting their environmental applications. Consequently, there is much interest in combining other adsorbents with the GO/CS hydrogels. For example, metal–organic frameworks (MOFs), which consist of metal ions and polyfunctional organic ligands and have good adsorption potential, have been successfully integrated within GO/CS for the elimination of Cr(VI) [219]. Moreover, they have been employed as both an adsorbent and photocatalyst in the removal of methylene blue, reaching an adsorption capacity of approximately 357.15 mg g^{-1} [220]. The intercalated MOF can also provide the GO channels with molecular-sieving properties. Using this strategy, Chang et al. [221] prepared a GO/CS/MOF membrane for the purification of water. The GO/CS/MOF membrane exhibited very good water flux (14.62 L m^{-2} h^{-1} bar^{-1}), with high rejection (>99% for dyes) and good antifouling characteristics.

Hydroxyapatite (Hap) is another promising adsorbent with good bioactive, non-toxic and biocompatible properties. Its adsorption capacity can be enhanced by combining it with other suitable adsorbent materials and GO/CS with its π–π stacking, hydrogen bonding and electrostatic interactions is a suitable host material. GO/CS/Hap has shown good adsorption of dye molecules with adsorption capacities of 43.06, 41.32 and 40.03 mg g^{-1} for the removal of Congo Red, Acid Red 1 and Reactive Red 2 from water, respectively [129]. Montmorillonite is an alternative cost effective, sheet-like adsorbent material that has rather poor adsorbent capacity as an individual material, but it has been shown to enhance the stability of rGO/CS [199]. This porous hydrogel was formed without any cross-linking of chitosan and served as an efficient adsorbent for the uptake of Cr(VI) (87.03 mg g^{-1}) [199]. Another candidate is layered double hydroxides (LDHs). These 2D materials consist of layers of divalent and trivalent cations, with intercalating anions. These layered materials have been employed in the removal of heavy metal ions [222]. More recently, they have been integrated with GO/CS hydrogels to give improved adsorption performances [148,223]. In Figure 8, a schematic describing the formation of GO/CS combined with Fe-Al double layered hydroxide is shown illustrating its application in the removal of As(V) [223].

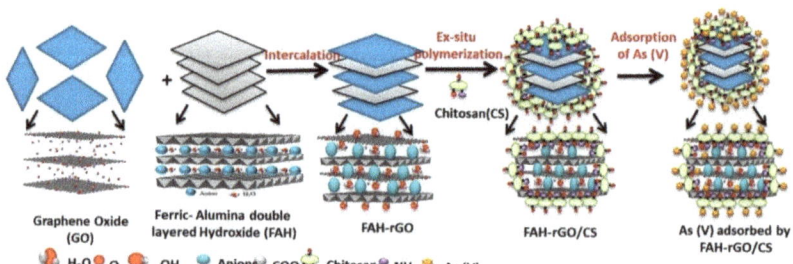

Figure 8. Schematic of the formation of GO/CS/LDH(Fe-Al) and illustration of the adsorption of As(V). Reproduced with permission from Priya et al. [223], Mater. Chem. Phys.; published by Elsevier, 2020.

Antifouling reagents, such as ZnO and silver ions, can also be easily integrated within the GO/CS hydrogels and used to increase antibacterial activity. For example, GO/CS/ZnO hybrid composites have shown impressive antibacterial activity against *E. coli* and *S. aureus* and very good adsorption of methylene blue [224], while GO/CS/MOF modified with silver ions, with good antibiofouling characteristics, was employed for the adsorption of uranium [225].

4.6. GO/CS and Polymer Blending and Hybrids

Although chitosan has several attractive properties as an immobilization matrix for GO, it can suffer from relatively poor hydrolytic stability. In an effort to overcome these technological challenges, chitosan copolymers and the blending of chitosan with other synthetic polymers has received considerable attention. A number of synthetic polymers, with well-defined structures, and various biopolymers, with less defined structures, have been combined with chitosan and GO. The polymers employed include poly(vinyl alcohol) (PVA) [226], polyacrylic acid (PAA) [227], polylactic acid (PLA) [201], cellulose (C) [228], carboxymethyl cellulose (CMC) [229] and alginate [169]. The performance of these chitosan blended polymer systems as a matrix for GO is summarized in Table 6, where it is readily evident that these systems have good adsorption capacity.

Table 6. Adsorption capacity of blended chitosan biopolymers combined with GO, where the solid adsorbent is added to the pollutant-containing solution phase.

Blended Polymer	Adsorbent/Adsorption Experiment	Adsorbate	Adsorption q_m (mg g^{-1})	Ref.
PVA (poly(vinyl alcohol))	GO/CS/PVA Batch, 30 °C, 160 rpm, 20 mg adsorbent, V = 40 mL	Cd(II) Ni(II)	172.11 70.37	[230]
PVA	GO/CS/PVA Batch, 140 rpm, 6 g/L adsorbent	Congo red dye	12.38	[226]
PVA	GO/CS/PVA Batch, 22.16 mg/L adsorbate, 0.5 g/L adsorbent	Sr(II)	17.48	[231]
PVA	GO/CS/PVA Batch, 30 °C, 150 rpm, 50 mg adsorbent, V = 100 mL	Cu(II)	162	[232]
PAA (polyacrylic acid)	GO/CS/PAA Batch, 25 °C, 0.2 g adsorbent, V = 150 mL	Rhodamine 6G Methyl violet Methyl orange	224.6 169.2 195.6	[227]
PAA	GO/CS/PAA/Fe$_3$O$_4$ Batch, 25 °C, 300 rpm, 10 mg adsorbent, V = 50 mL	Cu(II)	217.4	[187]
PLA (polylactic acid)	GO/CS/PLA Batch, 110 rpm, 25–45 °C, 30 mg adsorbent, V = 30 mL	Crystal violet	45	[201]
CMC (carboxymethyl cellulose)	GO/CMC/CS Batch, 25 °C, 200 rpm, 5 mg adsorbent, V = 20 mL	Sulfamethoxazole Sulfapyridine	312.2 161.8	[233]
CMC	GO/CS/CMC Batch, 25 °C, 0.1 0.6 g/L adsorbent	MB MO	655.98 404.52	[229]
C Cellulose	GO/CS/C Batch, 30 °C, 200 rpm, 5 mg adsorbent, V = 8 mL	Cu(II)	22.40	[228]
SA (sodium alginate)	GO/SA/CS/FeO Batch, 30 °C, 50 mg adsorbent	Cu(II) Cd(II) Pb(II)	55.96 86.28 189.04	[169]

Conducting polymers, such as polypyrrole [136,234,235], have also been added to the GO/CS hydrogel. This can be easily achieved through the in-situ polymerization or electropolymerization of the corresponding monomer within the GO/CS hydrogel. The addition of conducting polymers, such as polypyrrole, which have very good stability and a porous and high surface area, has the potential to enhance the adsorption capacity as the conducting polymers can capture charged contaminants as dopants, participate in hydrogen bonding and be involved in π–π interactions. Magnetic nanoparticles can also be easily deposited at the conducting polypyrrole to give hybrids which can be removed following adsorption using magnetic separation [236].

5. Adsorption and Regeneration Processes

The ideal adsorbent should possess a high adsorption capacity, but it should also be possible to desorb the adsorbate and regenerate the adsorbent. These processes are now described and discussed.

5.1. Adsorption Capacity

The adsorption process depends on a number of experimental parameters, such as the initial concentration of the adsorbate, the pH and ionic strength of the solution, temperature and adsorption contact time. The pH of the solution is one of the more significant parameters as it influences both the surface properties of the adsorbent and ionization state of the adsorbates, and therefore affects significantly the electrostatic interactions. Typically, the GO/CS adsorbents have a pH of zero charge (pH_{pzc}) ranging between about 2.2 [143] and 6.0 [237]. Accordingly, the GO/CS adopts an overall positive charge under acidic conditions and a more negative charge at higher pH values. Generally, the adsorption capacity for divalent metal ions, such as Cu(II), increases from a pH of about 2.0 to 7.0, but then decreases rapidly at higher pH values [187,238]. This can be explained in terms of the formation of protonated chitosan under acidic conditions, where the resulting NH_3^+ groups repel cations. However, as the pH increases and the GO/CS adopts an overall negative charge, with the generation of $-COO^-$ and $-NH_2$ groups, electrostatic interactions are favoured and the cations can be eliminated from the solution phase. As the pH is further increased, insoluble metal hydroxides are formed leading to precipitation in the solution phase. In terms of the removal of Cr(VI), the anionic dichromate ($Cr_2O_7^{2-}$) and chromate ($HCrO_4^-$) species are generated at pH values in the vicinity of 2.0 to 4.0, giving higher adsorption capacity under these acidic conditions [239]. These electrostatic interactions normally give poor selectivity in the adsorption step, and on increasing the ionic strength of the solution, the adsorption capacity of the targeted adsorbent is reduced [205].

In terms of the time-dependent adsorption, the kinetics of the adsorption process plays a significant role. The kinetics depends on different diffusional processes, including diffusion of the adsorbate from the bulk solution to the adsorbent–solution boundary, to the adsorbent surface, within the porous adsorbent and to the actual adsorption step. In most studies, the solution phase is agitated and this eliminates diffusional limitations within the bulk solution. Under these conditions, the intraparticle diffusion within the porous adsorbent tends to become the rate-determining step [198]. There is normally an initial rapid increase in the adsorption as the easily accessible adsorption sites are first reached and this is then followed by a more gradual increase until equilibrium adsorption is attained. The time required to achieve equilibrium can vary from about 200 min [198] to 58 h [139] for MB, and 500 min [240] to 20 h for MO [241], clearly highlighting the significant role of the diffusional processes within the adsorbents and the difficulty in comparing the different GO/CS adsorbents. Indeed, a direct comparison between the large number of GO/CS adsorbents employed, Tables 4–6, is very challenging given the significant variations in the properties of the chitosan used, including its DD levels, MW, porosity, particle size, the nature of the crosslinking agents employed during synthesis (Section 3) and the ratio of chitosan to GO and the presence of additional materials.

5.2. Regeneration Strategies

The successful regeneration of the GO/CS-based adsorbents is an important consideration in the final applications of these materials. Efficient regeneration, without significant loss in the adsorption capacity, is not only necessary in terms of operating costs, but is also essential in the recovery of the adsorbates and the prevention of secondary waste. The desorbing agents employed in the regeneration of the GO/CS-based adsorbents are normally acids [229], bases [134,229], chelating agents, such as EDTA [213], or organic solvents, such as ethanol [166,198], methanol [176] or acetone [92]. The selection of the desorbing agent depends on the nature of the adsorbate. For example, NaOH is a good choice for the desorption of anionic dyes and various anions, whereas acidic eluents are effective in the removal of cationic dyes and cations, and organic solvents are suitable for the desorption of organic molecules. When an acidic eluent, such as HCl, is used, the amino groups in chitosan become protonated to give NH_3^+ and the $-COO^-$ groups in GO become protonated to give –COOH. This favours the desorption of cationic species that were previously electrostatically bound to the $-COO^-$ groups on GO. On the other hand, anionic species bound to the $-NH_3^+$ groups can be desorbed with an increase in pH as the $-NH_3^+$ is converted to $-NH_2$ and the now negatively generated $-COO^-$ groups will favour desorption of the anions. Indeed, solutions of NaOH have been employed to desorb anionic species such as chromate [180] and methyl orange [242], while HCl solutions have been employed to desorb divalent cations such as Pb(II) [175] and Cd(II) [243]. Organic solvents such as methanol have been employed in the desorption of ciprofloxacin [176] and acetone has been used to desorb methyl violet and Alizarin yellow R [92].

While these strategies can be employed in the regeneration of the GO/CS based adsorbents, the adsorption efficiency of the regenerated adsorbents tend to decrease with each regeneration step. For example, Zhao et al. [244] observed a gradual loss in the adsorption capacity and removal of phenol and p-nitrophenol on regeneration of the GO/CS-based adsorbent with NaOH. Furthermore, concentrated acids and bases are not always the best options in the regeneration of the GO/CS adsorbents. Magnetic particles, such as Fe_3O_4, tend to corrode and dissolve in acidic media. Therefore, EDTA has been employed in the regeneration step to protect magnetic particles [213].

However, it is well known that chitosan undergoes hydrolysis reactions in the presence of acids [96]. These hydrolysis reactions, which attack the polymer chains, give rise to a lowering in the MW and a reduction in the mechanical strength of chitosan. This makes the regeneration of GO/CS more complex and challenging, as the structure of the GO/CS will alter with repetitive regeneration steps. Furthermore, these changes in the overall structure of chitosan are likely to give rise to some restacking of the GO sheets, reducing the adsorption capacity.

6. Conclusions

It is clearly evident from the increasing number of publications that focus on combining graphene and chitosan that these materials are emerging as interesting candidates for environmental applications. Both chitosan and graphene have good adsorption capacity. The addition of graphene to chitosan enhances its mechanical properties, while the chitosan hydrogel serves as an immobilization matrix for graphene. Fortunately, it is GO that has the best potential in the formulation of adsorbents, removing the need to use graphene in its pure form, which had been plagued with manufacturing issues. Other components, varying from supramolecular agents, magnetic nanoparticles and chelating agents, can also be added to the GO/CS hydrogel, to further enhance its properties.

However, this research is still in its early stages of development and a number of challenges exist and must be overcome before these materials can be employed as adsorbents in real applications. One of the more significant aspects from a health and well-being consideration is an evaluation of the toxic properties of GO. As increasing amounts of GO are used, it will eventually make its way into the aquatic environment, aided by its polar groups. There is increasing evidence to suggest that the build-up of GO within the aquatic environment may lead to bioaccumulation. Therefore, studies aimed at measuring the release and leaching of GO from the chitosan-based hydrogels are important. While chitosan can be employed as an immobilization matrix, it may be necessary to further modify the physiochemical properties of the chitosan hydrogel to trap more effectively the GO sheets.

Another aspect that requires consideration is the final costs associated with the fabrication of the adsorbents and its potential for regeneration. The costs associated with the production of GO and the GO/CS adsorbent must be competitive in terms of the competing technologies and especially the competing activated carbon, which is currently the preferred adsorbent material. Ideally, adsorbents should be readily regenerated with little or no additional costs. With GO and the GO/CS composites, the adsorption capacity tends to decrease with each adsorption–regeneration cycle. Consequently, new regeneration strategies that will give rise to more cost-effective and efficient regeneration are needed. Furthermore, more studies should be focussed on the recovery of the adsorbate and especially the recovery of valuable metals that have accumulated within the composite. The recovery of valuable metals for re-use could be used to offset some of the adsorbent fabrication and recovery costs and contribute to the development of the GO/CS adsorbents as a sustainable technology.

Selectivity in the adsorption process is another challenge as real water samples contain various ions, such as sodium, calcium and chlorides, that will consume the adsorption sites and thus reduce the removal capacity of the targeted pollutant or pollutants. Indeed, in many studies, an increase in the ionic strength gives rise to a reduction in the adsorption capacity of the targeted contaminant. Supramolecular chemistry, which not only includes cyclodextrins, but calixarenes and pillararenes, and imprinting technologies, has the potential to play a greater role in addressing these selectivity issues in the adsorption step.

Magnetic GO/CS composites have the capacity to be removed following adsorption. Nevertheless, these magnetic materials must be sufficiently anchored within the GO/CS composites to prevent their leaching or dissolution and escape into the aquatic environment. Consequently, more studies are required aimed at monitoring the release of the magnetic/iron particles, Fe^{2+} and/or Fe^{3+}, that could in turn lead to the development of sludge (insoluble iron hydroxide precipitates). Finally, batch experiments are normally employed with very few studies utilizing other techniques, such as fixed bed columns, that may be more applicable in real wastewater treatment applications.

In conclusion, while these GO/CS adsorbents require further research and development, these materials have a promising future as adsorbent materials. Besides, new 2D materials are continuously being developed and these could be relatively easily combined with GO/CS to generate a new family of high performing and cost-effective adsorbents.

Author Contributions: Conceptualization, C.B.B. and D.C.d.S.A.; writing—original draft preparation, C.B.B., D.C.d.S.A. and B.H.; writing—review and editing, C.B.B., D.C.d.S.A., T.Y. and B.H.; funding acquisition, T.Y. and C.B.B. All authors have read and agreed to the published version of the manuscript.

Funding: The authors would like to acknowledge funding provided by the Irish Research Council, award number GOIPG/2020/657 and Maynooth University.

Institutional Review Board Statement: Not applicable.

Informed Consent Statement: Not applicable.

Data Availability Statement: Not applicable.

Conflicts of Interest: The authors declare no conflict of interest.

References

1. Brindha, K.; Schneider, M. Impact of urbanization on groundwater quality. In *GIS and Geostatistical Techniques for Groundwater Science*; Elsevier Inc.: Amsterdam, The Netherlands, 2019; pp. 179–196. ISBN 9780128154137.
2. Brindha, K.; Paul, R.; Walter, J.; Tan, M.L.; Singh, M.K. Trace metals contamination in groundwater and implications on human health: Comprehensive assessment using hydrogeochemical and geostatistical methods. *Environ. Geochem. Health* **2020**, *42*, 3819–3839. [CrossRef] [PubMed]

3. Anjum, M.; Miandad, R.; Waqas, M.; Gehany, F.; Barakat, M.A. Remediation of wastewater using various nano-materials. *Arab. J. Chem.* **2016**, *12*, 4897–4919. [CrossRef]
4. Casagrande, T.; Lawson, G.; Li, H.; Wei, J.; Adronov, A.; Zhitomirsky, I. Electrodeposition of composite materials containing functionalized carbon nanotubes. *Mater. Chem. Phys.* **2008**, *111*, 42–49. [CrossRef]
5. Hansima, M.A.C.K.; Makehelwala, M.; Jinadasa, K.B.S.N.; Wei, Y.; Nanayakkara, K.G.N.; Herath, A.C.; Weerasooriya, R. Fouling of ion exchange membranes used in the electrodialysis reversal advanced water treatment: A review. *Chemosphere* **2021**, *263*, 127951. [CrossRef]
6. Kar, A.; Bahadur, V. Using excess natural gas for reverse osmosis-based flowback water treatment in US shale fields. *Energy* **2020**, *196*, 117145. [CrossRef]
7. Saffari, R.; Shariatinia, Z.; Jourshabani, M. Synthesis and photocatalytic degradation activities of phosphorus containing ZnO microparticles under visible light irradiation for water treatment applications. *Environ. Pollut.* **2020**, *259*, 113902. [CrossRef]
8. Vidotti, M.; Salvador, R.P.; Torresi, S.I.C.; Antonia, L.H.D. Electrocatalytic oxidation of urea by nanostructured nickel/cobalt hydroxide electrodes. *Electrochim. Acta* **2008**, *53*, 4030–4034. [CrossRef]
9. Yu, T.; Breslin, C.B. Graphene-modified composites and electrodes and their potential applications in the electro-fenton process. *Materials* **2020**, *13*, 2254. [CrossRef]
10. Daniele, C.; Alves, S.; Healy, B.; Pinto, L.A.D.A.; Sant, T.R.; Cadaval, A.; Breslin, C.B. Recent developments in chitosan-based adsorbents for the removal of pollutants from aqueous environments. *Molecules* **2021**, *26*, 594.
11. Subha, R.; Sridevi, O.A.; Anitha, D.; Sudha, D. Treatment methods for the removal of phenol from water—A Review. In Proceedings of the International Conference on Systems, Science, Control, Communication, Engineering and Technology, Coimbatore, India, 10–11 August 2015; pp. 199–203.
12. Chowdhury, S.; Balasubramanian, R. Recent advances in the use of graphene-family nanoadsorbents for removal of toxic pollutants from wastewater. *Adv. Colloid Interface Sci.* **2014**, *204*, 35–56. [CrossRef]
13. Travlou, N.A.; Kyzas, G.Z.; Lazaridis, N.K.; Deliyanni, E.A. Graphite oxide/chitosan composite for reactive dye removal. *Chem. Eng. J.* **2013**, *217*, 256–265. [CrossRef]
14. Zhang, Q.; Hou, Q.; Huang, G.; Fan, Q. Removal of heavy metals in aquatic environment by graphene oxide composites: A review. *Environ. Sci. Pollut. Res.* **2020**, *27*, 190–209. [CrossRef]
15. Mishra, A.K.; Ramaprabhu, S. Carbon dioxide adsorption in graphene sheets. *AIP Adv.* **2011**, *1*. [CrossRef]
16. Sui, Z.-Y.; Meng, Y.-N.; Xiao, P.-W.; Zhao, Z.-Q.; Wei, Z.-X.; Han, B.-H. Nitrogen-doped graphene aerogels as efficient supercapacitor electrodes and gas adsorbents. *ACS Appl. Mater. Interfaces* **2015**, *7*, 1431–1438. [CrossRef]
17. Liu, F.; Chung, S.; Oh, G.; Seo, T.S. Three-dimensional graphene oxide nanostructure for fast and efficient water-soluble dye removal. *ACS Appl. Mater. Interfaces* **2012**, *4*, 922–927. [CrossRef]
18. Guo, X.; Qu, L.; Tian, M.; Zhu, S.; Zhang, X.; Tang, X.; Sun, K. Chitosan/graphene oxide composite as an effective adsorbent for reactive red dye removal. *Water Environ. Res.* **2016**, *88*, 579–588. [CrossRef]
19. Xing-yu, H.; Yu-kui, T.; Fu-gui, Z.H.U.; Qin-fei, X.I.A.; Miao-miao, T. Graphene oxide-based magnetic boronate-affinity adsorbent for extraction of Horseradish Peroxidase. *Chin. J. Anal. Chem.* **2020**, *48*, 20158–20164. [CrossRef]
20. Sánchez-page, B.; Ana, M.P.; Jim, M.V.; Jesús, J.P.; Gonz, Z.; Fern, L.; Granda, M.; Men, R.; Blasco, J.; Subías, G.; et al. Influence of graphene sheet properties as supports of iridium-based N-heterocyclic carbene hybrid materials for water oxidation electrocatalysis. *J. Organomet. Chem.* **2020**, *919*, 121334. [CrossRef]
21. Li, A.; Lin, R.; Lin, C.; He, B.; Zheng, T.; Lu, L.; Cao, Y. An environment-friendly and multi-functional absorbent from chitosan for organic pollutants and heavy metal ion. *Carbohydr. Polym.* **2016**, *148*, 272–280. [CrossRef]
22. Karimi-Maleh, H.; Ayati, A.; Davoodi, R.; Tanhaei, B.; Karimi, F.; Malekmohammadi, S.; Orooji, Y.; Fu, L.; Sillanpää, M. Recent advances in using of chitosan-based adsorbents for removal of pharmaceutical contaminants: A review. *J. Clean. Prod.* **2021**, *291*, 125880. [CrossRef]
23. Yusuf, M.; Elfghi, F.M.; Zaidi, S.A.; Abdullah, E.C.; Khan, M.A. Applications of graphene and its derivatives as an adsorbent for heavy metal and dye removal: A systematic and comprehensive overview. *RSC Adv.* **2015**, *5*, 50392–50420. [CrossRef]
24. Kyzas, G.Z.; Deliyanni, E.A.; Matis, K.A. Graphene oxide and its application as an adsorbent for wastewater treatment. *J. Chem. Technol. Biotechnol.* **2014**, *89*, 196–205. [CrossRef]
25. Peng, W.; Li, H.; Liu, Y.; Song, S. A review on heavy metal ions adsorption from water by graphene oxide and its composites. *J. Mol. Liq.* **2017**, *230*, 496–504. [CrossRef]
26. Duru, İ.; Ege, D.; Kamali, A.R. Graphene oxides for removal of heavy and precious metals from wastewater. *J. Mater. Sci.* **2016**, *51*, 6097–6116. [CrossRef]
27. Liu, X.; Ma, R.; Wang, X.; Ma, Y.; Yang, Y.; Zhuang, L.; Zhang, S.; Jehan, R.; Chen, J.; Wang, X. Graphene oxide-based materials for efficient removal of heavy metal ions from aqueous solution: A review. *Environ. Pollut.* **2019**, *252*, 62–73. [CrossRef]
28. Sherlala, A.I.A.; Raman, A.A.A.; Bello, M.M.; Asghar, A. A review of the applications of organo-functionalized magnetic graphene oxide nanocomposites for heavy metal adsorption. *Chemosphere* **2018**, *193*, 1004–1017. [CrossRef]
29. Lai, K.C.; Lee, L.Y.; Hiew, B.Y.Z.; Thangalazhy-Gopakumar, S.; Gan, S. Environmental application of three-dimensional graphene materials as adsorbents for dyes and heavy metals: Review on ice-templating method and adsorption mechanisms. *J. Environ. Sci.* **2019**, *79*, 174–199. [CrossRef]

30. Lin, Y.; Tian, Y.; Sun, H.; Hagio, T. Progress in modifications of 3D graphene-based adsorbents for environmental applications. *Chemosphere* **2021**, *270*. [CrossRef]
31. Novoselov, K.S.; Geim, A.K.; Morozov, S.V.; Jiang, D.; Zhang, Y.; Dubonos, S.V.; Grigorieva, I.V.; Firsov, A.A. Electric field in atomically thin carbon films. *Science* **2004**, *306*, 666–669. [CrossRef] [PubMed]
32. Wang, B.; Ruan, T.; Chen, Y.; Jin, F.; Peng, L.; Zhou, Y.; Wang, D.; Dou, S. Graphene-based composites for electrochemical energy storage. *Energy Storage Mater.* **2020**, *24*, 22–51. [CrossRef]
33. Tyagi, D.; Wang, H.; Huang, W.; Hu, L.; Tang, Y.; Guo, Z.; Ouyang, Z.; Zhang, H. Recent advances in two-dimensional-material-based sensing technology toward health and environmental monitoring applications. *Nanoscale* **2020**, *12*, 3535–3559. [CrossRef] [PubMed]
34. Shu, R.; Zhang, J.; Guo, C.; Wu, Y.; Wan, Z.; Shi, J.; Liu, Y.; Zheng, M. Facile synthesis of nitrogen-doped reduced graphene oxide/nickel-zinc ferrite composites as high-performance microwave absorbers in the X-band. *Chem. Eng. J.* **2020**, *384*. [CrossRef]
35. Rathi, B.S.; Kumar, P.S.; Show, P.-L. A review on effective removal of emerging contaminants from aquatic systems: Current trends and scope for further research. *J. Hazard. Mater.* **2021**, *409*. [CrossRef] [PubMed]
36. Hussain, M.M.; Wang, J.; Bibi, I.; Shahid, M.; Niazi, N.K.; Iqbal, J.; Mian, I.A.; Shaheen, S.M.; Bashir, S.; Shah, N.S.; et al. Arsenic speciation and biotransformation pathways in the aquatic ecosystem: The significance of algae. *J. Hazard. Mater.* **2021**, *403*. [CrossRef]
37. Rout, P.R.; Zhang, T.C.; Bhunia, P.; Surampalli, R.Y. Treatment technologies for emerging contaminants in wastewater treatment plants: A review. *Sci. Total Environ.* **2021**, *753*. [CrossRef]
38. Somani, P.R.; Somani, S.P.; Umeno, M. Planer nano-graphenes from camphor by CVD. *Chem. Phys. Lett.* **2006**, *430*, 56–59. [CrossRef]
39. Gao, L.; Guest, J.R.; Guisinger, N.P. Epitaxial graphene on Cu(111). *Nano Lett.* **2010**, *10*, 3512–3516. [CrossRef]
40. Dreyer, D.R.; Park, S.; Bielawski, C.W.; Ruoff, R.S. The chemistry of graphene oxide. *Chem. Soc. Rev.* **2010**, *39*, 228–240. [CrossRef]
41. Hummers, W.S.; Offeman, R.E. Preparation of graphitic oxide. *J. Am. Chem. Soc.* **1958**, *80*, 1339. [CrossRef]
42. Eda, G.; Chhowalla, M. Chemically derived graphene oxide: Towards large-area thin-film electronics and optoelectronics. *Adv. Mater.* **2010**, *22*, 2392–2415. [CrossRef]
43. Adeel, M.; Bilal, M.; Rasheed, T.; Sharma, A.; Iqbal, H.M.N. Graphene and graphene oxide: Functionalization and nano-bio-catalytic system for enzyme immobilization and biotechnological perspective. *Int. J. Biol. Macromol.* **2018**, *120*, 1430–1440. [CrossRef]
44. Wang, J.; Zhang, J.; Han, L.; Wang, J.; Zhu, L.; Zeng, H. Graphene-based materials for adsorptive removal of pollutants from water and underlying interaction mechanism. *Adv. Colloid Interface Sci.* **2021**, *289*. [CrossRef]
45. Cong, H.-P.; Chen, J.-F.; Yu, S.-H. Graphene-based macroscopic assemblies and architectures: An emerging material system. *Chem. Soc. Rev.* **2014**, *43*, 7295–7325. [CrossRef]
46. Wang, S.; Li, X.; Liu, Y.; Zhang, C.; Tan, X.; Zeng, G.; Song, B.; Jiang, L. Nitrogen-containing amino compounds functionalized graphene oxide: Synthesis, characterization and application for the removal of pollutants from wastewater: A review. *J. Hazard. Mater.* **2018**, *342*, 177–191. [CrossRef]
47. Ai, L.; Zhang, C.; Chen, Z. Removal of methylene blue from aqueous solution by a solvothermal-synthesized graphene/magnetite composite. *J. Hazard. Mater.* **2011**, *192*, 1515–1524. [CrossRef]
48. Cheng, L.; Ji, Y.; Liu, X. Insights into interfacial interaction mechanism of dyes sorption on a novel hydrochar: Experimental and DFT study. *Chem. Eng. Sci.* **2021**, *233*. [CrossRef]
49. Ramesha, G.K.; Vijaya Kumara, A.; Muralidhara, H.B.; Sampath, S. Graphene and graphene oxide as effective adsorbents toward anionic and cationic dyes. *J. Colloid Interface Sci.* **2011**, *361*, 270–277. [CrossRef]
50. Xing, H.T.; Chen, J.H.; Sun, X.; Huang, Y.H.; Su, Z.B.; Hu, S.R.; Weng, W.; Li, S.X.; Guo, H.X.; Wu, W.B.; et al. NH_2-rich polymer/graphene oxide use as a novel adsorbent for removal of Cu(II) from aqueous solution. *Chem. Eng. J.* **2015**, *263*, 280–289. [CrossRef]
51. Li, D.; Huang, J.; Huang, L.; Tan, S.; Liu, T. High-performance three-dimensional aerogel based on hydrothermal pomelo peel and reduced graphene oxide as an efficient adsorbent for water/oil separation. *Langmuir* **2021**, *37*, 1521–1530. [CrossRef]
52. Ge, H.; Zou, W. Preparation and characterization of L-glutamic acid-functionalized graphene oxide for adsorption of Pb(II). *J. Dispers. Sci. Technol.* **2017**, *38*, 241–247. [CrossRef]
53. Basadi, N.; Ghanemi, K.; Nikpour, Y. l-Cystine-functionalized graphene oxide nanosheets for effective extraction and preconcentration of mercury ions from environmental waters. *Chem. Pap.* **2021**, *75*, 1083–1093. [CrossRef]
54. Verma, S.; Dutta, R.K. Development of cysteine amide reduced graphene oxide (CARGO) nano-adsorbents for enhanced uranyl ions removal from aqueous medium. *J. Environ. Chem. Eng.* **2017**, *5*, 4547–4558. [CrossRef]
55. Zhang, C.-Z.; Yuan, Y.; Guo, Z. Experimental study on functional graphene oxide containing many primary amino groups fast-adsorbing heavy metal ions and adsorption mechanism. *Sep. Sci. Technol.* **2018**, *53*, 1666–1677. [CrossRef]
56. Wang, Y.; Cui, X.; Wang, Y.; Shan, W.; Lou, Z.; Xiong, Y. A thiourea cross-linked three-dimensional graphene aerogel as a broad-spectrum adsorbent for dye and heavy metal ion removal. *New J. Chem.* **2020**, *44*, 16285–16293. [CrossRef]
57. Janik, P.; Zawisza, B.; Talik, E.; Sitko, R. Selective adsorption and determination of hexavalent chromium ions using graphene oxide modified with amino silanes. *Microchim. Acta* **2018**, *185*. [CrossRef]

58. Suddai, A.; Nuengmatcha, P.; Sricharoen, P.; Limchoowong, N.; Chanthai, S. Feasibility of hard acid-base affinity for the pronounced adsorption capacity of manganese(II) using amino-functionalized graphene oxide. *RSC Adv.* **2018**, *8*, 4162–4171. [CrossRef]
59. Wang, D.; Liu, L.; Jiang, X.; Yu, J.; Chen, X. Adsorption and removal of malachite green from aqueous solution using magnetic β-cyclodextrin-graphene oxide nanocomposites as adsorbents. *Colloids Surf. A Physicochem. Eng. Asp.* **2015**, *466*, 166–173. [CrossRef]
60. Wang, H.; Liu, Y.-G.; Zeng, G.-M.; Hu, X.-J.; Hu, X.; Li, T.-T.; Li, H.-Y.; Wang, Y.-Q.; Jiang, L.-H. Grafting of β-cyclodextrin to magnetic graphene oxide via ethylenediamine and application for Cr(VI) removal. *Carbohydr. Polym.* **2014**, *113*, 166–173. [CrossRef]
61. Liu, X.; Yan, L.; Yin, W.; Zhou, L.; Tian, G.; Shi, J.; Yang, Z.; Xiao, D.; Gu, Z.; Zhao, Y. A magnetic graphene hybrid functionalized with beta-cyclodextrins for fast and efficient removal of organic dyes. *J. Mater. Chem. A* **2014**, *2*, 12296–12303. [CrossRef]
62. Del Valle, E.M.M. Cyclodextrins and their uses: A review. *Process Biochem.* **2004**, *39*, 1033–1046. [CrossRef]
63. Crini, G. Review: A history of cyclodextrins. *Chem. Rev.* **2014**, *114*, 10940–10975. [CrossRef]
64. Challa, R.; Ahuja, A.; Ali, J.; Khar, R.K. Cyclodextrins in drug delivery: An updated review. *AAPS PharmSciTech* **2005**, *6*. [CrossRef]
65. Zhao, Y.; Zheng, X.; Wang, Q.; Zhe, T.; Bai, Y.; Bu, T.; Zhang, M.; Wang, L. Electrochemical behavior of reduced graphene oxide/cyclodextrins sensors for ultrasensitive detection of imidacloprid in brown rice. *Food Chem.* **2020**, *333*. [CrossRef]
66. Healy, B.; Yu, T.; da Silva Alves, D.C.; Okeke, C.; Breslin, C.B. Cyclodextrins as supramolecular recognition systems: Applications in the fabrication of electrochemical sensors. *Materials* **2021**, *14*, 1668. [CrossRef]
67. Crini, G.; Morcellet, M. Synthesis and applications of adsorbents containing cyclodextrins. *J. Sep. Sci.* **2002**, *25*, 789–813. [CrossRef]
68. Nie, Z.-J.; Guo, Q.-F.; Xia, H.; Song, M.-M.; Qiu, Z.-J.; Fan, S.-T.; Chen, Z.-H.; Zhang, S.-X.; Zhang, S.; Li, B.-J. Cyclodextrin self-assembled graphene oxide aerogel microspheres as broad-spectrum adsorbent for removing dyes and organic micropollutants from water. *J. Environ. Chem. Eng.* **2021**, *9*. [CrossRef]
69. Zhong, Y.; He, Y.; Ge, Y.; Song, G. β-Cyclodextrin protected Cu nanoclusters as a novel fluorescence sensor for graphene oxide in environmental water samples. *Luminescence* **2017**, *32*, 596–601. [CrossRef]
70. Yang, Z.; Liu, X.; Liu, X.; Wu, J.; Zhu, X.; Bai, Z.; Yu, Z. Preparation of β-cyclodextrin/graphene oxide and its adsorption properties for methylene blue. *Colloids Surf. B Biointerfaces* **2021**, *200*. [CrossRef]
71. Gupta, V.K.; Agarwal, S.; Sadegh, H.; Ali, G.A.M.; Bharti, A.K.; Hamdy Makhlouf, A.S. Facile route synthesis of novel graphene oxide-β-cyclodextrin nanocomposite and its application as adsorbent for removal of toxic bisphenol A from the aqueous phase. *J. Mol. Liq.* **2017**, *237*, 466–472. [CrossRef]
72. Tan, P.; Hu, Y. Improved synthesis of graphene/β-cyclodextrin composite for highly efficient dye adsorption and removal. *J. Mol. Liq.* **2017**, *242*, 181–189. [CrossRef]
73. Samuel, M.S.; Selvarajan, E.; Subramaniam, K.; Mathimani, T.; Seethappan, S.; Pugazhendhi, A. Synthesized β-cyclodextrin modified graphene oxide (β-CD-GO) composite for adsorption of cadmium and their toxicity profile in cervical cancer (HeLa) cell lines. *Process Biochem.* **2020**, *93*, 28–35. [CrossRef]
74. Tian, H.; Zeng, H.; Zha, F.; Tian, H.; Chang, Y. Synthesis of graphene oxide–supported β-cyclodextrin adsorbent for removal of p-nitrophenol. *Water. Air. Soil Pollut.* **2020**, *231*. [CrossRef]
75. Liu, J.; Liu, G.; Liu, W. Preparation of water-soluble β-cyclodextrin/poly(acrylic acid)/graphene oxide nanocomposites as new adsorbents to remove cationic dyes from aqueous solutions. *Chem. Eng. J.* **2014**, *257*, 299–308. [CrossRef]
76. Jiang, L.; Liu, Y.; Liu, S.; Hu, X.; Zeng, G.; Hu, X.; Liu, S.; Liu, S.; Huang, B.; Li, M. Fabrication of β-cyclodextrin/poly (L-glutamic acid) supported magnetic graphene oxide and its adsorption behavior for 17β-estradiol. *Chem. Eng. J.* **2017**, *308*, 597–605. [CrossRef]
77. Boulanouar, S.; Mezzache, S.; Combès, A.; Pichon, V. Molecularly imprinted polymers for the determination of organophosphorus pesticides in complex samples. *Talanta* **2018**, *176*, 465–478. [CrossRef] [PubMed]
78. Cheng, L.; Pan, S.; Ding, C.; He, J.; Wang, C. Dispersive solid-phase microextraction with graphene oxide based molecularly imprinted polymers for determining bis(2-ethylhexyl) phthalate in environmental water. *J. Chromatogr. A* **2017**, *1511*, 85–91. [CrossRef]
79. Xu, K.; Wang, X.; Lu, C.; Liu, Y.; Zhang, D.; Cheng, J. Toxicity of three carbon-based nanomaterials to earthworms: Effect of morphology on biomarkers, cytotoxicity, and metabolomics. *Sci. Total Environ.* **2021**, *777*. [CrossRef]
80. Malhotra, N.; Villaflores, O.B.; Audira, G.; Siregar, P.; Lee, J.-S.; Ger, T.-R.; Hsiao, C.-D. Toxicity studies on graphene-based nanomaterials in aquatic organisms: Current understanding. *Molecules* **2020**, *25*, 3618. [CrossRef]
81. Dasmahapatra, A.K.; Dasari, T.P.S.; Tchounwou, P.B. Graphene-Based Nanomaterials Toxicity in Fish. *Rev. Environ. Contam. Toxicol.* **2019**, *247*, 1–58.
82. Martínez-Álvarez, I.; Le Menach, K.; Devier, M.-H.; Barbarin, I.; Tomovska, R.; Cajaraville, M.P.; Budzinski, H.; Orbea, A. Uptake and effects of graphene oxide nanomaterials alone and in combination with polycyclic aromatic hydrocarbons in zebrafish. *Sci. Total Environ.* **2021**, *775*. [CrossRef]
83. Shi, Y.-C.; Wang, A.-J.; Wu, X.-L.; Chen, J.-R.; Feng, J.-J. Green-assembly of three-dimensional porous graphene hydrogels for efficient removal of organic dyes. *J. Colloid Interface Sci.* **2016**, *484*, 254–262. [CrossRef]
84. Weng, D.; Song, L.; Li, W.; Yan, J.; Chen, L.; Liu, Y. Review on synthesis of three-dimensional graphene skeletons and their absorption performance for oily wastewater. *Environ. Sci. Pollut. Res.* **2021**, *28*, 16–34. [CrossRef]

85. Arabkhani, P.; Asfaram, A. Development of a novel three-dimensional magnetic polymer aerogel as an efficient adsorbent for malachite green removal. *J. Hazard. Mater.* **2020**, *384*. [CrossRef]
86. Kabiri, S.; Tran, D.N.H.; Cole, M.A.; Losic, D. Functionalized three-dimensional (3D) graphene composite for high efficiency removal of mercury. *Environ. Sci. Water Res. Technol.* **2016**, *2*, 390–402. [CrossRef]
87. Zhao, Q.; Zhu, X.; Chen, B. Stable graphene oxide/poly(ethyleneimine) 3D aerogel with tunable surface charge for high performance selective removal of ionic dyes from water. *Chem. Eng. J.* **2018**, *334*, 1119–1127. [CrossRef]
88. Rashidi Nodeh, H.; Sereshti, H.; Gaikani, H.; Kamboh, M.A.; Afsharsaveh, Z. Magnetic graphene coated inorganic-organic hybrid nanocomposite for enhanced preconcentration of selected pesticides in tomato and grape. *J. Chromatogr. A* **2017**, *1509*, 26–34. [CrossRef]
89. Geng, Z.; Lin, Y.; Yu, X.; Shen, Q.; Ma, L.; Li, Z.; Pan, N.; Wang, X. Highly efficient dye adsorption and removal: A functional hybrid of reduced graphene oxide-Fe_3O_4 nanoparticles as an easily regenerative adsorbent. *J. Mater. Chem.* **2012**, *22*, 3527–3535. [CrossRef]
90. Deng, J.-H.; Zhang, X.-R.; Zeng, G.-M.; Gong, J.-L.; Niu, Q.-Y.; Liang, J. Simultaneous removal of Cd(II) and ionic dyes from aqueous solution using magnetic graphene oxide nanocomposite as an adsorbent. *Chem. Eng. J.* **2013**, *226*, 189–200. [CrossRef]
91. Gul, K.; Sohni, S.; Waqar, M.; Ahmad, F.; Norulaini, N.A.N.; AK, M.O. Functionalization of magnetic chitosan with graphene oxide for removal of cationic and anionic dyes from aqueous solution. *Carbohydr. Polym.* **2016**, *152*, 520–531. [CrossRef]
92. Kharissova, O.V.; Dias, H.V.R.; Kharisov, B.I. Magnetic adsorbents based on micro- and nano-structured materials. *RSC Adv.* **2015**, *5*, 6695–6719. [CrossRef]
93. Sahraei, R.; Sekhavat Pour, Z.; Ghaemy, M. Novel magnetic bio-sorbent hydrogel beads based on modified gum tragacanth/graphene oxide: Removal of heavy metals and dyes from water. *J. Clean. Prod.* **2017**, *142*, 2973–2984. [CrossRef]
94. Li, J.; Ma, J.; Chen, S.; Huang, Y.; He, J. Adsorption of lysozyme by alginate/graphene oxide composite beads with enhanced stability and mechanical property. *Mater. Sci. Eng. C* **2018**, *89*, 25–32. [CrossRef]
95. Szymańska, E.; Winnicka, K. Stability of chitosan—A challenge for pharmaceutical and biomedical applications. *Mar. Drugs* **2015**, *13*, 1819–1846. [CrossRef]
96. Wu, Z.; Huang, W.; Shan, X.; Li, Z. Preparation of a porous graphene oxide/alkali lignin aerogel composite and its adsorption properties for methylene blue. *Int. J. Biol. Macromol.* **2020**, *143*, 325–333. [CrossRef]
97. Kyzasa, G.Z.; Deliyannib, E.A.; Bikiarisb, D.N.; Mitropoulos, A.C. Graphene composites as dye adsorbents: Review. *Chem. Eng. Res. Des.* **2017**, *129*, 75–88. [CrossRef]
98. Dotto, G.L.; Pinto, L.A.A. General considerations about chitosan. In *Frontiers in Biomaterials*; Bentham Science: Sharjah, United Arab Emirates, 2017; pp. 3–33.
99. Rodrigues, D.A.S.; Moura, J.M.; Dotto, G.L.; Cadaval, T.R.S.; Pinto, L.A.A. Preparation, characterization and dye adsorption/reuse of chitosan-vanadate films. *J. Polym. Environ.* **2018**, *26*, 2917–2924. [CrossRef]
100. Alves, D.C.S.; Coseglio, B.B.; Pinto, L.A.A.; Cadaval, T.R.S. Development of Spirulina/chitosan foam adsorbent for phenol adsorption. *J. Mol. Liq.* **2020**, *309*, 113256. [CrossRef]
101. Gerhardt, E.; Farias, B.S.; Moura, J.M.; De Almeida, L.S.; Adriano, R.; Dias, D.; Cadaval, T.R.S.; Pinto, L.A.A. Development of chitosan/Spirulina sp. blend films as biosorbents for Cr^{6+} and Pb^{2+} removal. *Int. J. Biol. Macromol.* **2020**, *155*, 142–152. [CrossRef] [PubMed]
102. Li, Z.; Wang, X.; Zhang, X.; Yang, Y.; Duan, J. A high-efficiency and plane-enhanced chitosan film for cefotaxime adsorption compared with chitosan particles in water. *Chem. Eng. J.* **2020**. [CrossRef]
103. Firozjaee, T.T.; Mehrdadi, N.; Baghdadi, M.; Nabi Bidhendi, G.R.N. The removal of diazinon from aqueous solution by chitosan/carbon nanotube adsorbent. *Desalin. Water Treat.* **2017**, *79*, 291–300. [CrossRef]
104. Liu, L.; Li, C.; Bao, C.; Jia, Q.; Xiao, P.; Liu, X.; Zhang, Q. Preparation and characterization of chitosan/graphene oxide composites for the adsorption of Au(III) and Pd(II). *Talanta* **2012**, *93*, 350–357. [CrossRef]
105. Jiménez-Gómez, C.P.; Cecilia, J.A. Chitosan: A Natural Biopolymer with a Wide and Varied Range of Applications. *Molecules* **2020**, *25*, 3981. [CrossRef]
106. Affonso, L.N.; Marques, J.L.; Lima, V.V.C.; Gonçalves, J.O.; Barbosa, S.C.; Primel, E.G.; Burgo, T.A.L.; Dotto, G.L.; Pinto, L.A.A.; Cadaval, T.R.S. Removal of fluoride from fertilizer industry effluent using carbon nanotubes stabilized in chitosan sponge. *J. Hazard. Mater.* **2020**, *388*. [CrossRef]
107. Gonçalves, J.O.; Santos, J.P.; Rios, E.C.; Crispim, M.M.; Dotto, G.L.; Pinto, L.A.A. Development of chitosan based hybrid hydrogels for dyes removal from aqueous binary system. *J. Mol. Liq.* **2017**, *225*, 265–270. [CrossRef]
108. Crini, G.; Badot, P.M. Application of chitosan, a natural aminopolysaccharide, for dye removal from aqueous solutions by adsorption processes using batch studies: A review of recent literature. *Prog. Polym. Sci.* **2008**, *33*, 399–447. [CrossRef]
109. Akpan, E.I.; Gbenebor, O.P.; Adeosun, S.O.; Cletus, O. Solubility, degree of acetylation, and distribution of acetyl groups in chitosan. In *Handbook of Chitin and Chitosan*; Elsevier: Amsterdam, The Netherlands, 2020; pp. 131–164. ISBN 9780128179703.
110. Zhou, H.Y.; Guang, X.; Kong, M.; Sheng, C.; Su, D.; Kennedy, J.F. Effect of molecular weight and degree of chitosan deacetylation on the preparation and characteristics of chitosan thermosensitive hydrogel as a delivery system. *Carbohydrate* **2008**, *73*, 265–273. [CrossRef]
111. Nunthanid, J.; Puttipipatkhachorn, S.; Yamamoto, K.; Peck, G.E. Physical properties and molecular behavior of chitosan films. *Drug Dev. Ind. Pharm.* **2001**, *27*, 143–157. [CrossRef]

112. Bof, M.J.; Bordadgaray, V.C.; Locaso, D.E.; García, M.A. Chitosan molecular weight effect on starch-composite film properties. *Food Hydrocoll.* **2015**, *51*, 281–294. [CrossRef]
113. Habiba, U.; Chin, T.; Siddique, T.A.; Salleh, A.; Chin, B.; Afifi, A.M. Effect of degree of deacetylation of chitosan on adsorption capacity and reusability of chitosan / polyvinyl alcohol / TiO_2 nano composite. *Int. J. Biol. Macromol.* **2017**, *104*, 1133–1142. [CrossRef]
114. Moura, J.M.; Farias, B.S.; Rodrigues, D.A.S. Preparation of chitosan with different characteristics and its application for bofilms production. *J. Polym. Environ.* **2015**, *23*, 470–477. [CrossRef]
115. Tavares, L.; Emanuel, E.; Flores, E.; Rodrigues, R.C.; Hertz, P.F. Effect of deacetylation degree of chitosan on rheological properties and physical chemical characteristics of genipin-crosslinked chitosan beads. *Food Hydrocoll.* **2020**, *106*. [CrossRef]
116. Gupta, K.C.; Jabrail, F.H. Effects of degree of deacetylation and cross-linking on physical characteristics, swelling and release behavior of chitosan microspheres. *Carbohydr. Polym.* **2006**, *66*, 43–54. [CrossRef]
117. Gonçalves, J.O.; Duarte, D.A.; Dotto, G.L.; Pinto, L.A.A. Use of chitosan with different deacetylation degrees for the adsorption of food dyes in a binary system. *Clean Soil Air Water* **2013**, *9*, 767–774. [CrossRef]
118. Piccin, J.S.; Vieira, M.L.G.; Gonçalves, J.O.; Dotto, G.L.; Pinto, L.A.A. Adsorption of FD & C Red No. 40 by chitosan: Isotherms analysis. *J. Food Eng.* **2009**, *95*, 16–20. [CrossRef]
119. Zhu, H.; Fu, Y.; Jiang, R.; Yao, J.; Liu, L.; Chen, Y.; Xiao, L.; Zeng, G. Preparation, characterization and adsorption properties of chitosan modified magnetic graphitized multi-walled carbon nanotubes for highly effective removal of a carcinogenic dye from aqueous solution. *Appl. Surf. Sci.* **2013**, *285*, 865–873. [CrossRef]
120. Iamsamai, C.; Hannongbua, S.; Ruktanonchai, U. The effect of the degree of deacetylation of chitosan on its dispersion of carbon nanotubes. *Carbon* **2010**, *48*, 25–30. [CrossRef]
121. Gonçalves, J.O.; Esquerdo, V.M.; Cadaval, T.R.S.; Pinto, L.A.A. Chitosan-based hydrogels. In *Sustainable Agriculture Reviews 36*; Springer: Aix-en-Provence, France, 2019; pp. 49–123. ISBN 9783540228608.
122. Pillai, C.K.S.; Paul, W.; Sharma, C.P. Chitin and chitosan polymers: Chemistry, solubility and fiber formation. *Prog. Polym. Sci.* **2009**, *34*, 641–678. [CrossRef]
123. Kubota, N.; Eguchi, Y. Facile preparation of water-soluble N-acetylated chitosan and molecular weight dependence of its water-solubility. *Polym. J.* **1997**, *29*, 123–127. [CrossRef]
124. Chang, S.H.; Lin, H.T.V.; Wu, G.J.; Tsai, G.J. pH Effects on solubility, zeta potential, and correlation between antibacterial activity and molecular weight of chitosan. *Carbohydr. Polym.* **2015**, *134*, 74–81. [CrossRef]
125. Rinaudo, M.; Pavlov, G.; Desbrières, J. Solubilization of chitosan in strong acid medium. *Int. J. Polym. Anal. Charact.* **1999**, *5*, 267–276. [CrossRef]
126. Lu, S.; Song, X.; Cao, D.; Chen, Y.; Yao, K. Preparation of water-soluble chitosan. *J. Appl. Polym. Sci.* **2004**, *91*, 3497–3503. [CrossRef]
127. Esquerdo, V.M.; Cadaval, T.R.S.; Dotto, G.L.; Pinto, L.A.A. Chitosan scaffold as an alternative adsorbent for the removal of hazardous food dyes from aqueous solutions. *J. Colloid Interface Sci.* **2014**, *424*, 7–15. [CrossRef]
128. Zhang, D.; Wang, L.; Zeng, H.; Yan, P.; Nie, J.; Sharma, V.K. A three-dimensional macroporous network structured chitosan / cellulose biocomposite sponge for rapid and selective removal of mercury (II) ions from aqueous solution. *Chem. Eng. J.* **2019**, *363*, 192–202. [CrossRef]
129. Sirajudheen, P.; Karthikeyan, P.; Ramkumar, K.; Meenakshi, S. Effective removal of organic pollutants by adsorption onto chitosan supported graphene oxide-hydroxyapatite composite: A novel reusable adsorbent. *J. Mol. Liq.* **2020**, *318*. [CrossRef]
130. Zhang, H.; Dang, Q.; Liu, C.; Cha, D.; Yu, Z.; Zhu, W.; Fan, B. Uptake of Pb(II) and Cd(II) on chitosan microsphere surface successively grafted by methyl acrylate and diethylenetriamine. *ACS Appl. Mater. Interfaces* **2017**, *9*, 11144–11155. [CrossRef]
131. Dragan, E.S.; Dinu, M.V. Advances in porous chitosan-based composite hydrogels: Synthesis and applications Ecaterina. *React. Funct. Polym.* **2019**, *146*, 104372. [CrossRef]
132. Pakdel, P.M.; Peighambardoust, S.J. Review on recent progress in chitosan-based hydrogels for wastewater treatment application. *Carbohydr. Polym.* **2018**, *201*, 264–279. [CrossRef]
133. Saheed, O.I.; Oh, D.W.; Suah, M.B.F. Chitosan modifications for adsorption of pollutants—A review. *J. Hazard. Mater.* **2021**, *408*, 124889. [CrossRef]
134. Lai, K.C.; Lee, L.Y.; Hiew, B.Y.Z.; Yang, T.C.-K.; Pan, G.-T.; Thangalazhy-Gopakumar, S.; Gan, S. Utilisation of eco-friendly and low cost 3D graphene-based composite for treatment of aqueous Reactive Black 5 dye: Characterisation, adsorption mechanism and recyclability studies. *J. Taiwan Inst. Chem. Eng.* **2020**, *114*, 57–66. [CrossRef]
135. Yang, A.; Yang, P.; Huang, C.P. Preparation of graphene oxide–chitosan composite and adsorption performance for uranium. *J. Radioanal. Nucl. Chem.* **2017**, *313*, 371–378. [CrossRef]
136. Kamal, S.; Khan, F.; Kausar, H.; Khan, M.S.; Ahmad, A.; Ishraque Ahmad, S.; Asim, M.; Alshitari, W.; Nami, S.A.A. Synthesis, characterization, morphology, and adsorption studies of ternary nanocomposite comprising graphene oxide, chitosan, and polypyrrole. *Polym. Compos.* **2020**, *41*, 3758–3767. [CrossRef]
137. Mohseni Kafshgari, M.; Tahermansouri, H. Development of a graphene oxide/chitosan nanocomposite for the removal of picric acid from aqueous solutions: Study of sorption parameters. *Colloids Surf. B Biointerfaces* **2017**, *160*, 671–681. [CrossRef] [PubMed]
138. Rinaudo, M. Chitin and chitosan: Properties and applications. *Prog. Polym. Sci.* **2006**, *31*, 603–632. [CrossRef]

139. Chen, Y.; Chen, L.; Bai, H.; Li, L. Graphene oxide-chitosan composite hydrogels as broad-spectrum adsorbents for water purification. *J. Mater. Chem. A* **2013**, *1*, 1992–2001. [CrossRef]
140. Wu, K.; Liu, X.; Li, Z.; Jiao, Y.; Zhou, C. Fabrication of chitosan/graphene oxide composite aerogel microspheres with high bilirubin removal performance. *Mater. Sci. Eng. C* **2020**, *106*. [CrossRef]
141. Luo, J.; Fan, C.; Xiao, Z.; Sun, T.; Zhou, X. Novel graphene oxide/carboxymethyl chitosan aerogels via vacuum-assisted self-assembly for heavy metal adsorption capacity. *Colloids Surf. A Physicochem. Eng. Asp.* **2019**, *578*. [CrossRef]
142. Zhang, M.; Ma, G.; Zhang, L.; Chen, H.; Zhu, L.; Wang, C.; Liu, X. Chitosan-reduced graphene oxide composites with 3D structures as effective reverse dispersed solid phase extraction adsorbents for pesticides analysis. *Analyst* **2019**, *144*, 5164–5171. [CrossRef]
143. Yang, X.; Tu, Y.; Li, L.; Shang, S.; Tao, X.-M. Well-dispersed chitosan/graphene oxide nanocomposites. *ACS Appl. Mater. Interfaces* **2010**, *2*, 1707–1713. [CrossRef]
144. Zhang, Y.; Zhang, M.; Jiang, H.; Shi, J.; Li, F.; Xia, Y.; Zhang, G.; Li, H. Bio-inspired layered chitosan/graphene oxide nanocomposite hydrogels with high strength and pH-driven shape memory effect. *Carbohydr. Polym.* **2017**, *177*, 116–125. [CrossRef]
145. Wang, Y.; Xia, G.; Wu, C.; Sun, J.; Song, R.; Huang, W. Porous chitosan doped with graphene oxide as highly effective adsorbent for methyl orange and amido black 10B. *Carbohydr. Polym.* **2015**, *115*, 686–693. [CrossRef]
146. Yan, M.; Huang, W.; Li, Z. Chitosan cross-linked graphene oxide/lignosulfonate composite aerogel for enhanced adsorption of methylene blue in water. *Int. J. Biol. Macromol.* **2019**, *136*, 927–935. [CrossRef]
147. Han Lyn, F.; Tan, C.P.; Zawawi, R.M.; Nur Hanani, Z.A. Enhancing the mechanical and barrier properties of chitosan/graphene oxide composite films using trisodium citrate and sodium tripolyphosphate crosslinkers. *J. Appl. Polym. Sci.* **2021**, *138*. [CrossRef]
148. Wang, R.; Zhang, X.; Zhu, J.; Bai, J.; Gao, L.; Liu, S.; Jiao, T. Facile preparation of self-assembled chitosan-based composite hydrogels with enhanced adsorption performances. *Colloids Surf. A Physicochem. Eng. Asp.* **2020**, *598*. [CrossRef]
149. Vlasceanu, G.M.; Crica, L.E.; Pandele, A.M.; Ionita, M. Graphene oxide reinforcing genipin crosslinked chitosan-gelatin blend films. *Coatings* **2020**, *10*, 189. [CrossRef]
150. Liu, Y.; Liu, R.; Li, M.; Yu, F.; He, C. Removal of pharmaceuticals by novel magnetic genipin-crosslinked chitosan/graphene oxide-SO$_3$H composite. *Carbohydr. Polym.* **2019**, *220*, 141–148. [CrossRef]
151. Yan, N.; Capezzuto, F.; Lavorgna, M.; Buonocore, G.G.; Tescione, F.; Xia, H.; Ambrosio, L. Borate cross-linked graphene oxide-chitosan as robust and high gas barrier films. *Nanoscale* **2016**, *8*, 10783–10791. [CrossRef]
152. Ruan, J.; Wang, X.; Yu, Z.; Wang, X.; Xie, Q.; Zhang, D.; Huang, Y.; Zhou, H.; Bi, X.; Xiao, C.; et al. Enhanced Physiochemical and Mechanical Performance of Chitosan-Grafted Graphene Oxide for Superior Osteoinductivity. *Adv. Funct. Mater.* **2016**, *26*, 1085–1097. [CrossRef]
153. Salzano de Luna, M.; Ascione, C.; Santillo, C.; Verdolotti, L.; Lavorgna, M.; Buonocore, G.G.; Castaldo, R.; Filippone, G.; Xia, H.; Ambrosio, L. Optimization of dye adsorption capacity and mechanical strength of chitosan aerogels through crosslinking strategy and graphene oxide addition. *Carbohydr. Polym.* **2019**, *211*, 195–203. [CrossRef]
154. Sharma, P.; Singh, A.K.; Shahi, V.K. Selective Adsorption of Pb(II) from aqueous medium by cross-linked chitosan-functionalized graphene oxide adsorbent. *ACS Sustain. Chem. Eng.* **2019**, *7*, 1427–1436. [CrossRef]
155. Kong, D.; He, L.; Li, H.; Zhang, F.; Song, Z. Preparation and characterization of graphene oxide/chitosan composite aerogel with high adsorption performance for Cr(VI) by a new crosslinking route. *Colloids Surf. A Physicochem. Eng. Asp.* **2021**, *625*. [CrossRef]
156. Sreeprasad, T.S.; Maliyekkal, S.M.; Lisha, K.P.; Pradeep, T. Reduced graphene oxide-metal/metal oxide composites: Facile synthesis and application in water purification. *J. Hazard. Mater.* **2011**, *186*, 921–931. [CrossRef]
157. Yang, K.; Hu, L.; Ma, X.; Ye, S.; Cheng, L.; Shi, X.; Li, C.; Li, Y.; Liu, Z. Multimodal imaging guided photothermal therapy using functionalized graphene nanosheets anchored with magnetic nanoparticles. *Adv. Mater.* **2012**, *24*, 1868–1872. [CrossRef]
158. Rebekah, A.; Bharath, G.; Naushad, M.; Viswanathan, C.; Ponpandian, N. Magnetic graphene/chitosan nanocomposite: A promising nano-adsorbent for the removal of 2-naphthol from aqueous solution and their kinetic studies. *Int. J. Biol. Macromol.* **2020**, *159*, 530–538. [CrossRef]
159. Singh, N.; Riyajuddin, S.; Ghosh, K.; Mehta, S.K.; Dan, A. Chitosan-graphene oxide hydrogels with embedded magnetic iron oxide nanoparticles for dye removal. *ACS Appl. Nano Mater.* **2019**, *2*, 7379–7392. [CrossRef]
160. Xu, L.; Suo, H.; Wang, J.; Cheng, F.; Liu, X.; Qiu, H. Magnetic graphene oxide decorated with chitosan and Au nanoparticles: Synthesis, characterization and application for detection of trace rhodamine B. *Anal. Methods* **2019**, *11*, 3837–3843. [CrossRef]
161. Foroughi, M.; Azqhandi, M.H.A. A biological-based adsorbent for a non-biodegradable pollutant: Modeling and optimization of Pb (II) remediation using GO-CS-Fe$_3$O$_4$-EDTA nanocomposite. *J. Mol. Liq.* **2020**, *318*. [CrossRef]
162. Sarkar, A.K.; Bediako, J.K.; Choi, J.-W.; Yun, Y.-S. Functionalized magnetic biopolymeric graphene oxide with outstanding performance in water purification. *NPG Asia Mater.* **2019**, *11*. [CrossRef]
163. Anush, S.M.; Chandan, H.R.; Gayathri, B.H.; Asma; Manju, N.; Vishalakshi, B.; Kalluraya, B. Graphene oxide functionalized chitosan-magnetite nanocomposite for removal of Cu(II) and Cr(VI) from waste water. *Int. J. Biol. Macromol.* **2020**, *164*, 4391–4402. [CrossRef] [PubMed]
164. Tran, H.V.; Hoang, L.T.; Huynh, C.D. An investigation on kinetic and thermodynamic parameters of methylene blue adsorption onto graphene-based nanocomposite. *Chem. Phys.* **2020**, *535*. [CrossRef]
165. Tasmia; Shah, J.; Jan, M.R. Microextraction of selected endocrine disrupting phenolic compounds using magnetic chitosan biopolymer graphene oxide nanocomposite. *J. Polym. Environ.* **2020**, *28*, 1673–1683. [CrossRef]

166. Shafaati, M.; Miralinaghi, M.; Shirazi, R.H.S.M.; Moniri, E. The use of chitosan/Fe$_3$O$_4$ grafted graphene oxide for effective adsorption of rifampicin from water samples. *Res. Chem. Intermed.* **2020**. [CrossRef]
167. Jiang, X.; Pan, W.; Chen, M.; Yuan, Y.; Zhao, L. The fabrication of a thiol-modified chitosan magnetic graphene oxide nanocomposite and its adsorption performance towards the illegal drug clenbuterol in pork samples. *Dalt. Trans.* **2020**, *49*, 6097–6107. [CrossRef]
168. Tang, T.; Cao, S.; Xi, C.; Li, X.; Zhang, L.; Wang, G.; Chen, Z. Chitosan functionalized magnetic graphene oxide nanocomposite for the sensitive and effective determination of alkaloids in hotpot. *Int. J. Biol. Macromol.* **2020**, *146*, 343–352. [CrossRef]
169. Wu, Z.; Deng, W.; Zhou, W.; Luo, J. Novel magnetic polysaccharide/graphene oxide @Fe$_3$O$_4$ gel beads for adsorbing heavy metal ions. *Carbohydr. Polym.* **2019**, *216*, 119–128. [CrossRef]
170. Le, T.T.N.; Le, V.T.; Dao, M.U.; Nguyen, Q.V.; Vu, T.T.; Nguyen, M.H.; Tran, D.L.; Le, H.S. Preparation of magnetic graphene oxide/chitosan composite beads for effective removal of heavy metals and dyes from aqueous solutions. *Chem. Eng. Commun.* **2019**, *206*, 1337–1352. [CrossRef]
171. Jiang, X.; Pan, W.; Xiong, Z.; Zhang, Y.; Zhao, L. Facile synthesis of layer-by-layer decorated graphene oxide based magnetic nanocomposites for β-agonists/dyes adsorption removal and bacterial inactivation in wastewater. *J. Alloys Compd.* **2021**, *870*. [CrossRef]
172. Rebekah, A.; Navadeepthy, D.; Bharath, G.; Viswanathan, C.; Ponpandian, N. Removal of 1-napthylamine using magnetic graphene and magnetic graphene oxide functionalized with Chitosan. *Environ. Nanotechnol. Monit. Manag.* **2021**, *15*. [CrossRef]
173. Fan, L.; Luo, C.; Li, X.; Lu, F.; Qiu, H.; Sun, M. Fabrication of novel magnetic chitosan grafted with graphene oxide to enhance adsorption properties for methyl blue. *J. Hazard. Mater.* **2012**, *215–216*, 272–279. [CrossRef]
174. Debnath, S.; Maity, A.; Pillay, K. Magnetic chitosan-GO nanocomposite: Synthesis, characterization and batch adsorber design for Cr(VI) removal. *J. Environ. Chem. Eng.* **2014**, *2*, 963–973. [CrossRef]
175. Samuel, M.S.; Shah, S.S.; Bhattacharya, J.; Subramaniam, K.; Pradeep Singh, N.D. Adsorption of Pb(II) from aqueous solution using a magnetic chitosan/graphene oxide composite and its toxicity studies. *Int. J. Biol. Macromol.* **2018**, *115*, 1142–1150. [CrossRef]
176. Wang, F.; Yang, B.; Wang, H.; Song, Q.; Tan, F.; Cao, Y. Removal of ciprofloxacin from aqueous solution by a magnetic chitosan grafted graphene oxide composite. *J. Mol. Liq.* **2016**, *222*, 188–194. [CrossRef]
177. Fan, L.; Luo, C.; Sun, M.; Qiu, H.; Li, X. Synthesis of magnetic β-cyclodextrin-chitosan/graphene oxide as nanoadsorbent and its application in dye adsorption and removal. *Colloids Surf. B Biointerfaces* **2013**, *103*, 601–607. [CrossRef] [PubMed]
178. Sherlala, A.I.A.; Raman, A.A.A.; Bello, M.M.; Buthiyappan, A. Adsorption of arsenic using chitosan magnetic graphene oxide nanocomposite. *J. Environ. Manage.* **2019**, *246*, 547–556. [CrossRef] [PubMed]
179. Shahzad, A.; Miran, W.; Rasool, K.; Nawaz, M.; Jang, J.; Lim, S.-R.; Lee, D.S. Heavy metals removal by EDTA-functionalized chitosan graphene oxide nanocomposites. *RSC Adv.* **2017**, *7*, 9764–9771. [CrossRef]
180. Zhang, L.; Luo, H.; Liu, P.; Fang, W.; Geng, J. A novel modified graphene oxide/chitosan composite used as an adsorbent for Cr(VI) in aqueous solutions. *Int. J. Biol. Macromol.* **2016**, *87*, 586–596. [CrossRef]
181. Li, L.; Fan, L.; Sun, M.; Qiu, H.; Li, X.; Duan, H.; Luo, C. Adsorbent for chromium removal based on graphene oxide functionalized with magnetic cyclodextrin-chitosan. *Colloids Surf. B Biointerfaces* **2013**, *107*, 76–83. [CrossRef]
182. Li, L.; Fan, L.; Sun, M.; Qiu, H.; Li, X.; Duan, H.; Luo, C. Adsorbent for hydroquinone removal based on graphene oxide functionalized with magnetic cyclodextrin-chitosan. *Int. J. Biol. Macromol.* **2013**, *58*, 169–175. [CrossRef]
183. Wang, Y.; Li, L.; Luo, C.; Wang, X.; Duan, H. Removal of Pb^{2+} from water environment using a novel magnetic chitosan/graphene oxide imprinted Pb^{2+}. *Int. J. Biol. Macromol.* **2016**, *86*, 505–511. [CrossRef]
184. Kong, D.; Wang, N.; Qiao, N.; Wang, Q.; Wang, Z.; Zhou, Z.; Ren, Z. Facile preparation of ion-imprinted chitosan microspheres enwrapping Fe$_3$O$_4$ and graphene oxide by inverse suspension cross-linking for highly selective removal of copper(II). *ACS Sustain. Chem. Eng.* **2017**, *5*, 7401–7409. [CrossRef]
185. Jamali, M.; Akbari, A. Facile fabrication of magnetic chitosan hydrogel beads and modified by interfacial polymerization method and study of adsorption of cationic/anionic dyes from aqueous solution. *J. Environ. Chem. Eng.* **2021**, *9*. [CrossRef]
186. Subedi, N.; Lähde, A.; Abu-Danso, E.; Iqbal, J.; Bhatnagar, A. A comparative study of magnetic chitosan (Chi@Fe$_3$O$_4$) and graphene oxide modified magnetic chitosan (Chi@Fe$_3$O$_4$GO) nanocomposites for efficient removal of Cr(VI) from water. *Int. J. Biol. Macromol.* **2019**, *137*, 948–959. [CrossRef]
187. Wu, X.-L.; Xiao, P.; Zhong, S.; Fang, K.; Lin, H.; Chen, J. Magnetic ZnFe$_2$O$_4$@chitosan encapsulated in graphene oxide for adsorptive removal of organic dye. *RSC Adv.* **2017**, *7*, 28145–28151. [CrossRef]
188. Zeng, W.; Liu, Y.-G.; Hu, X.-J.; Liu, S.-B.; Zeng, G.-M.; Zheng, B.-H.; Jiang, L.-H.; Guo, F.-Y.; Ding, Y.; Xu, Y. Decontamination of methylene blue from aqueous solution by magnetic chitosan lignosulfonate grafted with graphene oxide: Effects of environmental conditions and surfactant. *RSC Adv.* **2016**, *6*, 19298–19307. [CrossRef]
189. Hosseinzadeh, H.; Ramin, S. Effective removal of copper from aqueous solutions by modified magnetic chitosan/graphene oxide nanocomposites. *Int. J. Biol. Macromol.* **2018**, *113*, 859–868. [CrossRef]
190. Asadabadi, S.; Merati, Z. A tailored magnetic composite synthesized by graphene oxide, chitosan and aminopolycarboxylic acid for diminishing dye contaminant. *Cellulose* **2021**, *28*, 2327–2351. [CrossRef]
191. Tang, T.; Cao, S.; Xi, C.; Chen, Z. Multifunctional magnetic chitosan-graphene oxide-ionic liquid ternary nanohybrid: An efficient adsorbent of alkaloids. *Carbohydr. Polym.* **2021**, *255*. [CrossRef]

192. Taher, F.A.; Kamal, F.H.; Badawy, N.A.; Shrshr, A.E. Hierarchical magnetic/chitosan/graphene oxide 3D nanostructure as highly effective adsorbent. *Mater. Res. Bull.* **2018**, *97*, 361–368. [CrossRef]
193. Gong, Y.; Su, J.; Li, M.; Zhu, A.; Liu, G.; Liu, P. Fabrication and adsorption optimization of novel magnetic core-shell chitosan/graphene oxide/β-cyclodextrin composite materials for bisphenols in aqueous solutions. *Materials* **2020**, *13*, 5408. [CrossRef]
194. Fan, L.; Luo, C.; Sun, M.; Qiu, H. Synthesis of graphene oxide decorated with magnetic cyclodextrin for fast chromium removal. *J. Mater. Chem.* **2012**, *22*, 24577–24583. [CrossRef]
195. Wang, D.; Liu, L.; Jiang, X.; Yu, J.; Chen, X. Adsorbent for p-phenylenediamine adsorption and removal based on graphene oxide functionalized with magnetic cyclodextrin. *Appl. Surf. Sci.* **2015**, *329*, 197–205. [CrossRef]
196. Sun, Z.; Fang, S.; Hu, Y.H. 3D graphene materials: From understanding to design and synthesis control. *Chem. Rev.* **2020**, *120*, 10336–10453. [CrossRef] [PubMed]
197. Ma, Y.; Chen, Y. Three-dimensional graphene networks: Synthesis, properties and applications. *Natl. Sci. Rev.* **2015**, *2*, 40–53. [CrossRef]
198. Liu, Y.; Huang, S.; Zhao, X.; Zhang, Y. Fabrication of three-dimensional porous β-cyclodextrin/chitosan functionalized graphene oxide hydrogel for methylene blue removal from aqueous solution. *Colloids Surf. A Physicochem. Eng. Asp.* **2018**, *539*, 1–10. [CrossRef]
199. Yu, P.; Wang, H.-Q.; Bao, R.-Y.; Liu, Z.; Yang, W.; Xie, B.-H.; Yang, M.-B. Self-assembled sponge-like chitosan/reduced graphene oxide/montmorillonite composite hydrogels without cross-linking of chitosan for effective Cr(VI) sorption. *ACS Sustain. Chem. Eng.* **2017**, *5*, 1557–1566. [CrossRef]
200. Nasiri, R.; Arsalani, N.; Panahian, Y. One-pot synthesis of novel magnetic three-dimensional graphene/chitosan/nickel ferrite nanocomposite for lead ions removal from aqueous solution: RSM modelling design. *J. Clean. Prod.* **2018**, *201*, 507–515. [CrossRef]
201. Zhou, G.; Wang, K.P.; Liu, H.W.; Wang, L.; Xiao, X.F.; Dou, D.D.; Fan, Y.B. Three-dimensional polylactic acid@graphene oxide/chitosan sponge bionic filter: Highly efficient adsorption of crystal violet dye. *Int. J. Biol. Macromol.* **2018**, *113*, 792–803. [CrossRef]
202. Yu, R.; Shi, Y.; Yang, D.; Liu, Y.; Qu, J.; Yu, Z.-Z. Graphene oxide/chitosan aerogel microspheres with honeycomb-cobweb and radially oriented microchannel structures for broad-spectrum and rapid adsorption of water contaminants. *ACS Appl. Mater. Interfaces* **2017**, *9*, 21809–21819. [CrossRef]
203. Kovtun, A.; Campodoni, E.; Favaretto, L.; Zambianchi, M.; Salatino, A.; Amalfitano, S.; Navacchia, M.L.; Casentini, B.; Palermo, V.; Sandri, M.; et al. Multifunctional graphene oxide/biopolymer composite aerogels for microcontaminants removal from drinking water. *Chemosphere* **2020**, *259*. [CrossRef]
204. Zhang, D.; Li, N.; Cao, S.; Liu, X.; Qiao, M.; Zhang, P.; Zhao, Q.; Song, L.; Huang, X. A layered chitosan/grapheneoxide sponge as reusable adsorbent for removal of heavy metal ions. *Chem. Res. Chin. Univ.* **2019**, *35*, 463–470. [CrossRef]
205. Yu, B.; Xu, J.; Liu, J.-H.; Yang, S.-T.; Luo, J.; Zhou, Q.; Wan, J.; Liao, R.; Wang, H.; Liu, Y. Adsorption behavior of copper ions on graphene oxide-chitosan aerogel. *J. Environ. Chem. Eng.* **2013**, *1*, 1044–1050. [CrossRef]
206. Mei, J.; Zhang, H.; Mo, S.; Zhang, Y.; Li, Z.; Ou, H. Prominent adsorption of Cr(VI) with graphene oxide aerogel twined with creeper-like polymer based on chitosan oligosaccharide. *Carbohydr. Polym.* **2020**, *247*. [CrossRef]
207. Zhao, L.; Dong, P.; Xie, J.; Li, J.; Wu, L.; Yang, S.-T.; Luo, J. Porous graphene oxide-chitosan aerogel for tetracycline removal. *Mater. Res. Express* **2014**, *1*. [CrossRef]
208. Javadi, E.; Baghdadi, M.; Taghavi, L.; Ahmad Panahi, H. Removal of 4-nonylphenol from surface water and municipal wastewater effluent using three-dimensional graphene oxide–chitosan aerogel beads. *Int. J. Environ. Res.* **2020**, *14*, 513–526. [CrossRef]
209. Zhang, Y.; Chen, S.; Feng, X.; Yu, J.; Jiang, X. Self-assembly of sponge-like kaolin/chitosan/reduced graphene oxide composite hydrogels for adsorption of Cr(VI) and AYR. *Environ. Sci. Pollut. Res.* **2019**, *26*, 28898–28908. [CrossRef]
210. Bernsmann, F.; Ball, V.; Addiego, F.; Ponche, A.; Michel, M.; Gracio, J.J.D.A.; Toniazzo, V.; Ruch, D. Dopamine-melanin film deposition depends on the used oxidant and buffer solution. *Langmuir* **2011**, *27*, 2819–2825. [CrossRef]
211. Li, L.; Wei, Z.; Liu, X.; Yang, Y.; Deng, C.; Yu, Z.; Guo, Z.; Shi, J.; Zhu, C.; Guo, W.; et al. Biomaterials cross-linked graphene oxide composite aerogel with a macro–nanoporous network structure for efficient Cr (VI) removal. *Int. J. Biol. Macromol.* **2020**, *156*, 1337–1346. [CrossRef]
212. Cao, N.; Lyu, Q.; Li, J.; Wang, Y.; Yang, B.; Szunerits, S.; Boukherroub, R. Facile synthesis of fluorinated polydopamine/chitosan/reduced graphene oxide composite aerogel for efficient oil/water separation. *Chem. Eng. J.* **2017**, *326*, 17–28. [CrossRef]
213. Nekouei Marnani, N.; Shahbazi, A. A novel environmental-friendly nanobiocomposite synthesis by EDTA and chitosan functionalized magnetic graphene oxide for high removal of Rhodamine B: Adsorption mechanism and separation property. *Chemosphere* **2019**, *218*, 715–725. [CrossRef]
214. Croitoru, A.-M.; Ficai, A.; Ficai, D.; Trusca, R.; Dolete, G.; Andronescu, E.; Turculet, S.C. Chitosan/graphene oxide nanocomposite membranes as adsorbents with applications in water purification. *Materials* **2020**, *13*, 1687. [CrossRef]
215. Li, T.; Liu, X.; Li, L.; Wang, Y.; Ma, P.; Chen, M.; Dong, W. Polydopamine-functionalized graphene oxide compounded with polyvinyl alcohol/chitosan hydrogels on the recyclable adsorption of cu(II), Pb(II) and cd(II) from aqueous solution. *J. Polym. Res.* **2019**, *26*. [CrossRef]
216. Liao, Y.; Wang, M.; Chen, D. Preparation of polydopamine-modified graphene oxide/chitosan aerogel for uranium(VI) adsorption. *Ind. Eng. Chem. Res.* **2018**, *57*, 8472–8483. [CrossRef]

217. Zhang, Y.; Bian, T.; Jiang, R.; Zhang, Y.; Zheng, X.; Li, Z. Bionic chitosan-carbon imprinted aerogel for high selective recovery of Gd(III) from end-of-life rare earth productions. *J. Hazard. Mater.* **2021**, *407*. [CrossRef] [PubMed]
218. Luo, J.; Fan, C.; Zhou, X. Functionalized graphene oxide/carboxymethyl chitosan composite aerogels with strong compressive strength for water purification. *J. Appl. Polym. Sci.* **2021**, *138*. [CrossRef]
219. Samuel, M.S.; Subramaniyan, V.; Bhattacharya, J.; Parthiban, C.; Chand, S.; Singh, N.D.P. A GO-CS@MOF [Zn(BDC)(DMF)] material for the adsorption of chromium(VI) ions from aqueous solution. *Compos. Part B Eng.* **2018**, *152*, 116–125. [CrossRef]
220. Samuel, M.S.; Suman, S.; Venkateshkannan; Selvarajan, E.; Mathimani, T.; Pugazhendhi, A. Immobilization of $Cu_3(btc)_2$ on graphene oxide-chitosan hybrid composite for the adsorption and photocatalytic degradation of methylene blue. *J. Photochem. Photobiol. B Biol.* **2020**, *204*. [CrossRef]
221. Chang, R.; Ma, S.; Guo, X.; Xu, J.; Zhong, C.; Huang, R.; Ma, J. Hierarchically assembled graphene oxide composite membrane with self-healing and high-efficiency water purification performance. *ACS Appl. Mater. Interfaces* **2019**, *11*, 46251–46260. [CrossRef]
222. Ma, L.; Wang, Q.; Islam, S.M.; Liu, Y.; Ma, S.; Kanatzidis, M.G. Highly selective and efficient removal of heavy metals by layered double hydroxide intercalated with the MoS_4^{2-} Ion. *J. Am. Chem. Soc.* **2016**, *138*, 2858–2866. [CrossRef]
223. Priya, V.N.; Rajkumar, M.; Magesh, G.; Mobika, J.; Sibi, S.P.L. Chitosan assisted Fe-Al double layered hydroxide/reduced graphene oxide composites for As(V) removal. *Mater. Chem. Phys.* **2020**, *251*. [CrossRef]
224. Sanmugam, A.; Vikraman, D.; Park, H.J.; Kim, H.-S. One-pot facile methodology to synthesize chitosan-ZnO-graphene oxide hybrid composites for better dye adsorption and antibacterial activity. *Nanomaterials* **2017**, *7*, 363. [CrossRef]
225. Guo, X.; Yang, H.; Liu, Q.; Liu, J.; Chen, R.; Zhang, H.; Yu, J.; Zhang, M.; Li, R.; Wang, J. A chitosan-graphene oxide/ZIF foam with anti-biofouling ability for uranium recovery from seawater. *Chem. Eng. J.* **2020**, *382*. [CrossRef]
226. Das, L.; Das, P.; Bhowal, A.; Bhattachariee, C. Synthesis of hybrid hydrogel nano-polymer composite using graphene oxide, chitosan and PVA and its application in waste water treatment. *Environ. Technol. Innov.* **2020**, *18*, 100664. [CrossRef]
227. Tang, H.; Liu, Y.; Li, B.; Zhu, L.; Tang, Y. Preparation of chitosan graft polyacrylic acid/graphite oxide composite and the study of its adsorption properties of cationic dyes. *Polym. Sci. Ser. A* **2020**, *62*, 272–283. [CrossRef]
228. Zhao, L.; Yang, S.; Yilihamu, A.; Ma, Q.; Shi, M.; Ouyang, B.; Zhang, Q.; Guan, X.; Yang, S.-T. Adsorptive decontamination of Cu^{2+}-contaminated water and soil by carboxylated graphene oxide/chitosan/cellulose composite beads. *Environ. Res.* **2019**, *179*. [CrossRef]
229. Mittal, H.; Al Alili, A.; Morajkar, P.P.; Alhassan, S.M. GO crosslinked hydrogel nanocomposites of chitosan/carboxymethyl cellulose—A versatile adsorbent for the treatment of dyes contaminated wastewater. *Int. J. Biol. Macromol.* **2021**, *167*, 1248–1261. [CrossRef]
230. Li, C.; Yan, Y.; Zhang, Q.; Zhang, Z.; Huang, L.; Zhang, J.; Xiong, Y.; Tan, S. Adsorption of Cd^{2+} and Ni^{2+} from aqueous single-metal solutions on gaphene oxide-chitosan-poly(vinyl alcohol) hydrogels. *Langmuir* **2019**, *35*, 4481–4490. [CrossRef]
231. Huo, J.; Yu, G.; Wang, J. Adsorptive removal of Sr(II) from aqueous solution by polyvinyl alcohol/graphene oxide aerogel. *Chemosphere* **2021**, *278*, 130492. [CrossRef]
232. Li, L.; Wang, Z.; Ma, P.; Bai, H.; Dong, W.; Chen, M. Preparation of polyvinyl alcohol/chitosan hydrogel compounded with graphene oxide to enhance the adsorption properties for Cu(II) in aqueous solution. *J. Polym. Res.* **2015**, *22*. [CrossRef]
233. Liu, Y.; Nie, P.; Yu, F. Enhanced adsorption of sulfonamides by a novel carboxymethyl cellulose and chitosan-based composite with sulfonated graphene oxide. *Bioresour. Technol.* **2021**, *320*, 124373. [CrossRef]
234. Salahuddin, N.; EL-Daly, H.; El Sharkawy, R.G.; Nasr, B.T. Synthesis and efficacy of PPy/CS/GO nanocomposites for adsorption of ponceau 4R dye. *Polymer* **2018**, *146*, 291–303. [CrossRef]
235. Klongklaew, P.; Naksena, T.; Kanatharana, P.; Bunkoed, O. A hierarchically porous composite monolith polypyrrole/octadecyl silica/graphene oxide/chitosan cryogel sorbent for the extraction and pre-concentration of carbamate pesticides in fruit juices. *Anal. Bioanal. Chem.* **2018**, *410*, 7185–7193. [CrossRef]
236. Salahuddin, N.A.; EL-Daly, H.A.; El Sharkawy, R.G.; Nasr, B.T. Nano-hybrid based on polypyrrole/chitosan/grapheneoxide magnetite decoration for dual function in water remediation and its application to form fashionable colored product. *Adv. Powder Technol.* **2020**, *31*, 1587–1596. [CrossRef]
237. Lai, K.C.; Hiew, B.Y.Z.; Lee, L.Y.; Gan, S.; Thangalazhy-Gopakumar, S.; Chiu, W.S.; Khiew, P.S. Ice-templated graphene oxide/chitosan aerogel as an effective adsorbent for sequestration of metanil yellow dye. *Bioresour. Technol.* **2019**, *274*, 134–144. [CrossRef]
238. Li, X.; Zhou, H.; Wu, W.; Wei, S.; Xu, Y.; Kuang, Y. Studies of heavy metal ion adsorption on chitosan/sulfydryl-functionalized graphene oxide composites. *J. Colloid Interface Sci.* **2015**, *448*, 389–397. [CrossRef]
239. Samuel, M.S.; Shah, S.S.; Subramaniyan, V.; Qureshi, T.; Bhattacharya, J.; Pradeep Singh, N.D. Preparation of graphene oxide/chitosan/ferrite nanocomposite for chromium(VI) removal from aqueous solution. *Int. J. Biol. Macromol.* **2018**, *119*, 540–547. [CrossRef]
240. Ouyang, A.; Wang, C.; Wu, S.; Shi, E.; Zhao, W.; Cao, A.; Wu, D. Highly porous core-shell structured graphene-chitosan beads. *ACS Appl. Mater. Interfaces* **2015**, *7*, 14439–14445. [CrossRef]
241. Wang, Y.; Liu, X.; Wang, H.; Xia, G.; Huang, W.; Song, R. Microporous spongy chitosan monoliths doped with graphene oxide as highly effective adsorbent for methyl orange and copper nitrate ($Cu(NO_3)_2$) ions. *J. Colloid Interface Sci.* **2014**, *416*, 243–251. [CrossRef]

242. Zhang, C.; Chen, Z.; Guo, W.; Zhu, C.; Zou, Y. Simple fabrication of chitosan/graphene nanoplates composite spheres for efficient adsorption of acid dyes from aqueous solution. *Int. J. Biol. Macromol.* **2018**, *112*, 1048–1054. [CrossRef]
243. Sharififard, H.; Shahraki, Z.H.; Rezvanpanah, E.; Rad, S.H. A novel natural chitosan/activated carbon/iron bio-nanocomposite: Sonochemical synthesis, characterization, and application for cadmium removal in batch and continuous adsorption process. *Bioresour. Technol.* **2018**, *270*, 562–569. [CrossRef]
244. Zhao, R.; Li, Y.; Ji, J.; Wang, Q.; Li, G.; Wu, T.; Zhang, B. Efficient removal of phenol and p-nitrophenol using nitrogen-doped reduced graphene oxide. *Colloids Surf. A Physicochem. Eng. Asp.* **2021**, *611*. [CrossRef]

Article

A Strategy to Synthesize Multilayer Graphene in Arc-Discharge Plasma in a Semi-Opened Environment

Hai Tan [1], Deguo Wang [1,2] and Yanbao Guo [1,2,*]

1 College of Mechanical and Transportation Engineering, China University of Petroleum, Beijing 102249, China
2 Beijing Key Laboratory of Process Fluid Filtration and Separation, Beijing 102249, China
* Correspondence: gyb@cup.edu.cn

Received: 17 May 2019; Accepted: 5 July 2019; Published: 16 July 2019

Abstract: Graphene, as the earliest discovered two-dimensional (2D) material, possesses excellently physical and chemical properties. Vast synthetic strategies, including chemical vapor deposition, mechanical exfoliation, and chemical reduction, are proposed. In this paper, a method to synthesize multilayer graphene in a semi-opened environment is presented by introducing arc-discharge plasma technology. Compared with previous technologies, the toxic gases and hazardous chemical components are not generated in the whole process. The synthesized carbon materials were characterized by transmission electron microscopy, atomic force microscopy, X-ray diffraction, and Raman spectra technologies. The paper offers an idea to synthesize multilayer graphene in a semi-opened environment, which is a development to produce graphene with arc-discharge plasma.

Keywords: multilayer graphene; synthetic strategies; arc-discharge plasma; semi-opened environment

1. Introduction

Since the graphene was isolated in 2004 by A.K. Geim and K.S. Novoselov via the 'Scotch tape' method [1], research on graphene has attracted a deluge of interests from scholars due to its extraordinary properties (excluding large available specific surface areas) and potential applications [2–5], especially in lithium ion batteries [5]. Graphene is the strongest discovered material in the world, whose elastic modulus reaches about 1.0 TPa [6]. In a word, graphene possesses a wide range of applications in our daily life and industrial manufacturing.

After a slow start, there has been a rapid increase of the amount of research on graphene in recent years [7]. For example, less than 1000 patent applications were lodged before 2008. However, the number of patent publications lodged on graphene was more than 24,000 from 2008 to 2014. In these patent publications, various graphene synthetic methods have been presented. Current methods require an ultra-low pressure (vacuum in the quartz tube is always lower than 9.75 Torr) or toxic oxidation and reduction reagents [8–11]. Graphite oxide (GO) is the essential material for the oxidation-reduction method, and sometimes even for the atmospheric plasma method [12]. When synthesizing graphene by the oxidation-reduction method, strong oxidizing reagents with a pungent odor and strong causticity are barriers to operators [13,14]. Shahriary's team [15] used a modified Hummers method to oxidize graphite powder, and harmful sulphuric acid (H_2SO_4) and potassium permanganate ($KMnO_4$) were used. In addition, mechanical exfoliation of highly oriented pyrolytic graphite (HOPG) is a method that is suited to use in laboratory investigations due to its low yield [16,17]. Moreover, the cost of the epitaxial growth method is unacceptable in commercial production [18]. Chemical vapor deposition (CVD) is considered the most potential method to synthesize graphene, with high quality and large-scale in industrial manufacture. However, substrates (such as copper foils) are essential to be used to deposit and separate out the carbon atoms [19–21]. For further use, the graphene film should be transformed to the target material surface. The transformation processes—regardless of whether it is the traditional 'wet etching method [22]' or other advanced methods [23,24]—introduces defects on the surface of graphene. Lin et al. [25] indicated poly (methyl methacrylate) (PMMA) residues on the graphene surface as a barrier to decrease its performance, and the PMMA residues could only be decreased by the annealing process but could not be eliminated. Recently, Zhang et al. [26] reported a novel method to transform graphene without any polymer. The graphene was etched in the hexane and ammonium persulfate solution and then transformed to target substrates with the aid of a Si/SiO$_2$ plate. This method depends on the quality of graphene. If the quality of graphene is good, the transformation process is not complex. Among these synthetic strategies, mechanical exfoliation gives the best graphene to date. Graphene synthesis by the epitaxial growth method is non-uniform in thickness and the chemical reduction introduces functionalized organic groups into this 2D structure. Nowadays, some new technologies, such as roll-to-roll technology [27], have been studied to enlarge the size of graphene film based on the CVD method. However, substrates with a special surface structure are expensive and suffer from time costs.

In 1990, fullerene (C$_{60}$) was firstly synthesized using the direct current (DC) arc-discharge method by Kratschmer and Huffman with a high yield [28]. Moreover, Iijima and Toshinari [29] found single-shell carbon nanotubes in cathodic products when they used the arc-discharge method to synthesize fullerene. Arc-discharge plasma technology, with different sizes of graphite rods as the cathode and anode, is widely carried out to synthesize carbon nano-materials in a relatively confined space in a special atmosphere. Until now, multilayer graphene has been successfully synthesized by using this technology. Wu et al. [30] synthesized large-scaled few-layered graphene in the condition of carbon dioxide (CO$_2$) and helium (He) with an optimized current (about 150 A) in a closed water-cooling stainless steel chamber. Chen et al. [31] synthesized graphene in a mixture of hydrogen (H$_2$) and buffer gases at 400 Torr in a relatively short time (about 20 min). This proves that the arc-discharge method has the potential to quickly synthesize graphene. However, to our knowledge, a strictly experimental environment, such as whole ambient pressure, is needed in the synthetic process by arc-discharge plasma and the mechanism has seldom been discussed.

Herein, we will describe a novel methodology based on the plasma technique that can synthesize multilayer graphene in a semi-opened environment, which lowers the requirement of experimental conditions. A semi-opened environment means that there is no strict requirement of a sealed environment for the set-up. The local pressure around the two electrodes is maintained at 400 Torr. In addition, the method does not generate by-products of toxic gases or use hazardous chemical components, which are not friendly to operators. The synthesized material in the cathode and anode were characterized and analyzed with a number of techniques, including transmission electron microscopy (TEM), atomic force microscopy (AFM), Raman spectroscopy, and X-ray diffraction (XRD), respectively. The synthesized mechanism was further discussed depending on the dynamics of thermal plasma.

2. Experiment

2.1. Materials

Graphene was directly synthesized without catalysts using two different sizes of graphite rods as carbon sources by arc-discharge plasma in a semi-opened environment. The graphite rods used in the experiment of 99.99%. A six-millimeter-diameter graphite rod in a length of 10 mm was employed as the cathode and the anode was also a 10 cm long graphite rod but in a diameter of 10 mm as shown in Figure 1b. Prior to experiment, the graphite rods were rinsed in a sonic washer with deionized (DI) water for 10 min and then dried in the nitrogen flow. Argon (Ar) with a purity of 99.999% was used in the experiment.

2.2. Experimental Set-Up

As shown in Figure 1, this experimental set-up mainly consists of an arc-reaction system, a pressure control and test system, and an auxiliary vacuum system. As we can see in Figure 1a, the cathodic rod and anodic rod were in the chamber and the distance of these two rods was about 2 mm. Arc plasma was generated between these two rods in the arc-reaction chamber (Figure 1c). In order to ensure the sustainability of the experiment, the chamber was cooled by water. The cathodic rod was grounded through a cable. A welder power source (WS-400, Beijing Time Technologies Co., Ltd), which was triggered in high frequency, was selected to ensure the stability of the arc ignition process. The welder power source offered a constant current of 35 A. The local ambient pressure around the arc-charge area was maintained at 400 Torr by a mechanical screw vacuum pump in the experimental processes. An Ar flow was directly blown to the arc-discharge area with a speed of 21.6 slm. The synthesized materials in the anode and cathode were collected for further characterization and analysis, respectively.

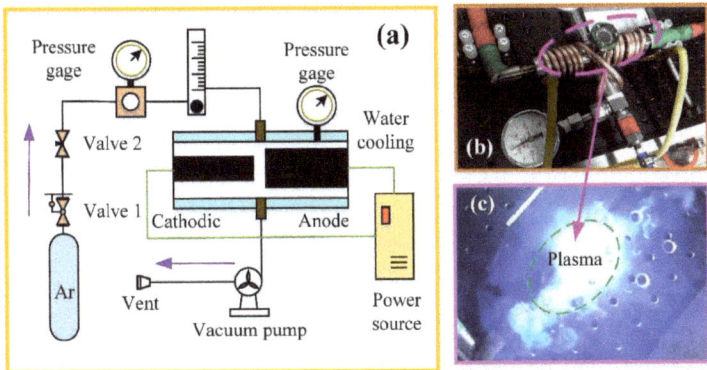

Figure 1. Arc-discharge plasma experimental set-up, (**a**) schematic diagram (**b**), real set-up, and (**c**) plasma generation area.

2.3. Characterizations

Raman spectroscopy (NRS-3000, JASCO Co., Tokyo, Japan) is one of the effective characterization methods to identify the product. The excited wavelength was 532 nm. A laser beam was focused through a micro-scope objective with a high numerical aperture (100×) and NA was set as 0.9. The incident laser power was 1 mW to prevent the pristine sample from being modified. X-ray diffraction (XRD) was measured by a D8 Focus (Bruker Co., Inc., Karlsruhe, Germany) with monochromatized Cu/Kα radiation (λ = 1.5418 Å). The tube voltage and current were 40 kV and 40 mA, respectively. To further determine the product, transmission electron microscopy (TEM, Tecnai G2 F20, FEI, Hillsboro, America) and electron diffraction images were used to get the information of topological and layer number of the product (Tecnai G^2 F20, FEI, operated at 200 kV). The thickness of the synthesized material was measured accurately in the tapping mode of an atomic force microscopy (AFM, Bruker Co., Inc., Karlsruhe, Germany).

3. Results and Discussions

The samples for Raman spectroscopy analysis were sonicated in ethyl alcohol for 10 min and then the solution was dropped on an SiO$_2$/Si wafer. Figure 2a shows the Raman spectra of synthesized material in different electrodes. The D-peak, 2D-peak, and G-peak were used to identify the synthesized materials' type. For graphene, the D peak that stands for the activation by defects is around 1350 cm^{-1}. Graphene with defects presents the existence of other bands and these bands are activated due to the breaking of the crystal symmetry that relax the Raman fundamental selection rule [32]. Such a phenomenon leads to an appearance of a D peak. The intensity of the 2D peak (~2700 cm^{-1}), which is derived from inelastic scattering of two phonons, and the G peak (~1580 cm^{-1}), which is related to the ordered in-plane sp^2 carbon structure, can be used to distinguish the number of graphene layer (mono-, bi-, or few-) [33]. For the double resonance mechanism, in the peak around 2700 cm^{-1}, only a resonant one can modify the Raman cross section. After the scattering process, the excited electron would be scattered back due to the phonon or a defect, and finally, a new photon is emitted [32]. As shown in Figure 2a, the synthesized material in the cathode (pink line) can be proven to be the graphene in a few layers according to Figure 2b,c. The shape and intensity of the 2D peak is sensitive to the number of graphene layers [34]. The insert partial enlarged drawing of the 2D peak (Figure 2b) appears to be symmetrical and the value of the symmetry axis is around 2690 cm^{-1} (a little bit less than 2700 cm^{-1}). With the increase of the graphene layers (from a monolayer to multilayers), the 2D peak upshifted [35]. For graphite, the shape of the 2D peak, which can be considered as an assembly of two smaller peaks (2D$_1$ and 2D$_2$ peaks), is not symmetrical and the 2D peak exceeded 2700 cm^{-1} [33,36]. The intensity ratio of 2D and G peaks of the synthesized products in the cathode is about 0.52. Tu and his team confirmed that the layer of graphene changes from three to seven layers when the ratio is around 0.5 as shown in Figure 2c [37]. It can also be found in Figure 2a that the synthesized material in the anode (green line) is the graphite in little layers due to the I_{2D}/I_G intensity ratio being less than 0.5. The I_D/I_G intensity ratio is an assessment method of the defect level. Compared with the cathode, the I_D/I_G ratio in the anode is greater than 0.7, thus showing more defects.

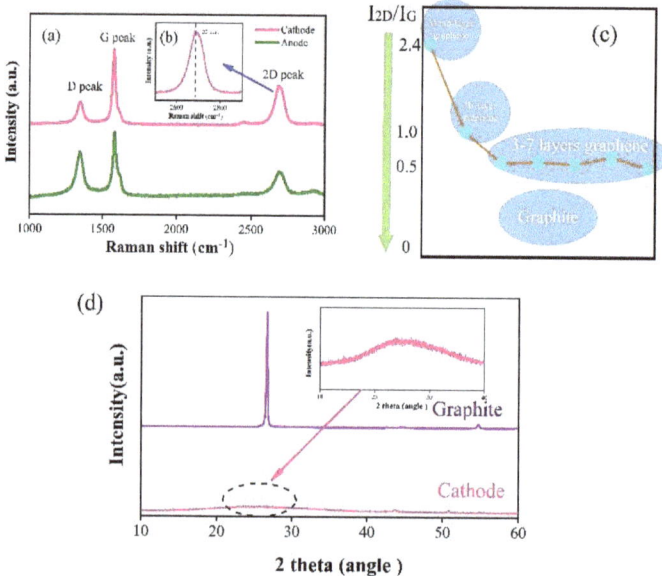

Figure 2. Raman spectra of the synthesized material (a) in different electrodes; (b) partial enlarged drawing of 2D peak (insert); (c) I_{2D}/I_G intensity ratio schematic diagram; and (d) XRD patterns.

Figure 2d shows the XRD patterns of the material in the cathode and the material collected from the original graphite rod. The strong peak in the XRD pattern of the material collected from the original graphite rod appears at 26.6°. For the material deposited in the cathode, there are weak and broad peaks at 24.8°, similar to [38].

To further identify the materials synthesized in the cathode and anode, the TEM and electron diffraction images were employed. The samples were first sonicated in dimethyl formamide (DMF) for 2 h, and then centrifuged. After that, the supernatant liquid was dropped on the micro-grid and dried by an infrared lamp. The information of the graphite and synthesized materials were detected by TEM as shown in Figure 3. The synthesized material in the cathode (Figure 3c,d) was more transparent and thinner after the arc-discharge process compared with the original graphite (Figure 3a,b). The layer information can be observed clearly in the high magnification TEM images as shown in Figure 3b,d.

The TEM image of the graphite (Figure 3a) showed a low level of transparency compared with the synthesized materials' TEM images. Additionally, the layer of graphite is more than 18 according to Figure 3b. It can be concluded from Figure 3d that the graphene synthesized in the cathode is four layers, and this conclusion is consistent with the Raman analysis presented above. The corresponding electron diffraction (ED) pattern of the synthesized material in the cathode is shown in Figure 4a. The carbon atoms are arranged in the hexagonal crystal mode marked by the purple circles. The TEM information proves that the graphene in a few layers has been successfully synthesized on the cathode.

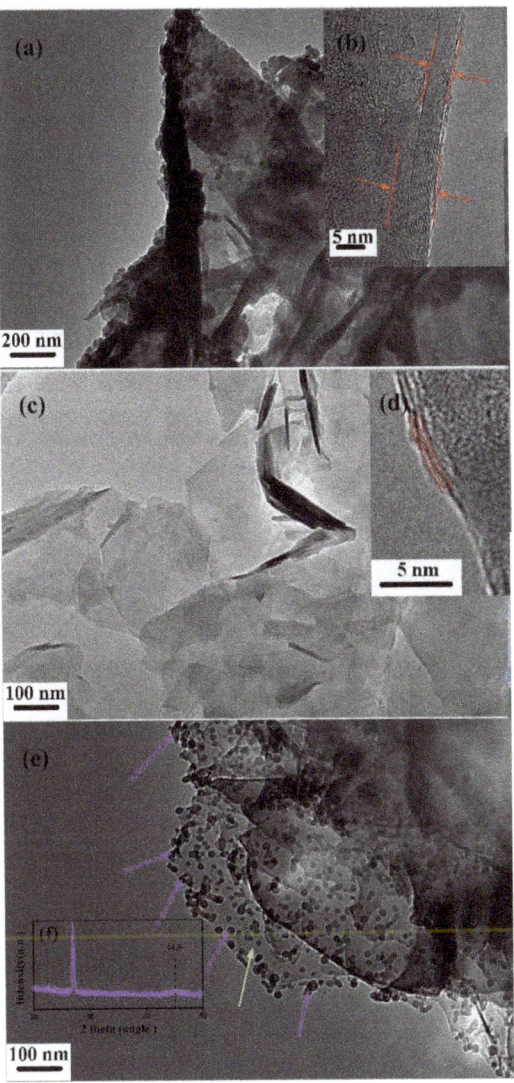

Figure 3. Transmission electron microscopies of (**a**) graphite at low magnification, (**b**) graphite at how magnification, (**c**) synthesized material in the cathode at low magnification, (**d**) synthesized material in the cathode at high magnification, (**e**) synthesized material in the anode, and (**f**) XRD of synthesized material in the anode.

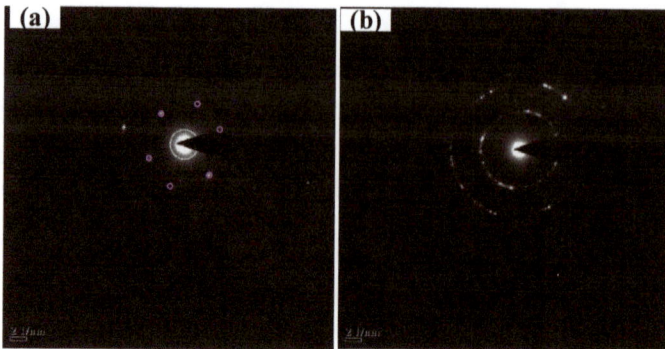

Figure 4. Electron diffraction images of (**a**) synthesized material in the cathode and (**b**) synthesized material in the anode.

Compared with the material synthesized in the cathode, the product deposited on the anode possessed a high defect as mentioned above. Figures 3e and 4b show the TEM and ED images of the material synthesized in the anode. We can see in Figure 3e that there are many black dots (pointed by the purple arrows) in the thin film (pointed by the green arrow). These black dots are expected to result in a sharp increase of the I_D/I_G ratio in the anode. To identify the composition of the material synthesized in the anode, XRD was employed. We found a slight peak around 44.6° in Figure 3f. This showed that Fe element was present. Beside the metal Fe, a sharp XRD peak was remarkably found at about 26.6° corresponding to the crystal structure of graphite according to the JCPDS card (No. 41-1487). The ED image of the synthesized material in the anode shows a more unordered arrangement. This phenomenon is also consistent with the TEM image.

The synthesized products in the cathode and anode were sonicated in the ethyl alcohol, respectively. Then, they were dropped on the mica surface for the AFM test to obtain the AFM images and the thickness information. The AFM testing of material collected from the original graphite rod was also detected. It can be seen that from the AFM image (Figure 5a,b) that the height of the brightest area is more than 8 nm, showing that the material in this area is not graphene, but is graphite. The carbon atoms were accumulated in this area that lead to the increase of the thickness. However, for most of the region in Figure 5a, the height is lower than 2 nm (Figure 5c,d). This indicates that the graphene in a few layers had been synthesized. The average particle size is about 1.0 μm. The AFM image and its height profiles of the material synthesized in the anode are shown in Figure 5e–g. It is obvious that the height of the anode material (about 6 nm) is much higher than the cathode due to the defects in the material as shown before. Figure 5h,i show the AFM and height profile of the material collected from the original graphite rod. The height is much higher than the synthesized materials. After the whole process of characterization and analysis, we can say that the graphene in a few layers has been successfully synthesized in the cathode. Moreover, the products deposited on the anode were a few layer graphite with Fe element.

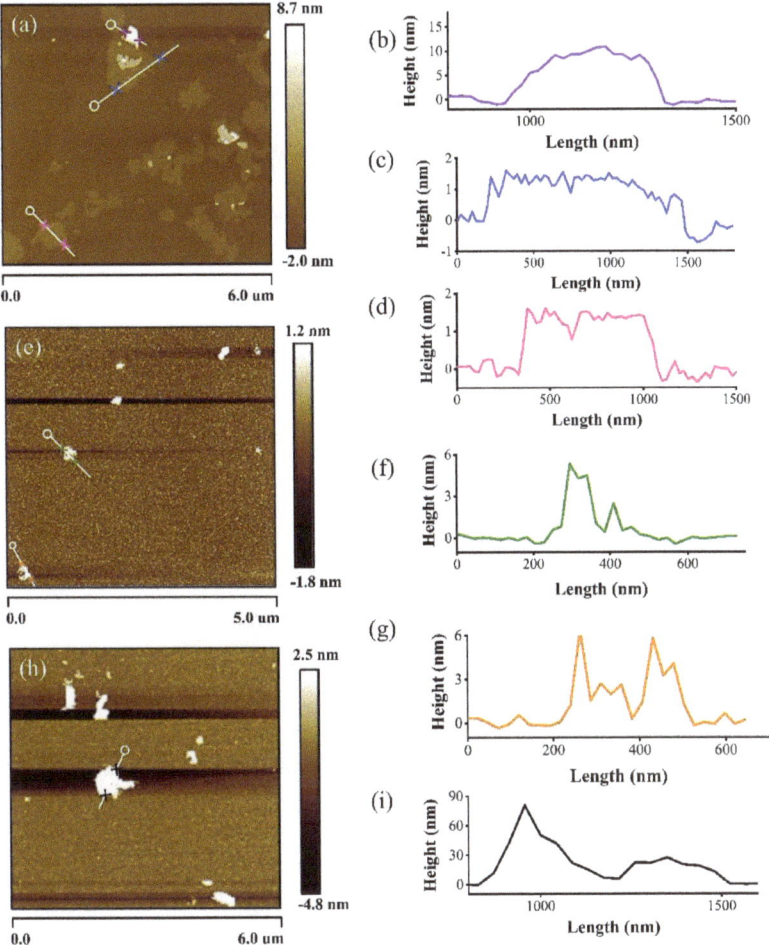

Figure 5. AFM (Atomic force microscopy) images of synthesized material in the cathode on mica surface, (**a**) AFM image (**b–d**) its height profiles, the material in the anode; (**e**) AFM image, (**f,g**) its height profiles, and the material collected from original graphite rod; (**h**) AFM image and (**i**) its height profile.

Now, the growth mechanism of graphene in the arc-discharge plasma condition is discussed. The growth pattern for synthesizing graphene without substrate is shown in Figure 6. Individual carbon atoms should be evaporated and then re-nucleated in the structure of a regular hexagon as shown in Figure 5a when the temperature between the anode and cathode electrodes is bigger than 3500 K. Figure 6a,b indicate the synthesized mechanism when using arc-discharge plasma technology without and with Ar flow, respectively. The type of synthesized carbon nanomaterial is limited by the temperature. For example, Keidar et al. [39] obtained single-wall carbon nanotube (SWNT) in the temperature range of about 1200–1800 K. When the gas between cathode and anode is triggered by high frequency and arc-discharge occurs, the temperature is more than 3500 K [40]. This high temperature lead to the evaporation of the carbon atom in the anode (Figure 6a). The carbon atom further changed to a carbon ion as shown in Equation (1). The graphene synthesized in the cathode in a special atmosphere as Chen et al. discussed [31]. The heat dissipation was achieved by the molecular heat conduction mechanism without gas flow in the chamber. Thus, the cathode is regarded as a deposited medium and graphene powder can be collected on the cathode. It is interesting that the different synthesized material grows in the cathode and anode in this experiment that is mentioned above. It can be seen in Figure 6d that the graphite rods of both the cathode and anode were consumed. Before the arc-discharge process, the graphite rods are complete and dense as shown in Figure 6c. However, after the experiment, the graphite rods become more porous and lots of hole can be seen through the eyes. When the Ar flow is introduced into the system, part of the heat may be taken away by Ar flow and bring the oscillation between the anode and cathode. Part of the arc column forms a closed system with the channel wall and thus some iron atoms evaporate to form iron ions (Equation (2)). These iron ions combine with carbon ions to nucleate and grow to the doped-graphite as shown in Figure 6b.

$$C + U_i \xrightarrow{Arc} C^+ + e \tag{1}$$

$$Fe + U_i \xrightarrow{Arc} Fe^+ + e \tag{2}$$

where C and Fe are the carbon atom, U_i is the foreign energy, Arc represents the changing condition, and C^+ and Fe^+ are the carbon ion and iron ion, respectively.

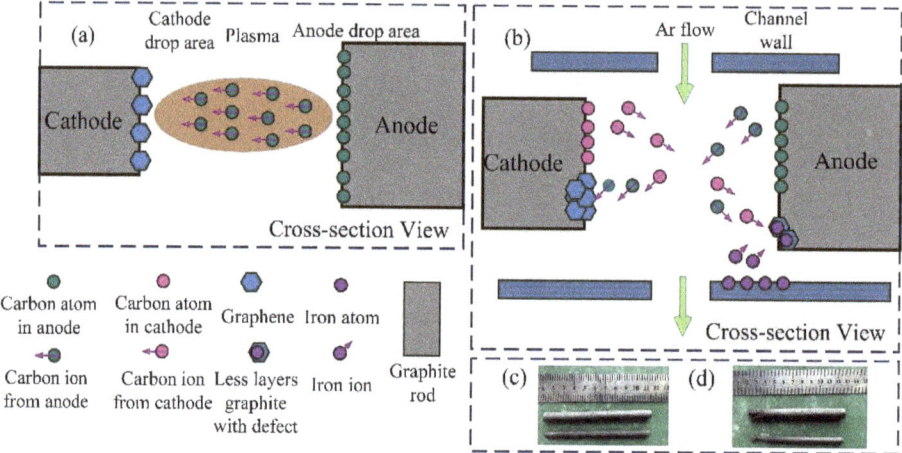

Figure 6. Synthetic mechanism of graphene (**a**) without Ar flow, (**b**) with Ar flow, and macro morphologies (**c**) before (**d**) after arc-discharge.

4. Conclusions

Multilayer graphene was successfully synthesized in the cathode in a semi-opened environment by introducing arc-discharge plasma. The synthesized material deposited on the cathode and anode was characterized and analyzed by Raman spectrum, TEM, XRD, and AFM. These results show that the graphene in a few layers was synthesized. Compared with the synthesized material in the cathode, the product in the anode showed a different structure. As we can see in the TEM images, the products deposited on the anode possess many black dots in the thin film, which is the reason why the I_D/I_G ratio (Raman spectrum) of the anode is high with a graphitic structure. Moreover, the synthesized mechanism was further discussed depending on the dynamics of the thermal plasma. When the gas between the cathode and anode was triggered and arc-discharge occurred, individual carbon atoms should evaporate from both the anode and cathode and then re-nucleate in the structure of a regular hexagon. In summary, we provided a method to synthesize multilayer graphene in a semi-opened environment by introducing arc-plasma technology.

Author Contributions: Conceptualization, D.W.; methodology, H.T.; validation, H.T. and Y.G.; formal analysis, H.T.; investigation, Y.G.; data curation, Y.G.; writing—original draft preparation, H.T.; writing—review and editing, Y.G.; visualization, H.T.; supervision, D.W.; project administration, Y.G.; funding acquisition, D.W.

Funding: This research is funded by the National Natural Science Foundation of China (No. 51875578), the Tribology Science Fund of State Key Laboratory of Tribology and the Science Foundation of China University of Petroleum, Beijing (No. 2462017BJB06).

Acknowledgments: The authors appreciate the assistance and critical discussion from Fan Yang, and Yongfeng Li in Department of Chemical Engineering at China University of Petroleum. This work was supported by the National Natural Science Foundation of China (No. 51875578), the Tribology Science Fund of State Key Laboratory of Tribology and the Science Foundation of China University of Petroleum, Beijing (No. 2462017BJB06).

Conflicts of Interest: The authors declare no conflict of interest.

References

1. Novoselov, K.S.; Geim, A.K.; Morozov, S.V.; Jiang, D.A.; Zhang, Y.; Dubonos, S.V.; Grigorieva, I.V.; Firsov, A.A. Electric field effect in atomically thin carbon films. *Science* **2004**, *306*, 666–669. [CrossRef] [PubMed]
2. Nair, R.R.; Blake, P.; Grigorenko, A.N.; Novoselov, K.S.; Booth, T.J.; Stauber, T.; Peres, N.M.R.; Geim, A.K. Fine structure constant defines visual transparency of graphene. *Science* **2008**, *320*, 1308. [CrossRef] [PubMed]
3. Zhang, J.; Yao, T.; Guan, C.; Zhang, N.; Huang, X.; Cui, T.; Wu, J.; Zhang, X. One-step preparation of magnetic recyclable quinary graphene hydrogels with high catalytic activity. *J. Colloid Interface Sci.* **2017**, *491*, 72–79. [CrossRef] [PubMed]
4. Bae, S.; Kim, H.; Lee, Y.; Xu, X.; Park, J.S.; Zheng, Y.; Balakrishnan, J.; Lei, T.; Kim, H.R.; Song, Y.; et al. Roll-to-roll production of 30-inch graphene films for transparent electrodes. *Nat. Nanotechnol.* **2010**, *5*, 574–578. [CrossRef] [PubMed]
5. Arsat, R.; Breedon, M.; Shafiei, M.; Spizziri, P.G.; Gilje, S.; Kaner, R.B.; Kalantar-zadeh, K.; Wlodarski, W. Graphene-like nano-sheets for surface acoustic wave gas sensor applications. *Chem. Phys. Lett.* **2009**, *467*, 344–347. [CrossRef]
6. Yi, M.; Shen, Z. A review on mechanical exfoliation for the scalable production of graphene. *J. Mater. Chem. A* **2015**, *3*, 11700–11715. [CrossRef]
7. Levchenko, I.; Ostrikov, K.K.; Zheng, J.; Li, X.; Keidar, M.; Teo, K.B. Scalable graphene production: Perspectives and challenges of plasma applications. *Nanoscale* **2016**, *8*, 10511–10527. [CrossRef] [PubMed]
8. Mueller, N.S.; Morfa, A.J.; Abou-Ras, D.; Oddone, V.; Ciuk, T.; Giersig, M. Growing graphene on polycrystalline copper foils by ultra-high vacuum chemical vapor deposition. *Carbon* **2014**, *78*, 347–355. [CrossRef]
9. Malesevic, A.; Vitchev, R.; Schouteden, K.; Volodin, A.; Zhang, L.; Tendeloo, G.V.; Vanhulsell, A.; Haesendonck, C.V. Synthesis of few-layer graphene via microwave plasma-enhanced chemical vapour deposition. *Nanotechnology* **2008**, *19*, 305604. [CrossRef] [PubMed]
10. Berger, C.; Song, Z.; Li, X.; Wu, X.; Brown, N.; Naud, C.; Mayou, D.; Li, T.; Hass, J.; Marchenkov, A.N.; et al. Electronic confinement and coherence in patterned epitaxial graphene. *Science* **2006**, *312*, 1191–1196. [CrossRef] [PubMed]
11. Stankovich, S.; Dikin, D.A.; Piner, R.D.; Kohlhaas, K.A.; Kleinhammes, A.; Jia, Y.; Wu, Y.; Nguyen, S.T.; Ruoff, R.S. Synthesis of graphene-based nanosheets via chemical reduction of exfoliated graphite oxide. *Carbon* **2007**, *45*, 1558–1565. [CrossRef]
12. Alotaibi, F.; Tung, T.T.; Nine, M.J.; Kabiri, S.; Moussa, M.; Tran, D.N.; Losic, D. Scanning atmospheric plasma for ultrafast reduction of graphene oxide and fabrication of highly conductive graphene films and patterns. *Carbon* **2018**, *127*, 113–121. [CrossRef]
13. Pei, S.; Cheng, H.M. The reduction of graphene oxide. *Carbon* **2012**, *50*, 3210–3228. [CrossRef]
14. Hummers, W.S., Jr.; Offeman, R.E. Preparation of graphitic oxide. *J. Am. Chem. Soc.* **1958**, *80*, 1339–1339. [CrossRef]
15. Shahriary, L.; Athawale, A.A. Graphene oxide synthesized by using modified hummers approach. *Int. J. Renew. Energy Environ. Eng.* **2014**, *2*, 58–63.
16. Tan, H.; Wang, D.; Guo, Y. Thermal Growth of Graphene: A Review. *Coatings* **2018**, *8*, 40. [CrossRef]
17. Sivudu, K.S.; Mahajan, Y. Mass production of high quality graphene: An analysis of worldwide patents. *Nanowerk*, 28 June 2015.

18. Wang, T.; Huang, D.; Yang, Z.; Xu, S.; He, G.; Li, X.; Hu, N.; Yin, G.; He, D.; Zhang, L. A review on graphene-based gas/vapor sensors with unique properties and potential applications. *Nano-Micro Lett.* **2016**, *8*, 95–119. [CrossRef]
19. Yamada, T.; Ishihara, M.; Kim, J.; Hasegawa, M.; Iijima, S. A roll-to-roll microwave plasma chemical vapor deposition process for the production of 294 mm width graphene films at low temperature. *Carbon* **2012**, *50*, 2615–2619. [CrossRef]
20. Liu, Y.; Wu, T.; Yin, Y.; Zhang, X.; Yu, Q.; Searles, D.J.; Ding, F.; Yuan, Q.; Xie, X. How Low Nucleation Density of Graphene on CuNi Alloy is Achieved. *Adv. Sci.* **2018**, *5*, 1700961. [CrossRef]
21. Reina, A.; Jia, X.; Ho, J.; Nezich, D.; Son, H.; Bulovic, V.; Dresselhaus, M.S.; Kong, J. Large area, few-layer graphene films on arbitrary substrates by chemical vapor deposition. *Nano Lett.* **2008**, *9*, 30–35. [CrossRef]
22. Li, X.; Cai, W.; An, J.; Kim, S.; Nah, J.; Yang, D.; Piner, R.; Velamakanni, A.; Jung, I.; Tutuc, E.; et al. Large-area synthesis of high-quality and uniform graphene films on copper foils. *Science* **2009**, *324*, 1312–1314. [CrossRef] [PubMed]
23. Wang, D.Y.; Huang, I.S.; Ho, P.H.; Li, S.S.; Yeh, Y.C.; Wang, D.W.; Chen, W.L.; Lee, Y.Y.; Chang, Y.M.; Chen, C.C.; et al. Clean-Lifting Transfer of Large-area Residual-Free Graphene Films. *Adv. Mater.* **2013**, *25*, 4521–4526. [CrossRef] [PubMed]
24. Gao, L.; Ren, W.; Xu, H.; Jin, L.; Wang, Z.; Ma, T.; Ma, L.; Zhang, P.Z.; Fu, Q.; Peng, L.M.; et al. Repeated growth and bubbling transfer of graphene with millimetre-size single-crystal grains using platinum. *Nat. Commun.* **2012**, *3*, 699. [CrossRef] [PubMed]
25. Lin, Y.C.; Lu, C.C.; Yeh, C.H.; Jin, C.; Suenaga, K.; Chiu, P.W. Graphene annealing: How clean can it be? *Nano Lett.* **2011**, *12*, 414–419. [CrossRef]
26. Zhang, G.; Güell Aleix, G.; Kirkman, P.M.; Lazenby, R.; Miller, T.S.; Unwin, P.R. Versatile polymer-free graphene transfer method and applications. *ACS Appl. Mater. Interface* **2016**, *8*, 8008–8016. [CrossRef] [PubMed]
27. Park, W.K.; Kim, T.; Kim, H.; Kim, Y.; Tung, T.T.; Lin, Z.; Jang, A.; Shin, H.S.; Han, J.H.; Yoon, D.H.; et al. Large-scale patterning by the roll-based evaporation-induced self-assembly. *J. Mater. Chem.* **2012**, *22*, 22844–22847. [CrossRef]
28. Krätschmer, W.; Lamb, L.D.; Fostiropoulos, K.; Huffman, D.R. Solid C60: A new form of carbon. *Nature* **1990**, *347*, 354. [CrossRef]
29. Iijima, S.; Ichihashi, T. Single-shell carbon nanotubes of 1-nm diameter. *Nature* **1993**, *363*, 603. [CrossRef]
30. Wu, Y.; Wang, B.; Ma, Y.; Huang, Y.; Li, N.; Zhang, F.; Chen, Y. Efficient and large-scale synthesis of few-layered graphene using an arc-discharge method and conductivity studies of the resulting films. *Nano Res.* **2010**, *3*, 661–669. [CrossRef]
31. Chen, Y.; Zhao, H.; Sheng, L.; Yu, L.; An, K.; Xu, J.; Ando, Y.; Zhao, X. Mass-production of highly-crystalline few-layer graphene sheets by arc discharge in various H_2–inert gas mixtures. *Chem. Phys. Lett.* **2012**, *538*, 72–76. [CrossRef]
32. Merlen, A.; Buijnsters, J.; Pardanaud, C. A guide to and review of the use of multiwavelength Raman spectroscopy for characterizing defective aromatic carbon solids: From graphene to amorphous carbons. *Coatings* **2017**, *7*, 153. [CrossRef]
33. Ferrari, A.C.; Meyer, J.C.; Scardaci, V.; Casiraghi, C.; Lazzeri, M.; Mauri, F.; Piscanec, S.; Jiang, D.; Novoselov, K.S.; Roth, S.; et al. Raman spectrum of graphene and graphene layers. *Phys. Rev. Lett.* **2006**, *97*, 187401. [CrossRef] [PubMed]
34. Gupta, A.; Chen, G.; Joshi, P.; Tadigadapa, S.; Eklund, P.C. Raman scattering from high-frequency phonons in supported n-graphene layer films. *Nano Lett.* **2006**, *6*, 2667–2673. [CrossRef] [PubMed]
35. Ni, Z.; Wang, Y.; Yu, T.; Shen, Z. Raman spectroscopy and imaging of graphene. *Nano Res.* **2008**, *1*, 273–291. [CrossRef]
36. Vidano, R.P.; Fischbach, D.B.; Willis, L.J.; Loehr, T.M. Observation of Raman band shifting with excitation wavelength for carbons and graphites. *Solid State Commun.* **1981**, *39*, 341–344. [CrossRef]
37. Tu, Z.; Liu, Z.; Li, Y.; Yang, F.; Zhang, L.; Zhao, Z.; Xu, C.; Wu, S.; Liu, H.; Yang, H.; et al. Controllable growth of 1–7 layers of graphene by chemical vapour deposition. *Carbon* **2014**, *73*, 252–258. [CrossRef]
38. Moon, I.K.; Lee, J.; Ruoff, R.S.; Lee, H. Reduced graphene oxide by chemical graphitization. *Nat. Commun.* **2010**, *1*, 73. [CrossRef] [PubMed]
39. Keidar, M.; Shashurin, A.; Volotskova, O.; Raitses, Y.; Beilis, I.I. Mechanism of carbon nanostructure synthesis in arc plasma. *Phys. Plasmas* **2010**, *17*, 56. [CrossRef]

40. Keidar, M.; Beilis, I.I. Modeling of atmospheric-pressure anodic carbon arc producing carbon nanotubes. *J. Appl. Phys.* **2009**, *106*, 103304. [CrossRef]

© 2019 by the authors. Licensee MDPI, Basel, Switzerland. This article is an open access article distributed under the terms and conditions of the Creative Commons Attribution (CC BY) license (http://creativecommons.org/licenses/by/4.0/).

Article

Catalyst-Less and Transfer-Less Synthesis of Graphene on Si(100) Using Direct Microwave Plasma Enhanced Chemical Vapor Deposition and Protective Enclosures

Rimantas Gudaitis, Algirdas Lazauskas, Šarūnas Jankauskas and Šarūnas Meškinis *

Institute of Materials Science, Kaunas University of Technology, K. Baršausko St. 59, LT-51423 Kaunas, Lithuania; rimantas.gudaitis@ktu.lt (R.G.); algirdas.lazauskas@ktu.edu (A.L.); sarunas.jankauskas@ktu.lt (Š.J.)
* Correspondence: sarunas.meskinis@ktu.lt

Received: 30 October 2020; Accepted: 8 December 2020; Published: 10 December 2020

Abstract: In this study, graphene was synthesized on the Si(100) substrates via the use of direct microwave plasma-enhanced chemical vapor deposition (PECVD). Protective enclosures were applied to prevent excessive plasma etching of the growing graphene. The properties of synthesized graphene were investigated using Raman scattering spectroscopy and atomic force microscopy. Synthesis time, methane and hydrogen gas flow ratio, temperature, and plasma power effects were considered. The synthesized graphene exhibited n-type self-doping due to the charge transfer from Si(100). The presence of compressive stress was revealed in the synthesized graphene. It was presumed that induction of thermal stress took place during the synthesis process due to the large lattice mismatch between the growing graphene and the substrate. Importantly, it was demonstrated that continuous horizontal graphene layers can be directly grown on the Si(100) substrates if appropriate configuration of the protective enclosure is used in the microwave PECVD process.

Keywords: graphene; direct plasma synthesis; microwave plasma enhanced chemical vapor deposition

1. Introduction

Graphene is a monolayer or several layers of hexagonally shaped carbon atoms [1]. This 2D carbon nanomaterial has achieved considerable interest due to the huge mobility of electrons and holes, optical transparency, flexibility, and chemical inertness [1–3]. Graphene is already considered to be a new transparent conductor [4,5], a monolayer alternative to the Schottky contact metals [6,7], and even an active layer of semiconductor devices [8–12]. Graphene-based transistors [8], diodes [6,7], photodetectors [9–11], and solar cells [12–14] are also meaningful in this context.

A complicated graphene transfer process is one of the main limitations preventing the broader use of graphene in semiconductor device technology. In this instance, graphene is grown on the copper of nickel catalytic foils [15]; followed by the complicated graphene transfer onto the required dielectric or semiconductor substrates. During the transfer process, different adsorbates can contaminate graphene [16]. Additionally, the transfer process can cause wrinkled or rippled surface morphology of graphene [17]. In this case, the control of the graphene film or graphene-semiconductor interface properties becomes a tricky task. Graphene can be synthesized on a silicon carbide (SiC) substrate if appropriate vacuum heating conditions are used [18]. No catalytic metals are necessary in this case [18]. However, the present use of SiC as a semiconductor is mainly limited by some segments of high-power electronics [19]. SiC apart, it was shown recently that direct graphene synthesis on the semiconductor or dielectric surfaces is possible via the use of plasma-enhanced chemical vapor deposition (PECVD) [20]. In this case, plasma activation of the chemical vapor deposition is mandatory. It ensures enhanced dissociation of the plasma species during the graphene synthesis process. However, plasma-related ion and electron bombardment of the growing graphene surface is detrimental. It results in the creation of defects and may even make the etching process prevail over the graphene growth [21]. Therefore, remote plasma is used for direct graphene synthesis.

Nevertheless, remote plasma mode is unavailable in the most conventional microwave and inductively coupled plasma-based PECVD units. However, there are few studies on catalyst-less and transfer-less horizontal graphene synthesis using direct PECVD. In this process, the growing graphene film is additionally protected by some plasma shielding. Notably, the [21] sample was enclosed in a metal cage with a honeycomb mesh shield, while in [22], a copper-foam-based Faraday cage was applied. Direct graphene synthesis on insulating substrates such as glass [22], sapphire [21], quartz [21] plates, as well as thermally deposited SiO_2, Al_2O_3, MnO_2, HfO_2, and TiO_2 films [21] was demonstrated. However, for many devices, graphene synthesis on semiconductor substrates is necessary. Monocrystalline Si(100) is still the most often used substrate for the fabrication of microelectronic devices, solar cells, and different photodiodes. Therefore, catalyst-less and transfer-less graphene synthesis on Si(100) using a direct microwave plasma system was considered in the present study.

In this paper, the samples were protected from direct plasma action using several different configurations of protective enclosures. The enclosures' design was varied, taking into account two processes: eliminating the unwanted direct plasma effects and flow of the reactive carbon, hydrocarbon, and hydrogen species towards the substrate. The protective enclosure should screen the substrate from direct plasma. At the same time, gas flows are changed due to the presence of the enclosure. Herein, we wanted to know to what extent we can further suppress excessive direct plasma action by reducing the enclosure's top hole size or removing the top holes above a substrate, and to what extent we can reduce protective enclosure design complexity. Synthesis parameters and their influence on the graphene structure were analyzed thoroughly. We have shown that graphene can be synthesized on the Si(100) substrate in a one-step process using a combination of the direct plasma and differently shaped enclosures. It was revealed that even a very simple enclosure design consisting of a single rectangular steel sheet without holes could be used as protective shielding.

2. Materials and Methods

The direct transfer-less synthesis of graphene was performed by the microwave PECVD system Cyrannus (Innovative Plasma Systems (Iplas) GmbH, Troisdorf, Germany). A methane and hydrogen gas mixture was used as a source of carbon and hydrogen. The hydrogen plasma was ignited until the heater reached the target temperature. Hydrogen gas flow and plasma power were the same as in the graphene growth process (Table 1). Methane gas was introduced when the temperature necessary for graphene synthesis was reached. The growth process was conducted in one-step without a separate nucleation stage.

Monocrystalline Si(100) (UniversityWafer Inc., South Boston, MA, USA) was applied as a substrate for the direct synthesis of graphene. No additional wet chemical cleaning of the substrate was performed. Special enclosures protected the sample from excessive plasma action. Four steel enclosures of different designs were used (Figure 1) to protect the sample from excessive plasma action. Three circular enclosures (1st–3rd) had holes of different diameters and pattern arrangements on the top. The 4th enclosure had a much simpler design and consisted of a rectangular steel sheet folded in two places.

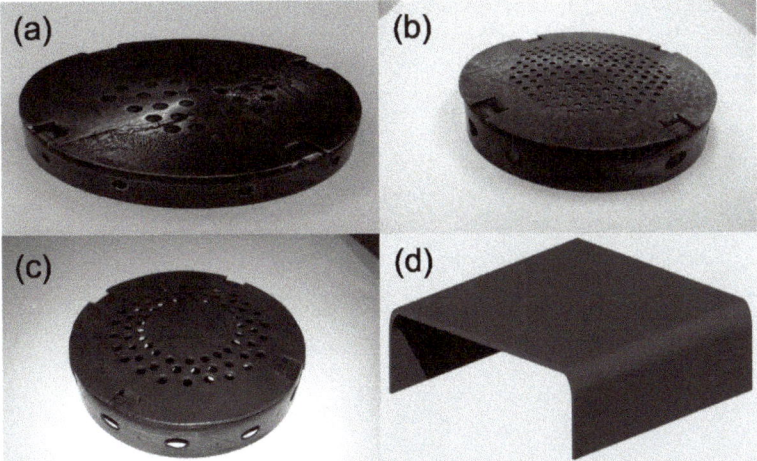

Figure 1. Protective enclosures used for direct synthesis of the graphene on Si(100): the first (1st) enclosure (top hole size 3.5 mm) (**a**), the second (2nd) enclosure (top hole size 2 mm) (**b**), the third (3rd) enclosure (top hole size 3.5 mm, no holes at the center) (**c**), the fourth (4th) enclosure (enclosure height 5 mm) (**d**).

Technological parameters such as plasma power, CH_4/H_2 gas flow ratio, pressure, temperature, and time were varied. The graphene direct synthesis conditions can be found in Table 1.

Raman scattering spectra of the synthesized samples were acquired using the Raman spectrometer inVia (Renishaw, Wotton-under-Edge, UK). The excitation wavelength was 532 nm. The excitation laser beam power was 1.5 mW. The ratio of 2D and G peak intensities (I_{2D}/I_G ratio) was estimated to evaluate the number of graphene layers [23]. The I_D/I_G peak intensity ratio was calculated to estimate the defect density of graphene [24,25]. Additionally, the positions of G and 2D peaks (Pos(G) and Pos(2D)) were also taken into account. Table S1 shows the possible relations between the Raman scattering spectra parameters mentioned above and the number of graphene layers, stress, doping, and defect density. The spectra were measured in several different places on the sample. The average values and standard deviation of the different Raman scattering spectra parameters were calculated for each sample.

The surface morphology of the selected graphene layers was investigated using atomic force microscopy (AFM) at several different places on a sample. The measurements were done at room temperature in ambient air. The NanoWizardIII atomic force microscope (JPK Instruments, Bruker Nano GmbH, Berlin, Germany) was used. A v-shaped silicon cantilever operating in a contact mode was applied, as this mode is less sensitive to the possible presence of adsorbed species. The cantilever's spring constant was 3 N/m, the tip curvature radius was 10.0 nm, and the cone angle was 20°. A 2 µm × 2 µm AFM scan area was chosen to reveal small graphene layer features. The SurfaceXplorer and JPKSPM Data Processing software (version spm-4.3.13, JPK Instruments) were applied for data analysis.

Table 1. Graphene synthesis conditions used in the present study.

Sample No.	Enclosure No.	P, kW	H₂, sccm	CH₄, sccm	p, mBar	t, °C	t, min
2E1	1	1.2	150	50	30	900	30
3E1	1	1.2	180	20	30	900	30
4E1	1	1.2	120	80	30	900	30
5E1	1	1.2	150	50	30	900	15
6E1	1	1.2	150	50	30	900	45
7E1	1	1.2	150	50	30	700	30
8E1	1	1.2	150	50	30	800	30
9E1	1	0.8	150	50	30	900	30
10E1	1	1.0	150	50	30	900	30
1E2	2	1.2	150	50	30	800	30
2E2	2	1.2	150	50	30	700	30
3E2	2	1.2	150	50	30	900	30
4E2	2	1.2	180	20	30	900	30
5E2	2	1.2	120	80	30	900	30
6E2	2	1.0	150	50	30	900	30
7E2	2	0.8	150	50	30	900	30
3E3	3	1.2	150	50	30	900	30
4E3	3	1.2	180	20	30	900	30
5E3	3	1.2	120	80	30	900	30
6E3	3	0.8	150	50	30	900	30
7E3	3	1.0	150	50	30	900	30
1E4	4	1.2	150	50	30	800	30
2E4	4	1.2	150	50	22	700	30
3E4	4	1.2	150	50	22	800	30
4E4	4	1.2	150	50	22	900	30

3. Results

3.1. Raman Spectra of Directly Synthesized Graphene

In the present study, graphene was synthesized on the Si(100) substrates using microwave PECVD. Firstly, it is important to note that the direct synthesis of graphene on the Si(100) was not possible when the protective enclosure was not used in the plasma discharge zone during the microwave PECVD process. In this case, only the direct plasma interacting with the substrate was obtainable, which suppressed the graphene's growth.

In another instance, the sample was protected by the enclosure to prevent excessive direct plasma interaction with the substrate. Protective enclosures with several different designs were used as plasma shielding (Figure 1). In this case, the direct synthesis of graphene on the Si(100) substrate was successful. The recorded Raman scattering spectra of the samples were typical for graphene (Figure 2) [23–43]. Characteristic G and 2D peaks as well as defects related peaks (D peak as well as less intensive D+D″ and D+D′ bands [39,40]) were observed. No separate D′ peak at ~1620 cm^{-1} was observed in all cases. It is noteworthy that, in the Raman spectra of directly synthesized graphene, the D peak is always observed [20–22]. That is the main difference from the graphene synthesized by chemical vapor deposition (CVD) on catalytic foil and afterwards transferred onto the target substrate. This difference is mainly related to the presence of many grain boundary defects related to the nanocrystalline nature of directly synthesized graphene.

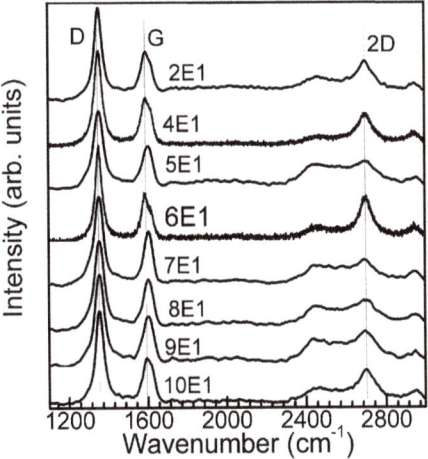

Figure 2. Typical Raman scattering spectra of graphene directly synthesized on Si(100).

3.2. Effect of Synthesis Conditions and Enclosure Design on the Graphene Structure

The influence of the several key technological synthesis parameters (i.e., protective enclosure design, plasma power, methane and hydrogen gas flow ratio, pressure, temperature, and synthesis time) on the growth and structure of synthesized graphene was investigated. The 1st enclosure had 3.5 mm size holes on the top, while for the 2nd enclosure, the hole size was decreased to 2 mm. The 3rd enclosure had 3.5 mm size holes on the top, but no holes at the center (Figure 1).

Firstly, the effect of the synthesis time was considered (Figure 3). The graphene was already formed after 15 min of the microwave PECVD process. It was determined that after this time mark graphene gets thinner, and more defects are promptly introduced into the graphene structure: the increase in I_{2D}/I_G from ~0.72 to ~0.93 (Figure 3a) and the increase in I_D/I_G from ~1.8 to ~2.1 (Figure 3b). The other experiments' synthesis time was chosen by taking into account these results (30 min).

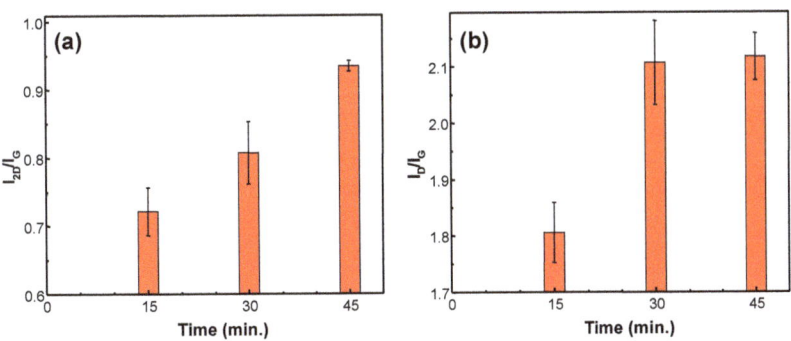

Figure 3. I_{2D}/I_G (a) and I_D/I_G (b) ratios of graphene synthesized at different process time (15 min, 30 min, 45 min). In all cases, every other process parameter is kept constant (H_2 gas flow 150 sccm sccm, CH_4 gas flow 50 sccm, power 1.2 kW, pressure 30 mBar, temperature 900 °C). The 1st enclosure was used.

Figure 4 shows the I_{2D}/I_G and I_D/I_G ratios of graphene synthesized using different plasma power and enclosures. As it can be seen in Figure 4a, plasma power effect varies for different enclosures. The I_{2D}/I_G ratio of graphene synthesized using the 1st enclosure increases with plasma power. This result implies that the number of graphene layers is reduced with an increase in plasma power. No clear dependence was observed for the 2nd and the 3rd enclosure. However, in all cases, the I_{2D}/I_G ratio of the graphene synthesized using 1.2 kW power was higher than the I_{2D}/I_G ratio of the graphene synthesized using 0.8 kW power. The I_D/I_G ratio increased with plasma power for the 1st and the 3rd enclosure (Figure 4b). No clear dependence of the I_D/I_G ratio on plasma power was observed for graphene synthesized using the 2nd enclosure.

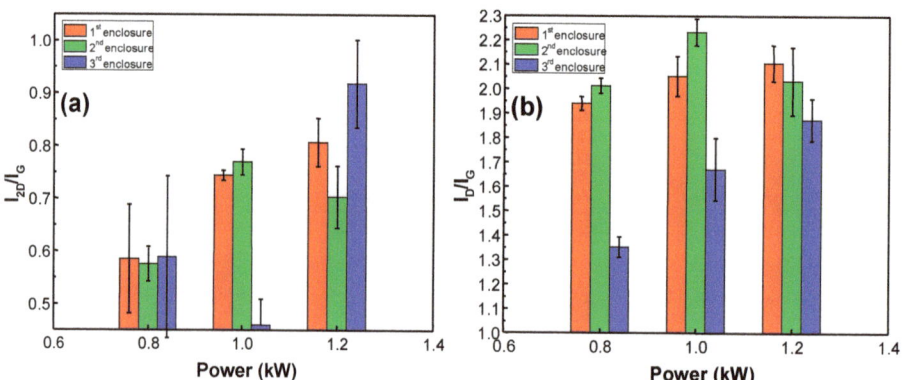

Figure 4. I_{2D}/I_G (a) and I_D/I_G (b) ratios of graphene synthesized using different plasma power (0.8 kW, 1.0 kW, 1.2 kW) and enclosures. In all cases, every other process parameter is kept constant (H_2 gas flow 150 sccm sccm, CH_4 gas flow 50 sccm, pressure 30 mBar, temperature 900 °C, time 30 min).

Figure 5 shows the I_{2D}/I_G and I_D/I_G ratios of graphene synthesized using different CH_4/H_2 gas flow ratio mixtures and enclosures. A too low CH_4/H_2 gas flow ratio (i.e., 0.11) was not sufficient to initiate the growth of graphene on the Si(100) substrate. The further increase in the methane flow and decrease in the hydrogen flow resulted in the increase in the number of graphene layers, as evident from the I_{2D}/I_G ratio decrease (Figure 5a). It was also found that the I_D/I_G ratio decreased with the increase in the CH_4/H_2 gas flow ratio for the 1st and the 2nd enclosure (Figure 5b).

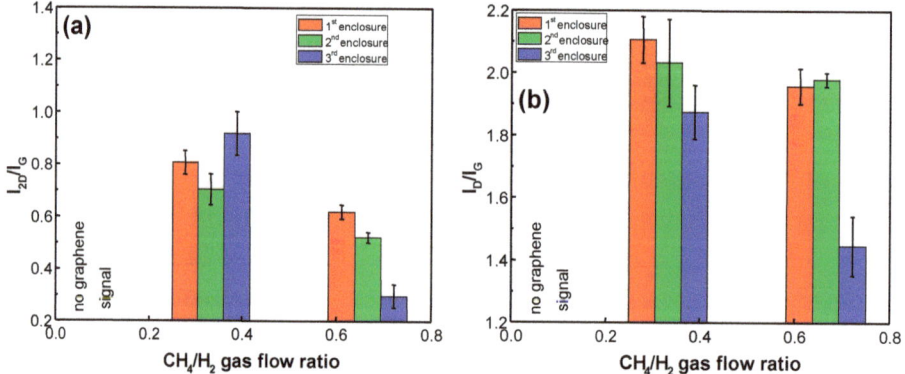

Figure 5. I_{2D}/I_G (a) and I_D/I_G (b) ratios of graphene synthesized using different flow ratio (0.11, 0.33, 0.67) CH_4/H_2 mixture of gas and enclosures. In all cases, every other process parameter is kept constant (power 1.2 kW, pressure 30 mBar, temperature 900 °C, time 30 min).

Figure 6 shows the I_{2D}/I_G and I_D/I_G ratios of graphene synthesized using different temperatures and enclosures. It is important to note that no graphene growth was observed at 600 °C. As it can be seen in Figure 6a, the I_{2D}/I_G ratio increases with the process temperature. Thus, the number of graphene layers grown decreases with the increase in temperature when the microwave PECVD process is performed in the range of 700–900 °C. For graphene synthesized using the 1st enclosure, the I_D/I_G ratio increases with temperature (Figure 6b). However, the highest density of defects was observed for the graphene synthesized using the 2nd enclosure at 700 °C.

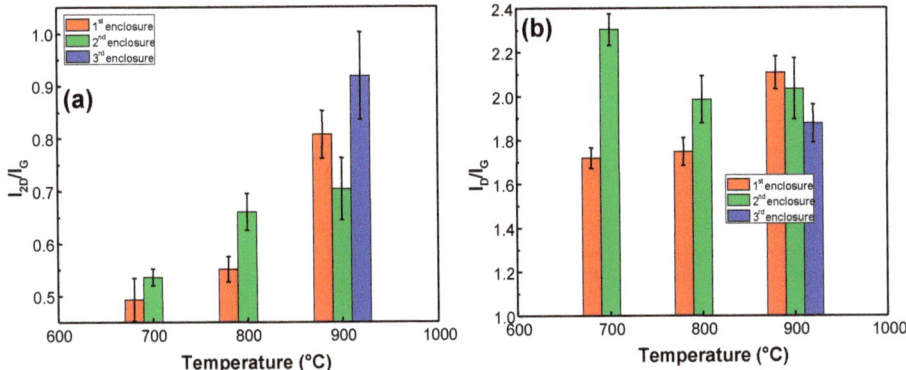

Figure 6. I_{2D}/I_G (a) and I_D/I_G (b) ratios of graphene synthesized using different temperatures (700 °C, 800 °C, 900 °C) and several enclosures. In all cases, every other process parameter is kept constant (H_2 gas flow 150 sccm sccm, CH_4 gas flow 50 sccm, power 1.2 kW, pressure 30 mBar, time 30 min).

Altogether, it was revealed that the most crucial graphene direct synthesis parameter in our case is the CH_4/H_2 gas flow ratio. A too low ratio results in no graphene synthesis. The number of graphene layers increases with methane flow when the ratio is substantially large for graphene synthesis. At the same time, defect density decreases. This is valid for all studied enclosures. When the temperature is too low, graphene does not grow. Subsequently, the temperature increase results in the decrease in graphene layer thickness. However, no typical behavior regarding defect density can be found. The plasma power effects are the least clear. The possible physical mechanisms hidden behind these results will be considered in Sections 4.2 and 4.3.

3.3. The Number of Graphene Layers and Defect Density

Figure 7 shows the I_{2D}/I_G vs. I_D/I_G ratio plot for all investigated samples (Table 1). It is considered that the I_{2D}/I_G vs. I_D/I_G ratio change for the 1st and the 3rd enclosure followed linear distribution pattern as the I_{2D}/I_G increased with I_D/I_G ratio. Such an outcome contradicts the results reported by [24] (Table S1), where I_{2D}/I_G decreased as a result of oxygen ion etching. It can be explained by the different nature of the defects in our study and [24] (boundary defects vs. irradiation defects).

No clear dependence of the I_{2D}/I_G ratio on the enclosure configuration was found (Figure 7). The lowest I_D/I_G ratios were observed for graphene synthesized using the 3rd enclosure, while for graphene samples synthesized using the 1st enclosure, the range of I_D/I_G ratio values was the broadest. Almost all I_D/I_G ratio values of the graphene synthesized using the 2nd enclosure were within the range typical for graphene grown using the 1st enclosure. Additionally, comparing the I_D/I_G ratios of the graphene samples synthesized using the 1st and the 2nd enclosures and the same other synthesis conditions, one can see that in some cases larger I_D/I_G ratio values were found for the 1st enclosure, and in other cases for the 3rd enclosure (Figures 4–6).

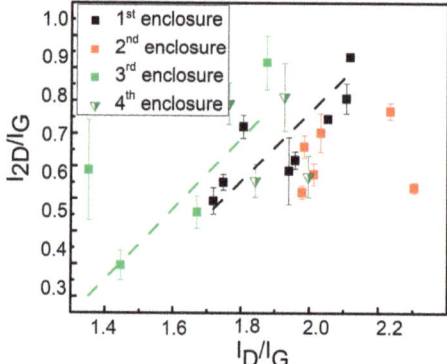

Figure 7. I_{2D}/I_G vs. I_D/I_G plot for all investigated samples. Dash lines represent the observed linear distributions.

The enclosure design was simplified even more, considering the results described above, though less graphene synthesis experiments were performed using the 3rd enclosure. No holes at the protective enclosure's center resulted in a decrease in the I_D/I_G ratio (see Figure 7). Therefore, holes were removed from the enclosure's entire surface to suppress the direct plasma effects further. The modified enclosure shape was simply a rectangular steel sheet folded in two places (Figure 1d). Graphene was synthesized at the temperature of 700 °C using a protective sheath of such a simple structure. The I_{2D}/I_G and I_D/I_G ratios of these graphene samples were within the typical values for the 3rd envelope, although no additional optimization of the deposition conditions was performed. Further detailed research on graphene synthesized using simplified enclosures is in progress.

3.4. Dopant Density and Stress

The 2D peak position dependence on the G peak position can provide information about the graphene's doping and stresses in graphene layers. This was shown in numerous studies investigating single-layer graphene synthesized by CVD on catalytic copper foil and transferred to the target substrate [28–33]. Some studies on multilayer transferred graphene have also been carried out [30]. The reported results from different authors follow similar dependencies. However, in directly synthesized graphene, the defect density is usually higher [20]. Therefore, the D peak is visible in the directly synthesized graphene's Raman scattering spectrum [20–22,38]. This makes the situation more complicated. Hence, an additional analysis was carried out regarding the possible influence of defects on other Raman scattering spectrum parameters. It should be noted that the I_{2D}/I_G ratio depends on the defect density, concentration of dopants, and the number of graphene layers. It was shown that the I_{2D}/I_G ratio decreases with the appearance of the defect-related D peak in the Raman spectrum and the subsequent increase in defect density [41]. Graphene doping also leads to a reduced I_{2D}/I_G ratio [41]. However, in our case, opposite dependence was found (Figure 7).

Another structural parameter that can influence the analysis of graphene doping and stress level is the different number of graphene layers. Both the 2D and G peak positions depend on the number of graphene layers (Table S1). In some studies, the upshift of the 2D peak position has been shown for few-layer graphene [30]. Following the latter research work, Raman spectra of the transferred graphene and the relationship between Pos(2D) vs. Pos(G) plots and stress as well as doping were analyzed [30].

Figure S1 shows a shift from Pos(2D) to the higher wavenumbers with increasing I_{2D}/I_G ratio. However, Pos(2D) should shift to the lower wavenumbers with increasing I_{2D}/I_G ratio due to the decrease in the number of graphene layers [39]. In our case, no clear dependence of the I_{2D}/I_G ratio on the G peak position was found (Figure S2). Thus, the Pos(2D) upshift should not be related to the increased number of graphene layers.

The dependence of Pos(2D) on Pos(G) is shown in Figure 8. Both the 2D and the G peak positions are shifted to the higher wavenumbers than the values typical for single-layer defect-free graphene. A vector analysis of the plot was performed according to the methodology presented in [28–33]. It can be seen that the 2D peak position is significantly shifted to the higher wavenumbers (by ~25–35 cm^{-1}) compared to a value typical for defect-free, undoped, transferred single-layer graphene. The graphene samples synthesized in the present study were determined to be in the range of 2–4 layers thick, according to the analysis of the I_{2D}/I_G ratio (see Figure 7 and Figure S3, for calculation method, see [23] and Table S1). However, the 2D peak position values are also shifted upwards compared to the position typical for two-layer defect-free undoped transferred graphene. Thus, it is considered that compressive stress was present in the investigated graphene samples (according to [28–33] and Table S1), in agreement with [42].

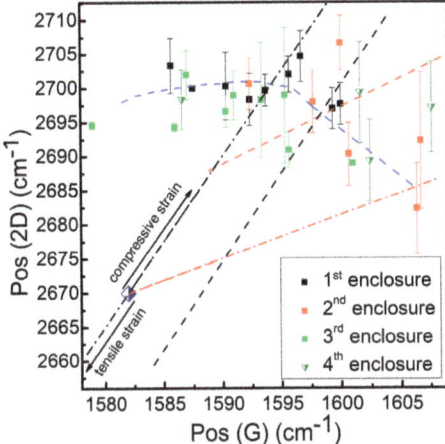

Figure 8. Pos(2D) vs. Pos(G) plot. The black dash-dot line refers to the undoped strained graphene (plotted according to the method [28]). The black dot line refers to the p-type doped strained graphene (constant hole concentration and different stress levels) (plotted according to [28]). The red dash-dot line refers to the unstrained p-type graphene (plotted according to [28]). The red dot line refers to the p-type doped strained graphene (constant stress level and different hole concentrations) (plotted according to the method [28]). The blue dash line refers to the strained n-type doped graphene (plotted according to [30], taking into account graphene layer number related shift of 2D peak position). The navy and white colored rhombus symbol refers to the unstrained and undoped graphene [28].

The analysis of the plot (Figure 8, Figure S4) regarding possible doping of the graphene revealed a less convenient picture. It should be emphasized that p-type graphene was investigated in most of the studies mentioned above [28–31,33]. The results reported in [28] revealed unintentional p-type doping of the graphene. However, the overall dependence of Pos(2D) on Pos(G) does not follow the vectors typical for p-type doped graphene (Figure 8, Figure S4). This behavior is rather typical for the n-type doped and strained graphene [30].

3.5. AFM Study

Several graphene samples synthesized using the 4th protective enclosure were studied by AFM to determine the number of the graphene layers as well as to evaluate the continuity of the graphene. AFM topographical images of the samples are presented in Figures S5–S7. The graphene samples' surface morphology is very different from silicon substrate morphology (Figure S8). In all cases,

AFM images revealed continuous horizontal graphene layers (Figures S5–S7). Similar AFM images were reported for directly synthesized graphene by other authors (e.g., [20,21,44,45]). One can notice some black features corresponding to the lowest surface points. It can be interpreted as holes in the graphene [46].

Previous studies [47–49] reported that step height of a single-layer graphene was found to be in the range of ~0.35–0.4 nm. The graphene AFM height profile was analyzed. Approximate graphene thickness was calculated by measuring the height from the zero points corresponding to the graphene holes to the profile maxima. A mixture of the single-layer and two-layer graphene was found for samples 1E4 and 2E4 (Figures S5–S7). In the case of the sample 3E4, up to three graphene layers can be found. An alternative graphene thickness evaluation method by using the AFM measurement data was performed following the histogram method [46,48]. Hence, the influence of the adsorbed contaminants can be minimized. According to the histogram method, the thickness of the graphene was between one and two layers for samples 1E4 and 3E4, while the thickness of sample 2E4 corresponded to single-layer graphene (Table S2, and Figures S5b, S6b, S7b and S8b). Thus, the average graphene thickness was found to be between one and two layers. In this case, graphene thickness evaluated via the use of AFM measurement data is in good agreement with graphene thickness calculated from the I_{2D}/I_G ratio according to [23] (Figure S3).

4. Discussion

4.1. Effect of the Deposition Conditions. Comparison with Results Reported Elsewhere

Our obtained results were further compared with previous studies. In most of the studies regarding direct graphene synthesis, graphene was grown on different dielectric substrates. There are substantially less studies reporting direct graphene synthesis on silicon. Therefore, graphene's direct growth on different substrates was considered in this comparison.

It must be pointed out that in [38], no clear dependence of the I_{2D}/I_G ratio on the process time was found for graphene directly synthesized on dielectric SiO_2, quartz, and sapphire substrates from a $CH_4/H_2/Ar$ gas mixture. Nevertheless, an I_D/I_G ratio increase with the process time was observed, in good agreement with the present research.

There are few studies on the effects of the hydrocarbon gas and hydrogen flow ratios on the structural properties of directly synthesized graphene. Analogously to our research, the I_{2D}/I_G and I_D/I_G ratios decreased with increasing C_2H_2/H_2 flow ratio for graphene directly synthesized on fused silica and quartz by electron cyclotron resonance (ECR) PECVD [44]. A similar tendency was found in [45], where direct graphene synthesis on quartz via the CVD process was performed. Following the latter work, a too low methane flow resulted in no graphene growth, in agreement with our results. However, the I_D/I_G ratio increased with the increase in CH_4/H_2 flow ratio [45].

Similarly to the graphene synthesized using the 1st enclosure, the I_{2D}/I_G ratio increased with plasma power for graphene directly synthesized via PECVD on SiO_2 [50]. However, the I_D/I_G ratio in that study decreased with plasma power [50].

Likewise to the graphene synthesized in our study using the 1st and the 2nd enclosure, the I_{2D}/I_G ratio increased with process temperature for graphene synthesized on Si(100) and glass from a C_2H_2/Ar gas mixture via microwave PECVD [51]. An opposite result was reported for graphene grown on quartz and fused silica substrates from a C_2H_2/H_2 gas mixture via ECR PECVD [44]. In in [50] and [21], I_{2D}/I_G ratio dependence on temperature was rather non-monotonic. The I_{2D}/I_G ratio was highest [21] or lowest [50] at the specific temperature range used for the synthesis of graphene. Similarly to the graphene synthesized using the 1st enclosure, the I_D/I_G ratio increased with temperature in [50]. A decrease in the I_D/I_G ratio with temperature was reported in [21,44,51], similar to the case of the graphene synthesized using the 2nd enclosure.

Thus, the different effects of the gas flow ratio, plasma power, and temperature on the graphene structure were found in various studies. The discrepancy of the results reported by other authors was more considerable than the results reported in the present study.

4.2. Effect of the Deposition Conditions. Physical and Chemical Phenomena Involved

The graphene structure's dependence on the technological synthesis process conditions found in the present research can be explained by considering the main graphene growth-related physical and chemical processes.

Notably, the decrease in the number of graphene layers with increasing process time can be explained by the hydrogen etching prevailing over the growth of additional graphene layers as it was reported in [52].

The increase in the graphene I_{2D}/I_G ratio with plasma power (Figure 4) can be related to the dependence of the methane and hydrogen dissociation rate on plasma power [50]. It can be considered that, in the case of graphene grown using the 1st enclosure, carbon-containing reactive species concentration increased with plasma power slower than hydrogen atoms and ion concentration, while in other cases, changes in that concentration with plasma power were non-monotonic. It seems that graphene defect density increased with plasma due to the decrease in the graphene crystallite size or the enhanced irradiation by ions and electrons.

The decrease in the number of graphene layers with increased CH_4/H_2 gas flow ratio was observed in Figure 5. It can be explained by competition between two processes: graphene growth due to the carbon-containing active species flux towards the surface [50,53] and etching of the carbon-carbon bonds by hydrogen [44,54,55]. If the CH_4/H_2 ratio is too low, the etching reaction is much faster than the growth of the graphene layers [56]. This is the reason why no graphene growth was observed for samples 3E1, 4E2, and 4E3. The increase in the methane to hydrogen gas flow ratio resulted in a suppression of graphene growth over the etching. In this case, a further increase in the active flux of CH_x and C species towards the substrate and the reduced amount of hydrogen atoms will result in the increase in the number of graphene layers. The increasing size of graphene nanocrystals cannot explain the reduction in the I_D/I_G ratio with the increasing CH_4/H_2 ratio presented in Figure 5. It is because the increase in graphene nucleus density with decreased H_2 content due to the hydrogen etching resulted in a smaller graphene grain size [44]. Thus, the possible lowering of the hydrogen plasma-induced defect density with decreased hydrogen gas flow should be considered [57].

Several phenomena should be taken into account to explain the lowering of the number of graphene layers with increased process temperature (Figure 6). It was reported in [44] that the rate of the graphene etching by hydrogen decreases with temperature [44]. Therefore, it should instead result in a decrease in the I_{2D}/I_G ratio with temperature. On the other hand, the desorption rate of carbon atoms increases with temperature, as reported in [52]. This can be a cause of the decrease in the number of graphene layers with increased deposition temperature.

4.3. Effect of Protective Enclosure Design

A few studies could be considered while analyzing the protective enclosure design's influence on the graphene structure. Notably, a study on graphene synthesized on glass using a copper foam-based protective Faraday cage revealed a tendency that a smaller aperture hole size can result in a better electric field shielding effect [22]. Thus, it can be considered that in our case, the lower I_D/I_G ratios for graphene synthesized using the 3rd protective enclosure can be explained by better suppression of the electric field due to the absence of the holes at the enclosure's center. On the other hand, no apparent difference between the I_D/I_G ratios of the graphene synthesized using the 1st and the 2nd protective enclosures is in good accordance with [22], where reduced protective cage hole size resulted in no further apparent decrease in electric field strength.

4.4. Mechanisms Responsible to the N-Type Self Doping of Graphene and Induction of Compressive Stress

According to Figure 8 and Figure S4, unintentional n-type doping was found for directly synthesized graphene. It is worth noting that self-induced doping was already observed for graphene directly synthesized on Ge(111) by CVD [58]. Possible sources of such behavior would be adsorbates or substrate-induced effects. Atmospheric adsorbates (oxygen, water) usually result in p-type doping of the graphene [59–61]. P-type self-doping for graphene transferred onto the SiO_2 substrate occurred due to various surface treatments and residual charges created on the substrate [62]. Unintentional graphene n-type doping was reported for epitaxial graphene directly synthesized on SiC via high-temperature annealing in a vacuum or inert ambient gas [63,64]. It was explained by the substrate-to-graphene charge transfer [64]. A study of the graphene transferred onto different substrates revealed that p-type self-doping, n-type self-doping, and no doping could be achieved via selection of the appropriate substrate [58,65]. Simulations revealed that when graphene is put onto SiO_2, the graphene's electronic structure strongly depends on the interface geometry and surface polarity [66]. In the case of the O-polar SiO2 surface with dangling bonds, graphene's p-type doping takes place [66]. Graphene placing on the Si-polar surface with dangling bonds results in graphene's n-type self-doping [66]. Considering the studies mentioned above, it is supposed in our case that charge transfer from the Si(100) substrate to the graphene took place during the microwave PECVD process. It resulted in the n-type self-doping of the graphene, similar to the cases of SiC and Si-polar SiO_2 substrates.

One can see in Figure 8 that directly synthesized graphene in the present study is found to be stressed. This stress is compressive in nature. It should be mentioned that, in the case of the exfoliated graphene transferred onto SiO_2, pristine graphene sheets may exhibit both compressive and tensile strain [28]. This native strain becomes compressive due to the annealing at 100 °C or higher temperatures [28]. Similar effects of the annealing were reported for CVD synthesized graphene transferred onto the SiO_2 substrate [67]. On the other hand, epitaxial graphene directly grown on SiC above 1100 °C exhibited substrate-induced compressive strain [28]. Compressive stress may be present in graphene directly synthesized on Si(100) [42], quartz [44], Ge(110) [58], and SiO_2 [68]. Thus, our results are in good agreement with the studies mentioned above. It is considered that, in our case, compressive stress was induced during a direct graphene synthesis as thermal stress due to the large lattice mismatch between graphene and Si, as was suggested in [20,69].

5. Conclusions

In conclusion, the transfer-less and catalyst-less synthesis of graphene on Si(100) substrates via a combination of direct microwave plasma-enhanced chemical vapor deposition and protective enclosures was performed.

A study of the effect of the CH_4/H_2 gas flow ratio, temperature, and plasma power on graphene structure revealed that the most significant technological parameter used in the present study was methane and hydrogen gas flow ratio. Plasma power effects were the least pronounced. It seems that if a temperature is sufficiently high for graphene synthesis, the crucial process is a competition between plasma etching by hydrogen and carbon-containing active species influx towards the surface. If hydrogen flow is too high and/or methane flow is too low, etching prevails against growth and no graphene is formed. Afterwards, the number of graphene layers increases with carbon species flow and/or with decreased hydrogen flow. Hydrogen species density is a much more critical etching factor than increased plasma power. The thermally stimulated desorption of carbon atoms is important, while the formation of the plasma-induced radiation defects has less influence on graphene growth and defect density.

A study of the enclosure effects revealed no top hole size effects for investigated enclosures. The absence of top holes in the middle of the enclosure reduced the plasma effect on the growing graphene and decreased defect density. The graphene was successfully synthesized using just a rectangular steel sheet folded in two places as a simplified protective enclosure, considering these results. Graphene was grown at a temperature of 700 °C using a protective sheath with such a simple structure.

Analysis of the positions of 2D and G peaks revealed unintentional n-type doping of the graphene. It was explained by charge transfer from the Si(100) substrate to the graphene. The presence of compressive stress was found in graphene. It was supposed that the large lattice mismatch between the growing graphene and the silicon induced thermal stress.

An atomic force microscopy study confirmed the growth of continuous horizontal graphene layers.

Supplementary Materials: The following are available online at http://www.mdpi.com/1996-1944/13/24/5630/s1, Table S1: Possible relations of the Raman scatterings spectra parameters mentioned above the number of graphene layers, stress, doping and defect density, Table S2: Graphene samples and silicon substrate surface roughness histogram peak maximums and graphene thickness values according to the histogram method, Figure S1: Pos(2D) Vs I2D/IG plot, Figure S2: I2D/IG Vs Pos(G) plot, Figure S3: I2D/IG ratio of samples 4E1, 4E2, 4E3 and number of the graphene layers calculated according to [23], Figure S4: Pos(2D) vs. Pos(G) plot for sample 1E4, Figure S5: AFM image (a), height distribution histogram (b) and height profile (c) of the graphene sample No 1E4, Figure S6: AFM image (a), height distribution histogram (b) and height profile (c) of the graphene sample No 2E4, Figure S7: AFM image (a), height distribution histogram (b) and height profile (c) of the graphene sample No 3E4.

Author Contributions: Conceptualization, R.G. and Š.M.; investigation, R.G., A.L. and Š.J.; writing—original draft preparation, Š.M., A.L. and Š.J.; writing—review and editing, Š.M.; visualization, Š.M. and Š.J.; project administration, Š.M.; funding acquisition, Š.M. All authors have read and agreed to the published version of the manuscript.

Funding: The research project No. 09.3.3-LMT-K-712-01-0183 is funded under the European Social Fund measure "Strengthening the Skills and Capacities of Public Sector Researchers for Engaging in High Level R&D Activities" administered by the Research Council of Lithuania.

Acknowledgments: The authors acknowledge other participants of the research project No. 09.3.3-LMT-K-712-01-0183—A. Vasiliauskas, A. Guobienė, K. Šlapikas, V. Stankus, D. Peckus, E. Rajackaitė, T. Tamulevičius, A. Jurkevičiūtė, and F. Kalyk.

Conflicts of Interest: The authors declare no conflict of interest.

References

1. Sattar, T. Current Review on Synthesis, Composites and Multifunctional Properties of Graphene. *Top. Curr. Chem.* **2019**, *377*, 10. [CrossRef]
2. Banszerus, L.; Schmitz, M.; Engels, S.; Dauber, J.; Oellers, M.; Haupt, F.; Stampfer, C. Ultrahigh-mobility graphene devices from chemical vapor deposition on reusable copper. *Sci. Adv.* **2015**, *1*, e1500222. [CrossRef] [PubMed]
3. Tielrooij, K.J.; Song, J.C.W.; Jensen, S.A.; Centeno, A.; Pesquera, A.; Zurutuza Elorza, A.; Bonn, M.; Levitov, L.S.; Koppens, F.H.L. Photoexcitation cascade and multiple hot-carrier generation in graphene. *Nat. Phys.* **2013**, *9*, 248–252. [CrossRef]
4. Song, Y.; Fang, W.; Brenes, R.; Kong, J. Challenges and opportunities for graphene as transparent conductors in optoelectronics. *Nano Today* **2015**, *10*, 681–700. [CrossRef]
5. Nayak, P.K. Pulsed-grown graphene for flexible transparent conductors. *Nanoscale Adv.* **2019**, *1*, 1215–1223. [CrossRef]
6. Song, L.; Yu, X.; Yang, D. A review on graphene-silicon Schottky junction interface. *J. Alloys Compd.* **2019**, *806*, 63–70. [CrossRef]
7. Di Bartolomeo, A. Graphene Schottky diodes: An experimental review of the rectifying graphene/semiconductor heterojunction. *Phys. Rep.* **2016**, *606*, 1–58. [CrossRef]
8. Donnelly, M.; Mao, D.; Park, J.; Xu, G. Graphene field-effect transistors: The road to bioelectronics. *J. Phys. D Appl. Phys.* **2018**, *51*, 493001. [CrossRef]
9. Geng, H.; Yuan, D.; Yang, Z.; Tang, Z.; Zhang, X.; Yang, G.; Su, Y. Graphene van der Waals heterostructures for high-performance photodetectors. *J. Mater. Chem. C* **2019**, *7*, 11056–11067. [CrossRef]
10. Rogalski, A. Graphene-based materials in the infrared and terahertz detector families: A tutorial. *Adv. Opt. Photon.* **2019**, *11*, 314–379. [CrossRef]

11. Shin, D.; Choi, S.-H. Graphene-Based Semiconductor Heterostructures for Photodetectors. *Micromachines* **2018**, *9*, 350. [CrossRef] [PubMed]
12. Kong, X.; Zhang, L.; Liu, B.; Gao, H.; Zhang, Y.; Yan, H.; Song, X. Graphene/Si Schottky solar cells: A review of recent advances and prospects. *RSC. Adv.* **2019**, *9*, 863–877. [CrossRef]
13. Patil, K.; Rashidi, S.; Wang, H.; Wei, W. Recent Progress of Graphene-Based Photoelectrode Materials for Dye-Sensitized Solar Cells. *Int. J. Photoenergy* **2019**, *2019*, 1–16. [CrossRef]
14. Iqbal, M.Z.; Rehman, A.-U. Recent progress in graphene incorporated solar cell devices. *Sol. Energy* **2018**, *169*, 634–647. [CrossRef]
15. Zhang, Y.; Zhang, L.; Zhou, C. Review of Chemical Vapor Deposition of Graphene and Related Applications. *Acc. Chem. Res.* **2013**, *46*, 2329–2339. [CrossRef] [PubMed]
16. Haigh, S.J.; Gholinia, A.; Jalil, R.; Romani, S.; Britnell, L.; Elias, D.C.; Novoselov, K.S.; Ponomarenko, L.A.; Geim, A.K.; Gorbachev, R. Cross-sectional Imaging of Individual Layers and Buried Interfaces of Graphene-Based Heterostructures and Superlattices. *Nat. Mater.* **2012**, *11*, 764–767. [CrossRef]
17. Deng, S.; Berry, V. Wrinkled, rippled and crumpled graphene: An overview of formation mechanism, electronic properties, and applications. *Mater. Today* **2016**, *19*, 197–212. [CrossRef]
18. Shtepliuk, I.; Khranovskyy, V.; Yakimova, R. Combining graphene with silicon carbide: Synthesis and properties—A review. *Semicond. Sci. Technol.* **2016**, *31*, 113004. [CrossRef]
19. She, X.; Huang, A.Q.; Lucia, O.; Ozpineci, B. Review of Silicon Carbide Power Devices and Their Applications. *IEEE Trans. Ind. Electron.* **2017**, *64*, 8193–8205. [CrossRef]
20. Khan, A.; Islam, S.M.; Ahmed, S.; Kumar, R.R.; Habib, M.R.; Huang, K.; Hu, M.; Yu, X.; Yang, D. Direct CVD Growth of Graphene on Technologically Important Dielectric and Semiconducting Substrates. *Adv. Sci.* **2018**, *5*, 1800050. [CrossRef]
21. Zheng, S.; Zhong, G.; Wu, X.; D'Arsiè, L.; Robertson, J. Metal-catalyst-free growth of graphene on insulating substrates by ammonia-assisted microwave plasma-enhanced chemical vapor deposition. *RSC Adv.* **2017**, *7*, 33185–33193. [CrossRef]
22. Qi, Y.; Deng, B.; Guo, X.; Chen, S.; Gao, J.; Li, T.; Dou, Z.; Ci, H.; Sun, J.; Chen, Z.; et al. Switching Vertical to Horizontal Graphene Growth Using Faraday Cage-Assisted PECVD Approach for High-Performance Transparent Heating Device. *Adv. Mater.* **2018**, *30*, 1704839. [CrossRef]
23. Hwang, J.-S.; Lin, Y.-H.; Hwang, J.-Y.; Chang, R.; Chattopadhyay, S.; Chen, C.-J.; Chen, P.; Chiang, H.-P.; Tsai, T.-R.; Chen, L.-C. Imaging layer number and stacking order through formulating Raman fingerprints obtained from hexagonal single crystals of few layer graphene. *Nanotechnology* **2012**, *24*, 015702. [CrossRef]
24. Childres, I.; Jauregui, L.A.; Tian, J.; Chen, Y.P. Effect of oxygen plasma etching on graphene studied using Raman spectroscopy and electronic transport measurements. *New J. Phys.* **2011**, *13*, 025008. [CrossRef]
25. Dresselhaus, M.S.; Jorio, A.; Souza Filho, A.G.; Saito, R. Defect characterization in graphene and carbon nanotubes using Raman spectroscopy. *Philos. Trans. R. Soc. A* **2010**, *368*, 5355–5377. [CrossRef] [PubMed]
26. Zhao, W.; Tan, P.H.; Liu, J.; Ferrari, A.C. Intercalation of Few-Layer Graphite Flakes with $FeCl_3$: Raman Determination of Fermi Level, Layer by Layer Decoupling, and Stability. *J. Am. Chem. Soc.* **2011**, *133*, 5941–5946. [CrossRef] [PubMed]
27. Szirmai, P.; Márkus, B.G.; Chacón-Torres, J.C.; Eckerlein, P.; Edelthalhammer, K.; Englert, J.M.; Mundloch, U.; Hirsch, A.; Hauke, F.; Náfrádi, B.; et al. Characterizing the maximum number of layers in chemically exfoliated graphene. *Sci. Rep.* **2019**, *9*, 19480. [CrossRef] [PubMed]
28. Lee, J.E.; Ahn, G.; Shim, J.; Lee, Y.S.; Ryu, S. Optical separation of mechanical strain from charge doping in graphene. *Nat. Commun.* **2012**, *3*, 1024. [CrossRef]
29. Sakavičius, A.; Astromskas, G.; Bukauskas, V.; Kamarauskas, M.; Lukša, A.; Nargelienė, V.; Niaura, G.; Ignatjev, I.; Treideris, M.; Šetkus, A. Long distance distortions in the graphene near the edge of planar metal contacts. *Thin Solid Films* **2020**, *698*, 137850. [CrossRef]
30. Kim, S.; Ryu, S. Thickness-dependent native strain in graphene membranes visualized by Raman spectroscopy. *Carbon* **2016**, *100*, 283–290. [CrossRef]
31. Armano, A.; Buscarino, G.; Cannas, M.; Gelardi, F.M.; Giannazzo, F.; Schilirò, E.; Agnello, S. Monolayer graphene doping and strain dynamics induced by thermal treatments in controlled atmosphere. *Carbon* **2018**, *127*, 270–279. [CrossRef]

32. Neumann, C.; Reichardt, S.; Venezuela, P.; Drögeler, M.; Banszerus, L.; Schmitz, M.; Watanabe, K.; Taniguchi, T.; Mauri, F.; Beschoten, B.; et al. Raman spectroscopy as probe of nanometre-scale strain variations in graphene. *Nat. Commun.* **2015**, *6*, 8429. [CrossRef] [PubMed]
33. Lee, U.; Han, Y.; Lee, S.; Kim, J.S.; Lee, Y.H.; Kim, U.J.; Son, H. Time Evolution Studies on Strain and Doping of Graphene Grown on a Copper Substrate Using Raman Spectroscopy. *ACS Nano* **2020**, *14*, 919–926. [CrossRef] [PubMed]
34. Wu, J.-B.; Lin, M.-L.; Cong, X.; Liu, H.-N.; Tan, P.-H. Raman spectroscopy of graphene-based materials and its applications in related devices. *Chem. Soc. Rev.* **2018**, *47*, 1822–1873. [CrossRef] [PubMed]
35. Zeng, Y.; Lo, C.-L.; Zhang, S.; Chen, Z.; Marconnet, A. Dynamically tunable thermal transport in polycrystalline graphene by strain engineering. *Carbon* **2020**, *158*, 63–68. [CrossRef]
36. Mohiuddin, T.M.G.; Lombardo, A.; Nair, R.R.; Bonetti, A.; Savini, G.; Jalil, R.; Bonini, N.; Basko, D.M.; Galiotis, C.; Marzari, N.; et al. Uniaxial strain in graphene by Raman spectroscopy: G peak splitting, Grüneisen parameters, and sample orientation. *Phys. Rev. B* **2009**, *79*, 205433. [CrossRef]
37. Ni, Z.H.; Yu, T.; Lu, Y.H.; Wang, Y.Y.; Feng, Y.P.; Shen, Z.X. Uniaxial Strain on Graphene: Raman Spectroscopy Study and Band-Gap Opening. *ACS Nano* **2008**, *2*, 2301–2305. [CrossRef]
38. Chugh, S.; Mehta, R.; Lu, N.; Dios, F.D.; Kim, M.J.; Chen, Z. Comparison of graphene growth on arbitrary non-catalytic substrates using low-temperature. *Carbon* **2015**, *93*, 393–399. [CrossRef]
39. Merlen, A.; Buijnsters, J.; Pardanaud, C. A Guide to and Review of the Use of Multiwavelength Raman Spectroscopy for Characterizing Defective Aromatic Carbon Solids: From Graphene to Amorphous Carbons. *Coatings* **2017**, *7*, 153. [CrossRef]
40. Thomsen, C.; Reich, S. Double Resonant Raman Scattering in Graphite. *Phys. Rev. Lett.* **2000**, *85*, 5214–5217. [CrossRef]
41. Zafar, Z.; Ni, Z.H.; Wu, X.; Shi, Z.X.; Nan, H.Y.; Bai, J.; Sun, L.T. Evolution of Raman spectra in nitrogen doped graphene. *Carbon* **2013**, *61*, 57. [CrossRef]
42. Soin, N.; Roy, S.S.; O'Kane, C.; McLaughlin, J.A.D.; Lim, T.H.; Hetherington, C.J.D. Exploring the fundamental effects of deposition time on the microstructure of graphene nanoflakes by Raman scattering and X-ray diffraction. *Cryst. Engl. Commun.* **2011**, *13*, 312–318. [CrossRef]
43. Ferrari, A.C.; Meyer, J.C.; Scardaci, V.; Casiraghi, C.; Lazzeri, M.; Mauri, F.; Piscanec, S.; Jiang, D.; Novoselov, K.S.; Roth, S.; et al. Raman Spectrum of Graphene and Graphene Layers. *Phys. Rev. Lett.* **2006**, *97*, 187401. [CrossRef] [PubMed]
44. Muñoz, R.; Munuera, C.; Martínez, J.I.; Azpeitia, J.; Gómez-Aleixandre, C.; García-Hernández, M. Low temperature metal free growth of graphene on insulating substrates by plasma assisted chemical vapor deposition. *2D Mater.* **2017**, *4*, 015009. [CrossRef] [PubMed]
45. Xu, S.; Man, B.; Jiang, S.; Yue, W.; Yang, C.; Liu, M.; Chen, C.; Zhang, C. Direct growth of graphene on quartz substrates for label-free detection of adenosine triphosphate. *Nanotechnology* **2014**, *25*, 165702. [CrossRef] [PubMed]
46. Zhou, L.; Fox, L.; Włodek, M.; Islas, L.; Slastanova, A.; Robles, E.; Bikondo, O.; Harniman, R.; Fox, N.; Cattelan, M.; et al. Surface structure of few layer graphene. *Carbon* **2018**, *136*, 255–261. [CrossRef]
47. Nemes-Incze, P.; Osváth, Z.; Kamarás, K.; Biró, L.P. Anomalies in thickness measurements of graphene and few layer graphite crystals by tapping mode atomic force microscopy. *Carbon* **2008**, *46*, 1435–1442. [CrossRef]
48. Yaxuan, Y.; Lingling, R.; Sitian, G.; Shi, L. Histogram method for reliable thickness measurements of graphene films using atomic force microscopy (AFM). *J. Mater. Sci. Technol.* **2017**, *33*, 815–820. [CrossRef]
49. Shearer, C.J.; Slattery, A.D.; Stapleton, A.J.; Shapter, J.G.; Gibson, C.T. Accurate thickness measurement of graphene. *Nanotechnology* **2016**, *27*, 125704. [CrossRef]
50. Kim, Y.S.; Joo, K.; Jerng, S.-K.; Lee, J.H.; Yoon, E.; Chun, S.-H. Direct growth of patterned graphene on SiO_2 substrates without the use of catalysts or Lithography. *Nanoscale* **2014**, *6*, 10100. [CrossRef]
51. Kalita, G.; Kayastha, M.S.; Uchida, H.; Wakita, K.; Umeno, M. Direct growth of nanographene films by surface wave plasma chemical vapor deposition and their application in photovoltaic devices. *RSC Adv.* **2012**, *2*, 3225–3230. [CrossRef]
52. Chaitoglou, S.; Bertran, E. Effect of temperature on graphene grown by chemical vapor deposition. *J. Mater. Sci.* **2017**, *52*, 8348–8356. [CrossRef]

53. Scaparro, A.M.; Miseikis, V.; Coletti, C.; Notargiacomo, A.; Pea, M.; De Seta, M.; Di Gaspare, L. Investigating the CVD Synthesis of Graphene on Ge(100): Toward Layer-by-Layer Growth. *ACS Appl. Mater. Interfaces* **2016**, *8*, 33083–33090. [CrossRef] [PubMed]
54. Kim, Y.S.; Lee, J.H.; Kim, Y.D.; Jerng, S.-K.; Joo, K.; Kim, E.; Jung, J.; Yoon, E.; Park, Y.D.; Seo, S.; et al. Methane as an effective hydrogen source for single-layer graphene synthesis on Cu foil by plasma enhanced chemical vapor deposition. *Nanoscale* **2013**, *5*, 1221–1226. [CrossRef]
55. Kaur, G.; Kavitha, K.; Lahiri, I. Transfer-Free Graphene Growth on Dielectric Substrates: A Review of the Growth Mechanism. *Crit. Rev. Solid State Mater. Sci.* **2019**, *44*, 157–209. [CrossRef]
56. Park, H.J.; Meyer, J.; Roth, S.; Skákalová, V. Growth and properties of few-layer graphene prepared by chemical vapor deposition. *Carbon* **2010**, *48*, 1088–1094. [CrossRef]
57. Hug, D.; Zihlmann, S.; Rehmann, M.K.; Kalyoncu, Y.B.; Camenzind, T.N.; Marot, L.; Watanabe, K.; Taniguchi, T.; Zumbühl, D.M. Anisotropic etching of graphite and graphene in a remote hydrogen plasma. *NPJ 2D Mater. Appl.* **2017**, *1*, 1–6. [CrossRef]
58. Kiraly, B.; Jacobberger, R.M.; Mannix, A.J.; Campbell, G.P.; Bedzyk, M.J.; Arnold, M.S.; Hersam, M.C.; Guisinger, N.P. Electronic and Mechanical Properties of Graphene–Germanium Interfaces Grown by Chemical Vapor Deposition. *Nano Lett.* **2015**, *15*, 7414–7420. [CrossRef]
59. Ryu, S.; Liu, L.; Berciaud, S.; Yu, Y.-J.; Liu, H.; Kim, P.; Flynn, G.W.; Brus, L.E. Atmospheric Oxygen Binding and Hole Doping in Deformed Graphene on a SiO_2 Substrate. *Nano Lett.* **2010**, *10*, 4944–4951. [CrossRef]
60. Casiraghi, C.; Pisana, S.; Novoselov, K.S.; Geim, A.K.; Ferrari, A.C. Raman fingerprint of charged impurities in graphene. *Appl. Phys. Lett.* **2007**, *91*, 233108. [CrossRef]
61. Kolesov, E.A.; Tivanov, M.S.; Korolik, O.V.; Kapitanova, O.O.; Fu, X.; Cho, H.D.; Kang, T.W.; Panin, G.N. The effect of atmospheric doping on pressure-dependent Raman scattering in supported graphene. *Beilstein J. Nanotechnol.* **2018**, *9*, 704–710. [CrossRef] [PubMed]
62. Goniszewski, S.; Adabi, M.; Shaforost, O.; Hanham, S.M.; Hao, L.; Klein, N. Correlation of p-doping in CVD Graphene with Substrate Surface Charges. *Sci. Rep.* **2016**, *6*, 22858. [CrossRef] [PubMed]
63. Eriksson, J.; Puglisi, D.; Vasiliauskas, R.; Lloyd Spetz, A.; Yakimova, R. Thickness uniformity and electron doping in epitaxial graphene on SiC. *Mater. Sci. Forum* **2013**, *740*, 153–156. [CrossRef]
64. Jee, H.-g.; Jin, K.-H.; Han, J.-H.; Hwang, H.-N.; Jhi, S.-H.; Kim, Y.D.; Hwang, C.-C. Controlling the self-doping of epitaxial graphene on SiC via Ar ion treatment. *Phys. Rev. B* **2011**, *84*, 075457. [CrossRef]
65. Banszerus, L.; Janssen, H.; Otto, M.; Epping, A.; Taniguchi, T.; Watanabe, K.; Beschoten, B.; Neumaier, D.; Stampfer, C. Identifying suitable substrates for high-quality graphene-based heterostructures. *2D Mater.* **2017**, *4*, 025030. [CrossRef]
66. Kang, Y.-J.; Kang, J.; Chang, K.J. Electronic structure of graphene and doping effect on SiO_2. *Phys. Rev. B.* **2008**, *78*, 115404. [CrossRef]
67. Armano, A.; Buscarino, G.; Cannas, M.; Gelardi, F.M.; Giannazzo, F.; Schiliro, E.; Lo Nigro, R.; Agnello, S. Graphene-SiO_2 Interaction from Composites to Doping. *Phys. Status Solidif.* **2019**, *216*, 1800540. [CrossRef]
68. Barbosa, A.N.; Ptak, F.; Mendoza, C.D.; Maia da Costa, M.E.H.; Freire, F.L., Jr. Direct synthesis of bilayer graphene on silicon dioxide substrates. *Diam. Relat. Mater.* **2019**, *95*, 71–76. [CrossRef]
69. Chen, Z.; Qi, Y.; Chen, X.; Zhang, Y.; Liu, Z. Direct CVD Growth of Graphene on Traditional Glass: Methods and Mechanisms. *Adv. Mater.* **2018**, *31*, 1803639. [CrossRef]

Publisher's Note: MDPI stays neutral with regard to jurisdictional claims in published maps and institutional affiliations.

© 2020 by the authors. Licensee MDPI, Basel, Switzerland. This article is an open access article distributed under the terms and conditions of the Creative Commons Attribution (CC BY) license (http://creativecommons.org/licenses/by/4.0/).

Article

Highly Water Dispersible Functionalized Graphene by Thermal Thiol-Ene Click Chemistry

Farzaneh Farivar [1,2], Pei Lay Yap [1,2], Tran Thanh Tung [1,2] and Dusan Losic [1,2,*]

[1] School of Chemical Engineering and Advanced Materials, The University of Adelaide, Adelaide, SA 5005, Australia; farzaneh.farivar@adelaide.edu.au (F.F.); peilay.yap@adelaide.edu.au (P.L.Y.); tran.tung@adelaide.edu.au (T.T.T.)
[2] ARC Hub for Graphene Enabled Industry Transformation, The University of Adelaide, Adelaide, SA 5005, Australia
* Correspondence: dusan.losic@adelaide.edu.au

Abstract: Functionalization of pristine graphene to achieve high water dispersibility remains as a key obstacle owing to the high hydrophobicity and absence of reactive functional groups on the graphene surface. Herein, a green and simple modification approach to prepare highly dispersible functionalized graphene via thermal thiol-ene click reaction was successfully demonstrated on pristine graphene. Specific chemical functionalities (–COO, –NH$_2$ and –S) on the thiol precursor (L-cysteine ethyl ester) were clicked directly on the sp^2 carbon of graphene framework with grafting density of 1 unit L-cysteine per 113 carbon atoms on graphene. This functionalized graphene was confirmed with high atomic content of S (4.79 at % S) as well as the presence of C–S–C and N–H species on the L-cysteine functionalized graphene (FG-CYS). Raman spectroscopy evidently corroborated the modification of graphene to FG-CYS with an increased intensity ratio of D and G band, I_D/I_G ratio (0.3 to 0.7), full-width at half-maximum of G band, FWHM [G] (20.3 to 35.5) and FWHM [2D] (64.8 to 90.1). The use of ethanol as the reaction solvent instead of common organic solvents minimizes the chemical hazards exposure to humans and the environment. This direct attachment of multifunctional groups on the surface of pristine graphene is highly demanded for graphene ink formulations, coatings, adsorbents, sensors and supercapacitor applications.

Keywords: graphene; functionalized graphene; thiol-ene click reaction; dispersible graphene

Citation: Farivar, F.; Yap, P.L.; Tung, T.T.; Losic, D. Highly Water Dispersible Functionalized Graphene by Thermal Thiol-Ene Click Chemistry. *Materials* **2021**, *14*, 2830. https://doi.org/10.3390/ma14112830

Academic Editors: Federico Cesano and Domenica Scarano

Received: 22 April 2021
Accepted: 24 May 2021
Published: 25 May 2021

Publisher's Note: MDPI stays neutral with regard to jurisdictional claims in published maps and institutional affiliations.

Copyright: © 2021 by the authors. Licensee MDPI, Basel, Switzerland. This article is an open access article distributed under the terms and conditions of the Creative Commons Attribution (CC BY) license (https://creativecommons.org/licenses/by/4.0/).

1. Introduction

Graphene materials have sparked enormous attraction and interest from both the scientific and industrial community owing to the fascinating properties of graphene and its widespread applications across many technological fields, such as nanoelectronics, sensors, composites, coatings, batteries, supercapacitors and hydrogen storage [1]. The ongoing demand for graphene and its products is reflected in the rapid wave of graphene technologies transferring from the laboratory to the marketplace [2]. Despite the outstanding optical, electrical, and thermal conductivity properties of graphene, it is no secret that there are still limitations associated with the poor water dispersibility and low intrinsic reactivity of graphene, which hamper its useful applications [2–5]. To address these limitations, functionalization of pristine graphene with organic functional groups to obtain stable dispersions in various solvents is a vital move to produce graphene materials with new properties [4]. Different strategies including covalent and non-covalent approaches were explored and successfully used for graphene functionalization [4].

Covalent attachment of organic functional groups to the graphene surface can be achieved either by chemical reaction of free radicals or dienophiles sp^2 carbons of pristine graphene, or by the formation of a covalent bond between organic molecules and the oxygen groups of graphene oxide, GO [4]. The radical-mediated thiol-ene reaction that directly attacks the sp² carbon in the graphene framework appears to be a promising covalent functionalization approach that offers several benefits. It is highly efficient, simple to execute, insensitive to oxygen and water, has no side products and proceeds rapidly with high yield [6]. Several examples in the literature have showcased the application of thiol-ene click functionalization on GO, including our recent studies that demonstrated the use of thiol-ene modified GO with cysteamine and pentaerythritol tetrakis(3-mercaptopropionate) through both thermal and photoinitiated approaches for the removal of water pollutants [7–12]. Thiol-ene click reaction has been proven for the chemical modification of graphene oxide (GO), and the thiol-ene clicked GO serves as an excellent host matrix for platinum nanoparticles for catalytic and sensing applications [12]. In another example, one-step thiol-ene click reaction by mercaptosuccinic acid was used to prepare graphene quantum dots (GQDs) decorated with a number of carboxyl groups on the surface which improved their dispersibility in water [13].

Surprisingly, up to now, the majority of the reports of functionalized graphene have primarily focused on GO and its reduced form, reduced graphene oxide (rGO), where the preparation methodologies usually involve the use of hazardous chemicals and toxic gases, with large quantities of chemical waste generated after the reactions [4]. To perform direct covalent functionalization on pristine graphene, having an inert surface and fewer defects, poses considerable challenges and requires high energetic species to break its sp² honeycomb framework. In fact, limited reports were found on the direct functionalization of pristine graphene via thiol-ene click chemistry. Castelain et al. used thiol-ene reaction for a direct functionalization of the graphene surface with short chain polyethylene (PE) brushes which were successfully used to prepare graphene based high-density polyethylene nanocomposites [14]. Peng et al. also used microwave-assisted thiol-ene click reaction for functionalization of graphene using thiol precursors with different functional groups [15]. In another study, cysteamine hydrochloride was bonded to the graphene surface via thiol-ene click chemistry to immobilize Au nanoparticles as an efficient electrochemical sensor [16]. Although these studies have shown that thiol-ene click functionalization can be successfully performed on pristine graphene, there is a research gap regarding the safe use of solvents, including highly volatile organic and non-environmentally friendly solvents such as N-methylpyrrolidone, N,N-dimethylformamide or ortho-dichlorobenzene. Ensuring dispersibility of graphene in these solvents during modification reactions remains a challenging issue. Moreover, the efficiency of thiol-ene click functionalization on pristine graphene is still not satisfactorily accomplished compared to the thiol-ene modification achieved by GO and rGO.

Herein, we present a simple, green and sustainable method to functionalize pristine graphene through thiol-ene click reaction to acquire water dispersible graphene materials with multiple functional groups that are important for many applications. The aims of this work are twofold: first, to improve an existing thiol-ene click reaction using more sustainable and green conditions for the functionalization of pristine graphene, and second, to demonstrate versatility of this method to generate graphene with multifunctional surface chemistry with different end groups including ester, amino, and thioether. The developed method is based on the covalent attachment of thiol molecules on the sp² carbon of graphene via thermal thiol-ene click reaction, which is schematically presented in Figure 1. The method is highly efficient, catalyst-free, simple, with mild reaction conditions for surface modifications, which may provide scalable production of the functional graphene materials. To eliminate the hazardous conditions commonly found in thiol-ene click reaction based on toxic solvents, herein, we introduce for the first time the use of ethanol/water mixture (70/30), which is less expensive, non-toxic and more environmentally friendly, to improve the scalability of highly dispersible functionalized graphene materials. A broad range of characterization techniques such as Raman spectroscopy, Fourier-transform infrared spectroscopy (FTIR), X-ray photoelectron spectroscopy (XPS), thermogravimetric analysis (TGA), and water dispersibility test were used to confirm the thiol-ene click functionalization.

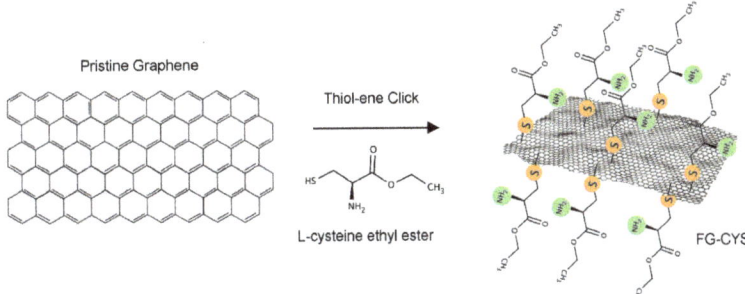

Figure 1. Schematic diagram of the green functionalization of pristine graphene using L-cysteine ethyl ester via thiol-ene click reaction.

2. Materials and Methods

2.1. Materials and Chemicals

Graphene was obtained from a local company (First Graphene pty ltd, Perth, WA, Australia). L-cysteine (Sigma-Aldrich, Sydney, NSW, Australia), thionyl chloride (Sigma-Aldrich, Sydney, NSW, Australia), diethyl ether (RCI Labscan, Bangkok, Thailand) 2,2′-azobis-(2-methylpropionitrile) (AIBN, Sigma-Aldrich, Sydney, NSW, Australia,), ethanol (Chem-Supply, Adelaide, SA, Australia) were used directly without prior purification. High purity milli-Q water (18.2 M$\Omega \cdot$cm^{-1}) was used throughout the work, unless otherwise stated.

2.2. Synthesis of L-Cysteine Ethyl Ester

Ethyl esters of cysteine were synthesized according to a reported method [17]. In brief, L-cysteine was first suspended in ethanol, and cooled to 5 °C. Thionyl chloride was added gradually, over a period of 20 min, and the reaction mixture was stirred at room temperature for 5 h. Dry ether was then added to the solution until an opaque solution appeared. The mixture was kept in the refrigerator for a few hours for crystallization of ethyl ester hydrochloride. The final product was collected by filtration.

2.3. Synthesis of Functionalized Graphene (FG-CYS)

Functionalized graphene with L-cysteine ethyl ester (FG-CYS), was prepared according to the adapted procedure reported by Yap et al. [8]. In a typical procedure, 50 mg graphene powder was first dispersed in 100 ml ethanol–water mixture 70% (v/v) via sonication for 1 h. Subsequently, the sonicated mixture was purged with nitrogen gas for 30 min to create an inert environment. Then, L-cysteine ethyl ester and 2,2′-azobis-(2-methylpropionitrile) (AIBN) were added into the mixture with a further nitrogen gas purge for 30 min, followed by a 30 min sonication. After that, the reaction mixture was poured into a round bottom flask with an additional 30 min of nitrogen gas purge and heated in an oil bath at 65 °C under reflux conditions overnight. After the reaction, the as-synthesized product was washed thoroughly with ethanol and deionized water using centrifuge, dried in oven at 65 °C overnight, and stored for further characterization.

2.4. Characterizations

FTIR spectra were collected at 500 to 4000 cm^{-1} on a Nicolet 6700 Fourier Transform Infrared (FTIR) Spectrometer (Thermo Fisher Sci, Sydney NSW, Australia). Morphology of the materials was imaged using a scanning electron microscope (FE-SEM, Quanta 450 FEG, FEI, USA) at an operating voltage of 10 kV and a transmission electron microscope at 120 kV (TEM, FEI Tecnai G2 Spirit, FEI, USA; Philips CM200, Japan at 200 kV). Chemical composition and chemical species were analyzed by X-ray Photoelectron Spectroscopy (XPS, AXIS Ultra DLD, Kratos, UK) equipped with a monochromatic Al Kα radiation source (hv = 1486.7 eV) at 225 W, 15 kV and 15 mA. XPS survey scans were performed at 0.5 eV step size over −10 to 1100 eV at 160 eV pass energy with peak fitting analysis executed using Casa XPSTM software. The core-level XPS spectra were calibrated at 284.8 eV. Raman spectrometer (LabRAM HR Evolution, Horiba Jvon Yvon Technology, Kyoto, Japan) with an excitation wavelength of 532 nm (mpc 3000 laser source) was applied from 500 to 3000 cm^{-1} with an integration time of 10 s for three accumulations to determine the vibrational features of pristine graphene, and thermogravimetric analysis (TGA) of functionalized graphene was performed at a heating rate of 10 °C/min in nitrogen atmosphere from 25 to 1000 °C on a METTLER TOLEDO TGA/DSC 2 instrument.

3. Results

The morphology of pristine graphene and the functionalized graphene was first examined using scanning (SEM) and transmission (TEM) electron microscopy, as shown in Figure 2a–c. Large and folded few-layer graphene (FLG) sheets, as depicted in Figure 2a, were observed under TEM analysis with an inset showing fewer than 10 layers of graphene used as the precursor in this work. A detailed statistical distribution analysis (Figure S1) indicates an average of six layers of graphene, confirming the FLG used in this study. Under the SEM, pristine graphene (Figure 2b) exhibited highly crumpled wrinkled thin sheets. After the modification using the thiol precursor, the primitive thin and crumpled graphene sheets still remained in FG-CYS, with its surface decorated with multiple fluffy clusters, as shown in Figure 2c. The presence of these fluffy clusters on the surface of graphene sheets preliminarily suggested the effective attachment of L-cysteine ethyl ester moieties on the surface of graphene.

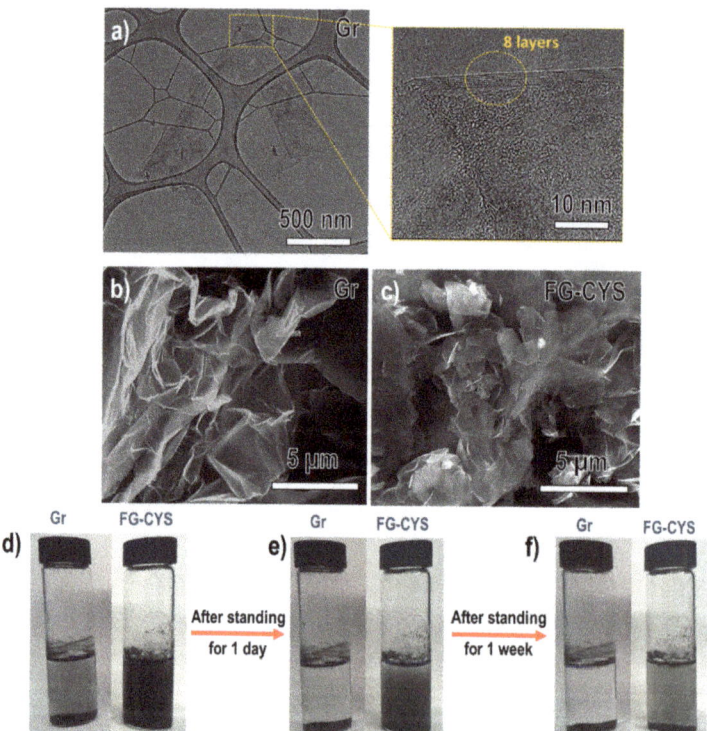

Figure 2. (a) TEM image of pristine graphene (Gr) with an enlarged image of HRTEM analysis showing 8 layers of graphene sheets, FESEM of (b) Gr and (c) FG-CYS. Dispersion test in water (0.5 mg/mL) from left to right: pristine graphene, functionalized graphene with cysteine ethyl ester (FG-CYS) for standing after (d) 2 h, (e) 1 day and (f) 1 week of free-standing.

A simple and rapid water dispersion test was performed to examine the effectiveness of thiol-ene functionalization (Figure 2d–f). As expected, pristine graphene showed poor dispersion, with its powder sedimented immediately after two hours of ultrasonication. The complete sedimentation of the graphene powder at the bottom of the bottle (Figure 2d–f) could be attributed to its low surface activity and weak bonding strength with the water matrix. In contrast, the thiol-ene functionalized graphene FG-CYS) visibly showed good dispersion in water after modification with thiol precursor, as depicted in Figure 2d–f. Remarkably, FG-CYS showed a stable aqueous dispersion in water even after standing for one week (Figure 2f). The enhanced dispersion of the functionalized graphene in water compared to pristine graphene suggested an effective attachment of polar functional groups, such as oxygen groups including ester and amino groups, from the thiol precursor to the graphene surface that rendered the interaction with the polar solvent (water) primarily through hydrogen bonds.

Chemical composition of the functionalized graphene was determined by XPS analysis (Table 1) with their survey spectra incorporated in Figure 3a. Significant appearance of S2p and N1s peaks, as well as the elevated intensity of O1s peaks of FG-CYS at binding energies of 164.0 eV, 401.5 eV and 531.0 eV, respectively, were observed in the survey scan of FG-CYS relative to pristine graphene. A notably increased atomic concentration of nitrogen (6.26 at %) and sulfur (4.79 at %) on FG-CYS, relative to pristine graphene, implied successful grafting of thiol precursor on the surface of graphene. As depicted in Figure 3b, the doublet peaks of $S2p_{3/2}$ and $S2p_{1/2}$ at around 164.1 eV and 165.2 eV, respectively, can be assigned to thioether sulfur bonded to carbon (C–S–C) species on the narrow XPS spectra of S2p of FG-CYS, and confirmed the successful attachment of the thiol groups from the thiol precursor via thiol-ene click reaction on the surface of the graphene [18,19]. Meanwhile, $S2p_{3/2}$ and $S2p_{1/2}$ peaks with lower intensities deconvoluted at 164.8 eV and 166.1 eV, respectively, could be ascribed to C–S or S–S species on the surface of graphene [20]. Note that S2p peak, associated with highly oxidized S species (>166 eV), was absent from the high-resolution S2p spectrum of FG-CYS, which implied good stability of the sulfur species formed on the surface of graphene sheets, despite the exposure of functionalized graphene material to the readily oxidized condition [21]. High-resolution N1s spectrum of FG-CYS (Figure 3c), on the other hand, exhibited several peak components including peaks at 400.0 eV, 401.1 eV, 401.7 eV and 402.5 eV, which can be allotted to $C-NH_2$, NH_3^+, N–H and $C-N^+$, respectively [8,18,22]. The presence of the nitrogen species from the N1s peak deconvolution analysis was in good correlation with the chemical structure of the thiol precursor, L-cysteine ethyl ester, that corroborated successful attachment of the thiol precursor on the surface of graphene [13,16,23]. These results were well-supported by the high-resolution C1s spectra of FG-CYS (Figure S2), with additional peak components of C–S, C–N, C=O and O–C=O identified at binding energies of 285.4 eV, 286.1, 287.7 eV and 288.6 eV, respectively, besides the typical C=C (284.5 eV) and C–C (284.8 eV) peaks.

Table 1. Normalized atomic percentage of elements determined from the XPS survey scan of Gr and FG-CYS.

Sample/Element	Atomic Composition (±0.3%)			
	C	O	S	N
Gr	91.60	8.40	N.A.	N.A.
FG-CYS	71.65	17.30	4.79	6.26

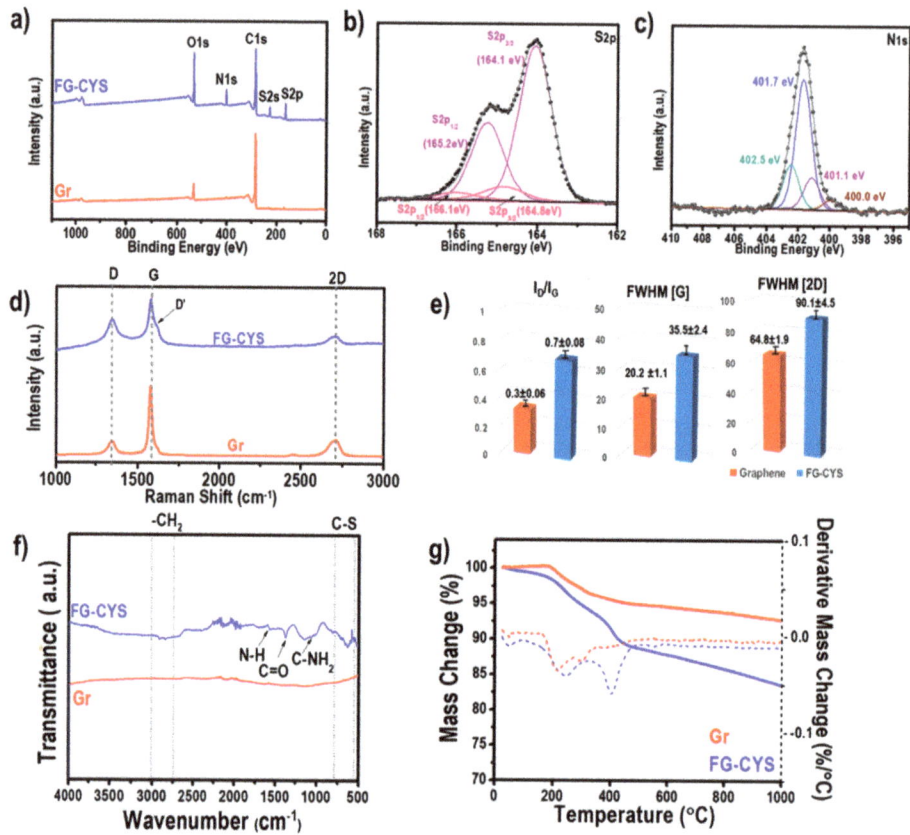

Figure 3. (a) XPS survey spectra of Gr and FG-CYS, deconvoluted high-resolution XPS spectra of (b) S2p and (c) N1s of Gr and FG-CYS, (d) Raman spectra, (e) I_D/I_G, FWHM of Raman G and FWHM of Raman 2D peak for Gr and FG-CYS, (f) FTIR spectra and (g) TGA-DTG thermograms in N_2 atmosphere of Gr and FG-CYS.

Detailed Raman analysis was performed to verify the modification created on the surface of pristine graphene and functionalized graphene, as illustrated in Figure 3d–e. Both of the samples showed the main features of G band at around 1580 cm^{-1}, D band at around 1350 cm^{-1}, and 2D band at 2700 cm^{-1} [24,25]. The presence of graphene can be confirmed with a sharp G peak (~1580 cm^{-1}) and a symmetrical 2D peak (~2700 cm^{-1}) as clearly observed on the Raman spectra of Gr and FG-CYS samples. As depicted in Figure 3d, an indicative D′ peak (indicated by the black arrow) for defective graphene positioned at ~1620 cm^{-1} was detected on the Raman spectrum of FG-CYS, suggesting a certain degree of defect found on the surface of FG-CYS. This finding was in consistent with the average value of I_D/I_G ratio (a measurand of the level of disorder of graphene) of FG-CYS (0.7), which was higher than Gr (0.3), indicating more defects were experienced on the surface of FG-CYS than Gr due to thiol-ene click modification using thiol precursor, as shown in Figure 3e [26]. The level of disorder after the modification using thiol precursor was also evidenced by the broadening of the 2D and G peaks, as attested by the significant increment of the full-width at half-maximum of the 2D peak and G peak, FWHM [2D] and FWHM [G] [27,28]. As shown in Figure 3e, the dramatic increase from 20.2 to 35.5 (FWHM [G]) and 64.8 to 90.1 (FWHM [2D]) for Gr to FG-CYS, respectively, clearly indicated that substantial defects were introduced to the surface of graphene that could have resulted from the thiol-ene click reaction.

The obtained FTIR characterization results presented in Figure 3f were further used to elucidate the functional groups attached on the surface of the functionalized graphene. A nearly flat FTIR curve with no obvious peak was detected for Gr sample due to the lack of chemical functional groups on its surface. Essentially, a band representing the C–S vibration was visible at around 600–800 cm^{-1} on the FTIR spectrum of the functionalized material, but not identified on the FTIR spectrum of pristine graphene [29]. This finding was consistent with the result from XPS analysis with a substantial amount of S detected and C–S–C species found in the functionalized sample, after the modification process using thiol precursor. The absence of a peak at around 2500 cm^{-1} (stretching of S–H group) in the functionalized graphene suggested that all the –SH groups in the thiol precursor had reacted with the sp^2 carbon in pristine graphene through thiol-ene click reaction [29]. Additionally, the successful modification of the pristine graphene was also evidenced by the appearance of the peak at approximately 1389 cm^{-1} for the stretching of C=O in the ester group on FG-CYS. Peaks at about 1640 cm^{-1} (N–H bending), 1090 cm^{-1} (C–N stretching) and 1054 cm^{-1} (C–NH2 vibration) were also found on the FTIR spectrum of FG-CYS, implying successful grafting of the functional groups on the pristine graphene. Two common peaks at 2913 and 2846 cm^{-1}, which can be attributed to the asymmetric and symmetric vibrations of C–H of –CH$_2$ group, respectively, were also observed in the functionalized graphene [29,30]. Based on the decoration of the functional groups on the functionalized graphene, as elucidated from the FTIR analysis, which was well-correlated with the XPS analysis discussed in the previous section, we can infer that the successful click of thiol precursor to the pristine graphene has occurred through the thiol-ene pathway.

Furthermore, TGA was also conducted on pristine and functionalized graphene to qualitatively and quantitatively determine the presence of attached functional groups on the pristine graphene after the thermal thiol-ene click reaction. As presented in Figure 3g, the TGA curves (solid blue and red lines) clearly showed a distinguishable mass loss pattern experienced by Gr and FG-CYS. Pristine graphene showed higher thermal stability with only 7.28% total mass loss, relative to its functionalized graphene that experienced an overall 16.41% mass loss when heated under inert atmosphere to 1000 °C. This result was consistent with the high thermal stability of the typical pristine graphene found in the literature [31–33]. From the DTG curve of the functionalized graphene, two major mass loss steps were identified at about 250 °C and 400 °C, apart from a small mass loss (0.52%) that was accountable for the elimination of water below 100 °C. The first mass loss step could be associated with the detachment of the oxygen functional groups, including ester from the graphene surface [8]. Meanwhile, the second mass loss at around 400 °C can be related to the cleavage of the remaining covalently bound groups, such as oxygen, sulfur, and amino moieties from the functionalized graphene surface. From the TGA curves of the control and functionalized graphene, the grafting density of the modified sample can be estimated by calculating the value of functional group coverage (average carbon atom number containing functional group) based on the mass loss difference of FG-CYS from Gr, as summarized in Table S1 [15,16]. Hence, it can be estimated that 1 cysteine molecule per 113 carbon atoms was grafted on the graphene sheets in this thiol-ene click modification, as quantitatively determined from the TGA.

Despite the key challenges of inertness of sp^2 carbon and the lack of reactive functional groups on the surface of pristine graphene, we successfully demonstrated an effective functionalization of pristine graphene through a green and scalable thiol-ene click modification to endow a highly dispersible graphene. A summary of our key findings with evidence supported by comprehensive techniques, as tabulated in Table 2, confirmed the proposed methodology as a powerful tool for the preparation of dispersible and functional graphene materials with promising applications across different arenas, including biomedical, coatings, energy storage and environment. In the next section, we compare and discuss the degree of functionalization of FG-CYS with similar functionalized graphene prepared in the literature.

Table 2. Summary of thiol-ene click functionalization of FG-CYS as evidenced by selected key techniques.

Sample	Water Dispersion	XPS (% S)	Raman (I_D/I_G Ratio)	FTIR (Attached Functional Groups)	TGA (Functional Group Coverage *)
Gr	Less than 2 h	N.A.	0.3	N.A.	N.A.
FG-CYS	More than a week	4.79	0.7	N–H, C–NH_2, C=O, C–S	113 *

* Detailed calculation can be found in the Supplementary Material.

4. Discussion

To date, functionalization of pristine graphene by covalent bonding has rarely been reported in relation to its derivative, GO, due to the lack of reactive functional groups on the pristine graphene surface [5]. It is undeniable that functionalized graphene has several limitations, including poor dispersion in water, low functional group coverage, and high toxicity of solvent used during the functionalization process. Click chemistry introduced by Sharpless et al. can be regarded as a promising solution to overcome the chemical inertness of C=C bonds in the graphene framework to afford a highly dispersible graphene, despite the fact that it is often accompanied by low reaction efficiency [6,15].

Numerous attempts have been made in the last decade to introduce organic functional groups onto the surface of pristine graphene through thiol-ene click chemistry. The degree of functionalization can be quantitatively gauged using both bulk (TGA) and surface (XPS) characterization techniques, as summarized in Table 3. To perform an impartial comparison, the value of the functional group coverage estimated from the mass loss through TGA method, and the chemical composition of specific element such as N or S in the thiol precursor determined through XPS analysis, were compared to determine the level of thiol-ene functionalization.

Table 3. Comparison of grafting density based on surface (XPS) and bulk (TGA) quantification techniques.

Sample	Modification Method	Solvent	Thiol Precursor	TGA (Mass Loss %); Functional Group Coverage	XPS at % (N; S)	References
G-LCHa	Microwave-assisted thiol-ene click (70 °C)	Tetrahydrofuran (THF)	L-cysteine	8.58; 137	0.89; 0.7	[15]
G-CHI	Microwave-assisted thiol-ene click (70 °C)	Tetrahydrofuran (THF)	cysteamine	7.60; 113	1.13; 1.66	[15]
GR-Cys	Thermal thiol-ene click (70 °C, 7 h)	N-methyl-2-pyrrolidone (NMP)	cysteamine	15.80; 96	4.78; 4.52	[16]
FG-CYS	Thermal thiol-ene click (65 °C, 24 h)	Ethanol–water	L-cysteine ethyl ester	8.61; 113 *	6.26; 4.79	This work

* Detailed calculations can be found in the Supplementary Material.

Based on the bulk TGA characterization technique, L-cysteine was found to be more effectively grafted on FG-CYS (present study) than G-LCHa, as reflected by its lower functional group coverage of 113, compared to 137 carbon atoms per unit of L-cysteine molecule, as tabulated in Table 3. Further surface elemental quantification analysis (XPS) showed that only 0.89 at % N and 0.7 at % S were detected on G-LCHa that was modified using microwave-assisted thiol-ene click reaction; while 6.26 at % N and 4.79 at % S were identified on the FG-CYS produced through the thermal thiol-ene click approach in our present study [15].

Another example of direct functionalization on pristine graphene was manifested in G-CHI and GR-Cys, with 1 cysteamine molecule per 113 and 96 carbon atoms, respectively, based on the value of functional group coverage. Meanwhile, 1.13 at % N and 1.66 at % S were found on G-CHI, while 4.78 at % N and 4.52 at % S were detected on GR-Cys after the thiol-ene click modification [15,16]. In comparison, FG-CYS (our present study) appeared to outperform G-CHI but slightly underperformed compared to GR-Cys. Note that several aspects including safety, scalability and economy feasibility should be considered when dealing with the use of a functionalization method. The majority of the thiol-ene functionalized graphene was produced using toxic solvents such as NMP and THF, which are not environmentally friendly and could potentially trigger chemical risk in the working environment. With respect to the environmental hazard, energy consumption, cost and scalability of producing dispersible graphene, the green thiol-ene click functionalization approach developed in the present work still surpassed the thiol-ene functionalized graphene reported in the literature based on the functionalization condition applied and the effective functionalization degree achieved by FG-CYS.

Despite the growing body of work recently reported to functionalize graphene directly using thiol-ene click strategy, there are still numerous open research opportunities to address the graphene dispersion issue, to endow a better processability of graphene materials (new functionalized graphene) with desired properties. Beyond the conventional chemical modification of graphene such as the thermal heating process, other methods including microwave irradiation, plasma etching, electrochemical modification or a combination of these methods could be explored as promising alternative routes to resolve the low reactivity of graphene. Based on our current work, functionalization parameters such as the ratio of graphene to thiol precursor, reaction time and temperature, could be tuned to accommodate a more effective grafting of functional groups on the graphene surface. The future direction of graphene functionalization may be structured towards room temperature modification of graphene, which can be feasibly achieved via the photoinitiated thiol-ene click approach. In addition, more research efforts should be dedicated to the optimization of functionalization methods to achieve not only the better dispersion of graphene, but also the scalability of processing of new graphene materials to be more economical.

5. Conclusions

In summary, the synthesis of functionalized pristine graphene by a green thermal thiol-ene click approach using L-cysteine ethyl ester was successfully demonstrated. The use of ethanol in replacing highly toxic organic solvents in this click reaction provides a generic, green, and scalable approach to graft graphene with desired functional groups, which can be achieved by selecting different types of thiol precursors bound with functional groups including hydroxyl, ester and amine groups. Successful modification of pristine graphene with the thiol precursor was confirmed by Raman, FTIR, TGA and XPS analyses. This green modification process provides a significant contribution towards the scalable production of functionalized graphene due to a simple and scalable chemical route which is beneficial for many applications, such as biomedical, sensing, graphene polymer composites, inks, supercapacitors, etc.

Supplementary Materials: The following are available online at https://www.mdpi.com/article/10.3390/ma14112830/s1, Figure S1: Histogram on the number of layers of Gr, showing an average six layers of graphene sheets with standard deviation determined from 20 measurements by HRTEM analysis, Figure S2: XPS high-resolution C1s plots of Gr and FG-CYS, Table S1: Results of TGA and the value of functional group coverage of FG-CYS estimated from TGA.

Author Contributions: Conceptualization, D.L., F.F., P.L.Y.; Methodology, F.F., P.L.Y., T.T.T.; Formal analysis, F.F., P.L.Y.; Investigation, F.F., P.L.Y.; Writing—Original Draft Preparation, F.F., P.L.Y.; Writing—Review and Editing, D.L., P.L.Y., F.F., T.T.T.; Resources, D.L.; Supervision, D.L.; Project Administration, D.L.; Funding acquisition, D.L. All authors have read and agreed to the published version of the manuscript.

Funding: This research was funded by the ARC Research Hub for Graphene Enabled Industry Transformation (IH150100003).

Institutional Review Board Statement: Not applicable.

Informed Consent Statement: Not applicable.

Data Availability Statement: The data presented in this study are available on reasonable request from the corresponding author.

Acknowledgments: The authors acknowledge the funding by the ARC Research for Graphene Enabled Industry Transformation Hub, (IH150100003) and thank Australian Microscopy and Microanalysis Research Facility (AMMRF) for the facilities access and technical support of HRTEM (Ramesh Karunagaran); and XPS (Chris Bassell) at the Microscopy Australia Facilities (NCRIS scheme) at the University of South Australia.

Conflicts of Interest: The authors declare no conflict of interest.

References

1. Geim, A.K.; Novoselov, K.S. The rise of graphene. *Nat. Mater.* **2007**, *6*, 183–191. [CrossRef]
2. Kong, W.; Kum, H.; Bae, S.-H.; Shim, J.; Kim, H.; Kong, L.; Meng, Y.; Wang, K.; Kim, C.; Kim, J. Path towards graphene commercialization from lab to market. *Nat. Nanotechnol.* **2019**, *14*, 927–938. [CrossRef]
3. Salavagione, H.J. Promising Alternative Routes for Graphene Production and Functionalization. *J. Mater. Chem. A* **2014**, *2*, 7138–7146. [CrossRef]
4. Georgakilas, V.; Otyepka, M.; Bourlinos, A.B.; Chandra, V.; Kim, N.; Kemp, K.C.; Hobza, P.; Zboril, R.; Kim, K.S. Functionalization of Graphene: Covalent and Non-Covalent Approaches, Derivatives and Applications. *Chem. Rev.* **2012**, *112*, 6156–6214. [CrossRef] [PubMed]
5. Liu, L.-H.; Lerner, M.M.; Yan, M. Derivitization of Pristine Graphene with Well-Defined Chemical Functionalities. *Nano Lett.* **2010**, *10*, 3754–3756. [CrossRef] [PubMed]
6. Hoyle, C.E.; Bowman, C.N. Thiol-Ene Click Chemistry. *Angew. Chem. Int. Ed.* **2010**, *49*, 1540–1573. [CrossRef] [PubMed]
7. Yap, P.L.; Hassan, K.; Auyoong, Y.L.; Mansouri, N.; Farivar, F.; Tran, D.N.; Losic, D. All-in-one bioinspired multifunctional graphene biopolymer foam for simultaneous removal of multiple water pollutants. *Adv. Mater. Interfaces* **2020**, 2000664. [CrossRef]
8. Yap, P.L.; Kabiri, S.; Tran, D.N.H.; Losic, D. Multifunctional binding chemistry on modified graphene composite for selective and highly efficient adsorption of mercury. *ACS Appl. Mater. Interfaces* **2019**, *11*, 6350–6362. [CrossRef] [PubMed]
9. Masteri-Farahani, M.; Modarres, M. Clicked graphene oxide supported venturello catalyst: A new hybrid nanomaterial as catalyst for the selective epoxidation of olefins. *Mater. Chem. Phys.* **2017**, *199*, 522–527. [CrossRef]
10. Liu, J.; Zhu, K.; Jiao, T.; Xing, R.; Hong, W.; Zhang, L.; Zhang, Q.; Peng, Q. Preparation of Graphene Oxide-Polymer Composite Hydrogels via Thiol-ene Photopolymerization as Efficient Dye Adsorbents for Wastewater Treatment. *Colloids Surf. A* **2017**, *529*, 668–676. [CrossRef]
11. Li, J.; Cheng, Y.; Zhang, S.; Li, Y.; Sun, J.; Qin, C.; Wang, J.; Dai, L. Modification of GO based on click reaction and its composite fibers with poly(vinyl alcohol). *Compos. Part A Appl. Sci. Manuf.* **2017**, *101*, 115–122. [CrossRef]
12. Luong, N.D.; Sinh, L.H.; Johansson, L.S.; Campell, J.; Seppälä, J. Functional Graphene by Thiol-ene Click Chemistry. *Chem. Eur. J.* **2015**, *21*, 3183–3186. [CrossRef] [PubMed]
13. Huang, H.; Liu, M.; Tuo, X.; Chen, J.; Mao, L.; Wen, Y.; Tian, J.; Zhou, N.; Zhang, X.; Wei, Y. A novel thiol-ene click reaction for preparation of graphene quantum dots and their potential for fluorescence imaging. *Mater. Sci. Eng. C* **2018**, *91*, 631–637. [CrossRef] [PubMed]
14. Castelaín, M.; Martínez, G.; Marco, C.; Ellis, G.; Salavagione, H.J. Effect of Click-Chemistry Approaches for Graphene Modification on the Electrical, Thermal, and Mechanical Properties of Polyethylene/Graphene Nanocomposites. *Macromolecules* **2013**, *46*, 8980–8987. [CrossRef]
15. Peng, Z.; Li, H.; Li, Q.; Hu, Y. Microwave-Assisted thiol-ene click chemistry of carbon nanoforms. *Colloids Surf. A* **2017**, *533*, 48–54. [CrossRef]
16. Li, Y.; Bao, L.; Zhou, Q.; Ou, E.; Xu, W. Functionalized Graphene Obtained via Thiol-Ene Click Reactions as an Efficient Electrochemical Sensor. *ChemistrySelect* **2017**, *2*, 9284–9290. [CrossRef]
17. Patel, R.P.; Price, S. Synthesis of Benzyl Esters of α-Amino Acids. *J. Org. Chem.* **1965**, *30*, 3575–3576. [CrossRef]
18. Yap, P.L.; Kabiri, S.; Auyoong, Y.L.; Tran, D.N.H.; Losic, D. Tuning the multifunctional surface chemistry of reduced graphene oxide via combined elemental doping and chemical modifications. *ACS Omega* **2019**, *4*, 19787–19798. [CrossRef]
19. Pang, Q.; Tang, J.; Huang, H.; Liang, X.; Hart, C.; Tam, K.C.; Nazar, L.F. A Nitrogen and Sulfur Dual-Doped Carbon Derived from Polyrhodanine@Cellulose for Advanced Lithium–Sulfur Batteries. *Adv. Mater.* **2015**, *27*, 6021–6028. [CrossRef]
20. Xing, X.; Li, Y.; Wang, X.; Petrova, V.; Liu, H.; Liu, P. Cathode electrolyte interface enabling stable Li–S batteries. *Energy Storage Mater.* **2019**, *21*, 474–480. [CrossRef]

21. Wang, M.; Zhai, D.D.; Liu, H.; Yang, X.M.; Chen, X.Y.; Zhang, Z.J. Design and synthesis of highly N, S co-doped 3D carbon materials with tunable porosity for supercapacitors. *Ionics* **2020**, *26*, 2031–2041. [CrossRef]
22. Salles, R.C.M.; Coutinho, L.H.; Veiga, A.G.d.; Sant'Anna, M.M.; Souza, G.G.B.d. Surface damage in cystine, an amino acid dimer, induced by keV ions. *J. Chem. Phys.* **2018**, *148*, 045107. [CrossRef] [PubMed]
23. Liu, S.; Tian, J.; Wang, L.; Zhang, Y.; Qin, X.; Luo, Y.; Asiri, A.M.; Al-Youbi, A.O.; Sun, X. Hydrothermal Treatment of Grass: A Low-Cost, Green Route to Nitrogen-Doped, Carbon-Rich, Photoluminescent Polymer Nanodots as an Effective Fluorescent Sensing Platform for Label-Free Detection of Cu(II) Ions. *Adv. Mater.* **2012**, *24*, 2037–2041. [CrossRef] [PubMed]
24. Ferrari, A.C. Raman spectroscopy of graphene and graphite: Disorder, electron–phonon coupling, doping and nonadiabatic effects. *Solid State Commun.* **2007**, *143*, 47–57. [CrossRef]
25. Venezuela, P.; Lazzeri, M.; Mauri, F. Theory of double-resonant Raman spectra in graphene: Intensity and line shape of defect-induced and two-phonon bands. *Phy. Rev. B* **2011**, *84*, 035433. [CrossRef]
26. ISO. Nanotechnologies—Structural characterization of graphene. In *Part 1: Graphene from Powders and Dispersions (ISO/TS 21356-1:2021)*; ISO: Geneva, Switzerland, 2021; p. 48.
27. Pollard, A.; Paton, K.; Clifford, C.; Legge, E. *Characterisation of the Structure of Graphene. Good Practice Guide No 145*; National Physical Laboratory (NPL) NPL: London, UK, 2017.
28. Ferrari, A.C.; Basko, D.M. Raman spectroscopy as a versatile tool for studying the properties of graphene. *Nat. Nanotechnol.* **2013**, *8*, 235–246. [CrossRef] [PubMed]
29. Devi, S.; Singh, B.; Paul, A.K.; Tyagi, S. Highly sensitive and selective detection of trinitrotoluene using cysteine-capped gold nanoparticles. *Anal. Methods* **2016**, *8*, 4398–4405. [CrossRef]
30. Țucureanu, V.; Matei, A.; Avram, A.M. FTIR Spectroscopy for Carbon Family Study. *Crit. Rev. Anal. Chem.* **2016**, *46*, 502–520. [CrossRef]
31. González-Domínguez, J.M.; León, V.; Lucío, M.I.; Prato, M.; Vázquez, E. Production of ready-to-use few-layer graphene in aqueous suspensions. *Nat. Protoc.* **2018**, *13*, 495–506. [CrossRef] [PubMed]
32. Farivar, F.; Yap, P.L.; Hassan, K.; Tung, T.T.; Tran, D.N.H.; Pollard, A.J.; Losic, D. Unlocking thermogravimetric analysis (TGA) in the fight against "Fake graphene" materials. *Carbon* **2021**, *179*, 505–513. [CrossRef]
33. Farivar, F.; Yap, P.L.; Karunagaran, R.U.; Losic, D. Thermogravimetric Analysis (TGA) of Graphene Materials: Effect of Particle Size of Graphene, Graphene Oxide and Graphite on Thermal Parameters. *C* **2021**, *72*, 41. [CrossRef]

Article

Electrochemically Exfoliated Graphene-Like Nanosheets for Use in Ceramic Nanocomposites

Rosalía Poyato [1,*], Reyes Verdugo [2], Carmen Muñoz-Ferreiro [2] and Ángela Gallardo-López [2]

1 Instituto de Ciencia de Materiales de Sevilla, ICMS (CSIC-US), Américo Vespucio 49, 41092 Sevilla, Spain
2 Departamento de Física de la Materia Condensada, ICMS, CSIC-Universidad de Sevilla, Apdo. 1065, 41080 Sevilla, Spain; ireyesvmz@gmail.com (R.V.); cmunoz7@us.es (C.M.-F.); angela@us.es (Á.G.-L.)
* Correspondence: rosalia.poyato@icmse.csic.es

Received: 21 May 2020; Accepted: 4 June 2020; Published: 11 June 2020

Abstract: In this work, the synthesis of graphene-like nanosheets (GNS) by an electrochemical exfoliation method, their microstructural characterization and their performance as fillers in a ceramic matrix composite have been assessed. To fabricate the composites, 3 mol % yttria tetragonal zirconia (3YTZP) powders with 1 vol % GNS were processed by planetary ball milling in tert-butanol to enhance the GNS distribution throughout the matrix, and densified by spark plasma sintering (SPS). According to a thorough Raman analysis and SEM observations, the electrochemically exfoliated GNS possessed less than 10 graphene layers and a lateral size lower than 1 µm. However, they contained amorphous carbon and vacancy-like defects. In contrast the GNS in the sintered composite exhibited enhanced quality with a lower number of defects, and they were wavy, semi-transparent and with very low thickness. The obtained nanocomposite was fully dense with a homogeneous distribution of GNS into the matrix. The Vickers hardness of the nanocomposite showed similar values to those of a monolithic 3YTZP ceramic sintered in the same conditions, and to the reported ones for a 3YTZP composite with the same content of commercial graphene nanosheets.

Keywords: graphene; electrochemical exfoliation method; 3YTZP; ceramic nanocomposites; planetary ball milling; SPS; Raman spectroscopy; electron microscopy; Vickers indentations

1. Introduction

Since the first isolation of single-layer graphene in 2004 by the mechanical exfoliation of graphite —the "Scotch tape" method [1]—its unique properties have motivated a continuous growth in research activity. It has been considered as a feasible candidate for applications in fuel cells, composites, electronic devices, sensors, and photodetectors [2].

In the last decade, graphene has mainly been synthesized using two different approaches: bottom-up, in which graphene is grown from small molecular carbon precursors, and top-down, in which graphene is exfoliated from graphite as parent material [2]. Among the bottom-up approaches, the chemical vapor deposition (CVD) technique is the most popular way of deposition of graphene films on metal foils or silicon substrates [3–5]. Together with epitaxial growth [6], these are methods that allow the formation of high-quality, large area graphene, encouraging its application in highly flexible and conducting films. However, these methods present drawbacks such as high manufacturing costs or the requirement of sophisticated equipment, high temperatures and expensive substrates [3]. On the other hand, when top-down approaches such as mechanochemical synthesis [7,8] or liquid phase exfoliation [9,10] are simple, cost-effective and easily scalable, they have been presented by different authors such as suitable methods for graphene mass production [7,11]. The synthesis of graphene oxide by the mechanochemical method has also been reported [12]. The main disadvantage of these techniques is that the obtained structures can have a greater number of defects than the ones that originate from bottom-up methods.

Electrochemical exfoliation has had a strong impact on the development of techniques to obtain graphene because it provides an economical, simple and fast way to produce it. It is easily reproducible because it can be performed under environmental conditions, and toxicity-free components that can be easily removed after the process are used. In addition, the use of graphite as starting material, together with the good results obtained from this process, reduce the cost of producing graphene, resulting in an efficient and affordable method for the scientific community [13]. Moreover, some studies have reported the production of high-quality thin graphene sheets with lateral sizes up to 30 µm by electrochemical exfoliation of graphite [10,11,14]. However, when this type of synthesis technique is used, it is not easy to generate single-layer graphene, and graphene nanosheets (GNS) are usually obtained. Thus, after the synthesis step, it is essential to characterize the nanostructures in order to assess the lateral dimension, the number of graphene layers and the possible presence of defects created during the synthesis process [7,8,10]. In recent years, new approaches for the electrochemical exfoliation technique have been suggested in order to improve the yield [11,14] and to promote the obtaining of mostly single- and few-layer graphene sheets [15].

Among the different applications of graphene, its use as a filler in composite materials has awakened the interest of the scientific community in the last years, owing to the relevant properties that these nanostructures impart to most materials [2,16]. In the case of ceramics, a strong interest has been generated in the development of advanced ceramics in which the presence of graphene as a second phase improves their fracture toughness and electrical conductivity [17,18]. However, these composite materials present a processing challenge due to graphene's strong tendency to agglomerate, as a consequence of its high surface area. This negatively affects the properties of the composite, so advanced processing techniques are usually needed [19,20]. Among the advanced ceramics, 3 mol % yttria tetragonal zirconia (3YTZP) presents a remarkable technological interest because of its excellent mechanical properties, such as Young's modulus, fracture toughness and hardness, as well as its chemical stability [21]. Recent studies about 3YTZP composites with graphene have reported enhancements on properties as fracture toughness or flexure strength for very low additions of graphene nanostructures [22,23].

Most of the published studies about graphene-ceramic composites generally use commercially acquired graphene nanosheets. Although promising results in terms of enhancement of mechanical and electrical properties have been reported for composites with cost-effective graphene nanoplatelets prepared using advanced powder processing techniques [20,23], the best results have been obtained in composites with thinner and more expensive graphene nanosheets or few-layer graphene [24–26]. This could hinder the industrial application of these composite materials due to the high manufacturing costs. In this context, the search for cost-effective synthesis techniques to obtain graphene nanosheets for its application in ceramic nanocomposites is very necessary.

In this work, the synthesis of graphene-like nanosheets has been assessed by means of a simple, cost-effective and fast electrochemical exfoliation technique, using graphite as parent material. After a detailed characterization of the as-synthesized nanosheets by Raman spectroscopy and electron microscopy observations, they were incorporated as filler in a 3YTZP matrix nanocomposite. Powders with 1 vol % GNS were processed by planetary ball milling in tert-butanol to enhance the GNS distribution throughout the matrix, and densified by spark plasma sintering (SPS). The quality and level of defects of the GNS in the composite were assessed by Raman spectroscopy. The microstructure and hardness of the obtained nanocomposite was analyzed and compared to the reported ones for 3YTZP composites prepared with commercial nanosheets.

2. Materials and Methods

2.1. Graphene Synthesis and Characterization

The graphene-like nanosheets were obtained by the electrochemical exfoliation method [10], using a graphite bar (1 cm diameter, 10 cm long, Goodfellow Cambridge Ltd., Huntingdon, UK) and a platinum wire acting as anode and cathode, respectively. The ionic solution was prepared by taking 1.3 mL of sulphuric acid (95–98%, Panreac, Castellar del Vallès, Spain) and diluting in 100 mL of DI water. The platinum wire and the graphite bar were immersed into the ionic solution with a separation of 5 cm, and the electrochemical exfoliation process was carried out by applying DC bias from 1 to 10 V, with steps of 1 V every ten minutes during a total time of 1.5 h. After this time, 10 V were applied for 30 min. Continuous magnetic agitation was applied during the whole exfoliation process.

After the exfoliation process, the suspensions were washed with DI water and isopropyl alcohol by vacuum filtration using 200 nm pore filter alumina membranes (Whatman, Maidstone, UK) and centrifuged (model SIGMA 3-30KS, Sigma Laboratory Centrifuges, Osterode am Harz, Germany) at 8500 r.p.m. for 15 min to remove graphite aggregates. The suspensions were frozen with liquid nitrogen and freeze-dried for 48 h at −80 °C in order to avoid re-agglomeration of the obtained nanosheets during drying (Cryodos-80, Telstar, Terrasa, Spain).

Raman spectroscopy and high-resolution scanning electron microscopy (HRSEM, S5200, Hitachi High-Technologies Corp., Tokyo, Japan) were used to characterize the number of layers, morphology and size distribution of the as-synthesized GNS. To that end, a few droplets of GNS suspension in isopropyl alcohol were deposited on a glass slide for Raman spectroscopy or on a Cu transmission grid with C coating for HRSEM inspection after drying. At least 10 Raman spectra were acquired on the electrochemically exfoliated GNS using a dispersive microscope Raman Horiba Jobin Yvon LabRam HR800 (ICMS), with a green laser He-Ne (532.1 nm) at 20 mW. The first-order (from 1000 to 2000 cm^{-1}) Raman spectra were fitted to a sum of five functions: two Gaussian and three pseudo-Voigt functions. In the second-order spectra (from 2250 to 3300 cm^{-1}) three Lorentz and three pseudo-Voigt functions were used. The fits were carried out using the OriginLab software (OriginPro 2019, OriginLab Corporation, Northampton, MA, USA).

2.2. Nanocomposite Processing and Characterization

Composite powders with 1 vol % GNS were prepared using the electrochemically exfoliated nanosheets and commercial 3YTZP powders (40 nm particle size, TZ-3YB-E, Tosoh Europe B.V, Amsterdam, The Netherlands), which were previously annealed at 850 °C for 30 min in air. Planetary ball milling (Pulverisette 7 classic line, Fritsch, Idar-Oberstein, Germany) was used to homogenize the powders in a 10 w/w% tert-butanol (t-BuOH)/water mixture at 700 r.p.m. for 15 min. A 45 mL zirconia jar and seven 15 mm diameter zirconia balls were used. After drying on a rotary evaporator, the composite powders were homogenized in an agatha mortar and spark plasma sintered at 1250 °C for 5 min, with an applied pressure of 75 MPa and heating and cooling ramps of 300 and 50 °C/min, respectively (SPS model 515 S, Dr. Sinter, Inc., Kanagawa, Japan). A sheet of graphite paper was placed between the powders and the die/punches to both ensure their electrical, mechanical and thermal contact and also for an easy removal. The temperature was continually monitored by means of an optical pyrometer focused on the side of the graphite die. Cylindrical samples with 10 mm diameter and 2 mm thickness were obtained. The surface graphite paper from the SPS molding system was manually eliminated by grinding.

To account for possible structural modifications of the graphene-like nanosheets after the composite powder processing and sintering, at least ten Raman spectra were acquired on the obtained powders after planetary ball milling, and on the fracture surface of the sintered composite. The first- and second-order Raman spectra were fitted to the functions described in Section 2.1. The density of the composite was determined with the Archimedes' method using distilled water as the immersion medium. The theoretical density was calculated by the rule of mixtures taking the density of the 3YTZP and the GNS as 6.05 g/cm^3 and 2.2 g/cm^3, respectively. Scanning electron microscopy (SEM) using backscattered electrons (BSE) for imaging (FEI-Teneo, FEI, Thermo Fisher, Cambridge, MA, USA) was used to analyze the dispersion of the GNS in the ceramic matrix. This microscope has two in-lens detectors which allow obtaining high resolution images at short work distances. Polished in-plane (i.p.) and cross-section (c.s.) surfaces were analyzed to account for the existence of any structural anisotropy on the composite. The grain size of the ceramic matrix was estimated from SEM images acquired on polished c.s. surfaces previously annealed in air for 15 min at 1150 °C. The planar equivalent diameter, $d = 2(area/\pi)^{1/2}$, namely the diameter corresponding to a circle with the same area as the measured grain, was taken as a measure of the grain size, averaging 200 to 300 grains, according to UNE-EN ISO 13383-1:2016 standard. The software packages ImageJ and OriginLab were used to determine the relevant parameters. The fracture surface of the composite was also examined by HRSEM (HRSEM, S5200, Hitachi High-Technologies Corp., Tokyo, Japan).

The hardness of the nanocomposite was estimated from standard Vickers micro-indentations (Vickers Duramin indenter, Struers, Copenhagen, Denmark) performed on the mirror polished i.p. and c.s. surfaces. These two orientations were evaluated to account for any possible anisotropy effects. Ten indentations were performed on each surface with 1.96 N applied load during 10 s. The hardness values were calculated following the equation: H_V (GPa) = 1854.4 P/D^2, where P is the applied load in N and D the average diagonal of the imprint in μm.

3. Results and Discussion

3.1. Microstructural Characterization of the Graphene Nanosheets

Typical HRSEM micrographs (acquired in Secondary Electron Image mode) for the electrochemically exfoliated nanosheets are shown in Figure 1. It can be observed that some nanosheets present a lateral size lower that 1 μm (Figure 1a). However, they show a strong tendency to agglomerate (Figure 1b), resulting in GNS interconnections with a lateral size of several microns.

Figure 1. High-resolution scanning electron microscopy (HRSEM) images for the electrochemically exfoliated graphene nanosheets, drop-casted on a Cu transmission grid (a) Isolated nanosheets; (b) agglomerated nanosheets.

The Raman spectrum acquired on the as-exfoliated nanosheets is presented in Figure 2a. It is very similar to the described ones in different works for graphene nanosheets, few-layer graphene or reduced graphene oxide (rGO) [10,25,27]. The typical bands described in literature for these nanomaterials are clearly observed at ~1350 (D), ~1585 (G) and ~2700 (2D) cm^{-1}. The G and 2D bands are always found in pristine graphene. The G band is due to the doubly degenerate zone center E2g mode and the 2D band is the second order of zone-boundary phonons [28–30]. On the other hand, the D band is the most prominent of the defect-induced bands. It has been reported that these bands arise from breathing-like modes of the carbon rings activated by defects via double-resonance Raman process [28–30]. Usually, the I_D/I_G intensity ratio is an indicative of the presence of defects on the graphene lattice [18,31,32].

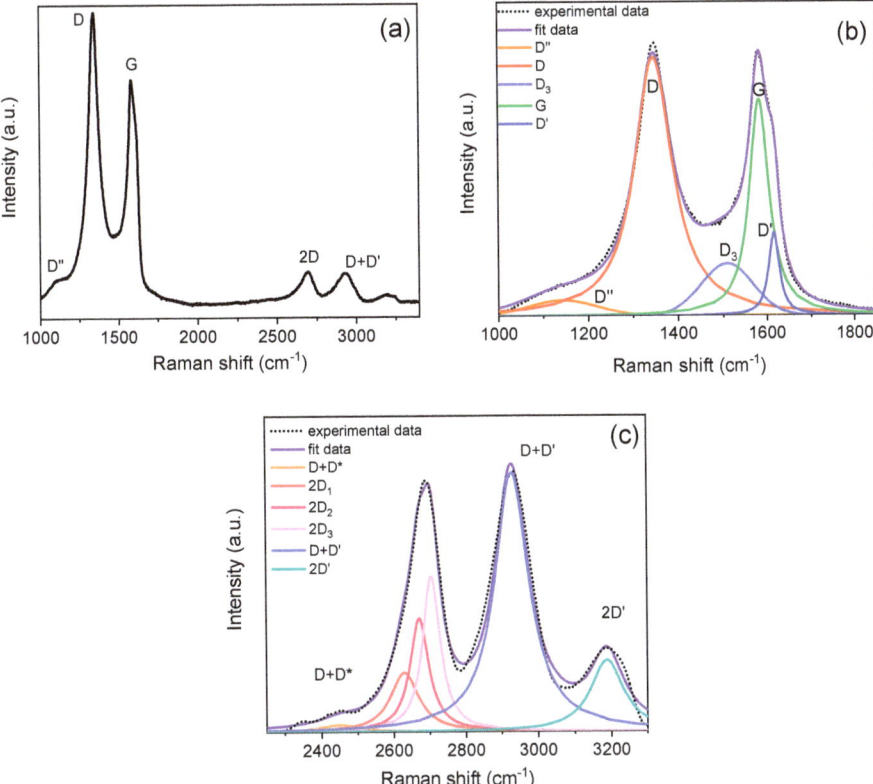

Figure 2. (a) Raman spectrum acquired on the as-synthesized graphene nanosheets; (b) Deconvolution of the first-order Raman spectrum using five functions (D″, D, D_3, G and D′ bands); (c) Deconvolution of the second-order Raman spectrum using six functions (D + D*, $2D_1$, $2D_2$, $2D_3$, D + D′ and 2D′ bands).

Along with the peaks that are clearly observed in the spectrum, other defect-induced bands are present at ~1100–1200 and ~1610–1620 cm^{-1}. These bands are detected in Figure 2a as a peak that overlaps with the left side of the D band and as a shoulder on the right side of the G peak, respectively. While the latter has been named in most of the published works as D′, the former has been named as T_1 [33], D_4 [34], D* [23,27,35] or D″ [29,30,36] depending on the authors and on the studied carbon-based material. Moreover, a broad shoulder between the D and G peaks can also be seen in Figure 2a. This feature has been related to a Raman band at ~1500 cm^{-1} in defected carbon-based materials, and has been named as T_2 [33], D_3 [34,37] or D″ [27,35] by different authors. This band has been related to the presence of amorphous carbon in graphene oxide [27,35], carbon nanotubes [33] or other carbon-based materials [34]. In the present work, we will assume the nomenclature D″, D_3 and D′ for the bands located at ~1100–1200, ~1500 and ~1610–1620 cm^{-1}, respectively.

Usually, the D″, D_3 and D′ bands are not described when analyzing the Raman spectra of graphene-based nanomaterials because they are very weak peaks. Nevertheless, when these bands present a remarkable intensity, they appear to overlap with the D and G peaks. This makes the deconvolution of the first-order spectrum (from 1000 to 2000 cm^{-1}) essential for the correct interpretation of the Raman spectrum, as it is suggested by different authors [27,33,35,38,39]. The fittings of the first- and second-order spectra allow the suitable obtaining of the position, intensity (integrated area) and band width of the different peaks. These parameters allow us to establish the presence and nature of defects in the electrochemically exfoliated nanosheets.

Figure 2b,c show examples of the fittings that have been carried out for all the Raman spectra acquired on the electrochemically exfoliated graphene nanosheets, using two Gaussian (D'' and D_3) and three pseudo-Voigt (D, G and D') functions for the first-order spectra, and three Lorentz (2D) and three pseudo-Voigt (D + D*, D + D' and 2D´) functions for the second-order spectra.

The high values of the I_D/I_G and $I_{D'}/I_G$ ratios (Table 1) point to the existence of defects and disorder in the exfoliated nanosheets. Moreover, the low value of I_{2D}/I_G supports this conclusion, as it has been published that the 2D band of highly disordered graphene reduces its intensity and increases its width [27,31]. Nevertheless, according to the terminology introduced by Ferrari et al. [29] regarding the Raman spectra of disordered graphene, the electrochemically exfoliated nanosheets obtained in the present work would correspond to low-defect graphene (stage I in the classification proposed by these authors). They established a transition between stages I (low-defect graphene) and II (disordered graphene) at I_D/I_G = 3.5, and the intensity ratio of the obtained GNS is lower than this value (2.24 ± 0.05). The existence of a pronounced D_3 band (see the high value of the I_{D3}/I_G ratio obtained after fitting, Table 1) is attributed to the presence of amorphous carbon in the nanosheets, as suggested by previous authors [27,34,35].

Table 1. Intensity ratios of the D, D_3, D' and 2D bands with respect to the G peak, obtained for the as-synthesized graphene-like nanosheets (GNS) and the GNS in the sintered composite after fitting the first- and second-order Raman spectra.

Sample	I_D/I_G	I_{D3}/I_G	$I_{D'}/I_G$	I_{2D}/I_G
As-synthesized GNS	2.24 ± 0.05	0.485 ± 0.021	0.23 ± 0.07	0.219 ± 0.011
Sintered 1 vol % GNS/3YTZP	1.93 ± 0.06	0.249 ± 0.021	0.18 ± 0.07	0.46 ± 0.05

The shape of the second-order spectrum (Figure 2c)—with a D+D' band with high intensity—is very similar to the reported one for monolayer graphene bombarded by low-energy argon ions in order to induce disorder in the system [31]. It has been shown that this ion bombardment promotes vacancy-type defects [31,32], so the disorder detected on the exfoliated nanosheets is very likely caused by vacancy-like defects.

Finally, the fitting of the 2D band could be carried out using three Lorentzian functions (Figure 2c), revealing that the GNS present a number of layers lower than 10, according to Ferrari et al. [28] and Malard et al. [40]. Thus, the electrochemical exfoliation technique used in this work allows the production of graphene-like nanosheets: reduced graphene oxide or few-layered defected graphene.

3.2. Microstructural Characterization of the Nanocomposite

A relative density of 99% was obtained for the sintered composite. This high-density value reveals the achievement of a high level of compaction and low porosity in the composite, as it is supported by the SEM micrographs of the composite polished surfaces annealed in air (Figure 3), where pores are not distinguished. It is possible to observe some voids closed to the ceramic grains; however, their size is very similar to that of the grains, which points to the fact that they are the consequence of grain pull-out during the grinding and polishing steps previous to the annealing. The full densification of this type of composites has been previously reported for composites with similar contents of other types of commercial graphene-based nanomaterials, prepared with similar processing and sintering routines [23,41].

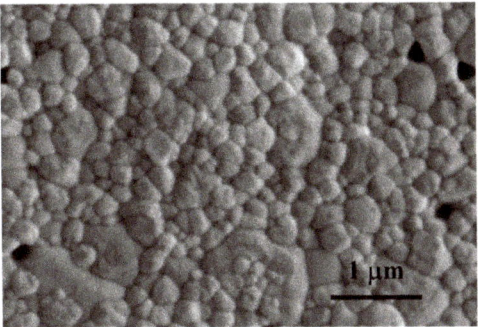

Figure 3. SEM micrograph of the polished c. s. surface of the 3YTZP composite after annealing in air.

A grain size of 0.17 ± 0.09 µm has been obtained for the nanocomposite, revealing a grain refinement with respect to a monolithic 3YTZP ceramic sintered using the same conditions (0.29 ± 0.02 µm [41]), in agreement with the grain growth inhibition effect previously reported for ceramic composites with commercial graphene-based nanomaterials [19,23,25,41]. The grain refinement shown by the composite in this work is more remarkable than the reported ones in previous works for 3YTZP composites with the same content of commercial graphene nanoplatelets [41] (0.27 µm) and graphene nanosheets obtained by mechanical exfoliation of commercial GNP [23] (0.25 µm). This could be related to the optimum GNS distribution throughout the matrix achieved in this work. This is a consequence, on the one hand, of the low dimensions of the electrochemically exfoliated GNS, and, on the other hand, of the adequate use of advanced powder processing and sintering techniques.

The Raman spectra of the composite powder after planetary ball milling and of the sintered ceramic composite are presented in Figure 4a. The characteristic peaks for graphene are clearly observed, revealing that neither the high-energy milling during powder processing nor the high temperature during sintering degraded the electrochemically exfoliated GNS. However, a peak with high intensity was detected at ~1000 cm^{-1}, which had not been observed in the spectrum of the as-exfoliated GNS (Figure 2a). In order to analyze the origin of this peak, the Raman spectra were acquired in an extended frequency range (inset in Figure 4a) revealing the existence of multiple peaks. Together with the peaks corresponding to the tetragonal (264, 320, 460, 643 cm^{-1}) and monoclinic (365, 488 cm^{-1}) phases of the zirconia matrix [42], sharp peaks in the range ~500–630 cm^{-1} and a broad band in the range ~700–1100 cm^{-1} were found. These bands can be attributed to the presence of a low percentage of an alumino-silicate (AS) glass [43] that could have been introduced as contamination into the composite powder during the high-energy ball milling. The percentage of this phase must be significantly low, as it was not detected by X-ray diffraction (results not shown). However, future efforts will be carried out to modify the planetary ball milling conditions in order to avoid the formation of this trace of AS glass. In order to perform the deconvolution of the first-order spectra to suitably analyze the defect-related peaks and the intensity ratios, we introduced a new peak—at ~1000 cm^{-1}—to the fittings.

Figure 4b,c shows examples of the fittings that have been carried out for all the Raman spectra acquired on the sintered ceramic composite using two Gaussian (D" and D_3) and four pseudo-Voigt (AS glass, D, G and D') functions for the first-order spectra, and three Lorentz (2D) and three pseudo-Voigt (D + D*, D + D' and 2D') functions for the second-order spectra.

A decrease of the defect-related D and D' peaks intensity, along with an increase of the intensity of the 2D band, is observed for the GNS in the sintered composite, in comparison with the as-exfoliated GNS (Table 1). Also, a decrease of the D_3 band is found, pointing to a lower amount of amorphous carbon in the GNS after sintering, in agreement with published results that the I_{D3}/I_G ratio decreases as the crystallinity increases [27]. All of this reveals a decrease of the number of defects and a restoration of the graphene network during the high-temperature sintering process [27,31,32,39].

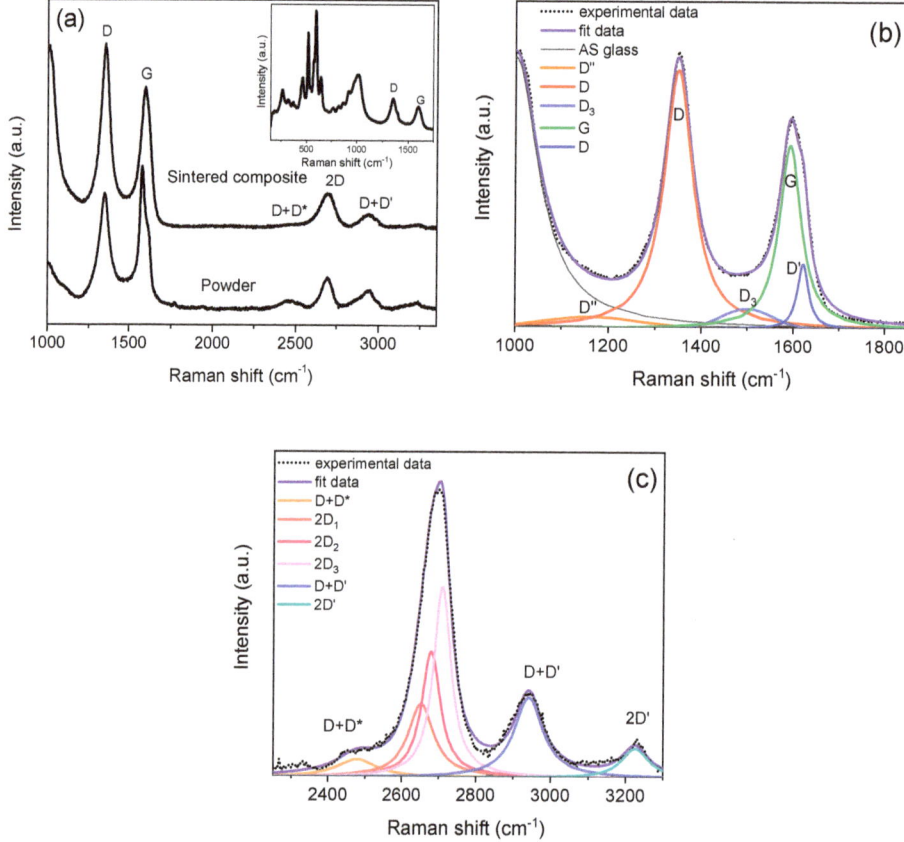

Figure 4. (a) Raman spectra acquired on the composite powders after high-energy planetary ball milling and on the sintered composite, inset: detail of the Raman spectrum acquired on the sintered composite in the range 150–1800 cm^{-1}; (b) Deconvolution of the first-order Raman spectrum of the sintered composite; (c) Deconvolution of the second-order Raman spectrum of the sintered composite (D + D*, 2D$_1$, 2D$_2$, 2D$_3$, D + D' and 2D' bands).

Another parameter that can give information about defects in graphene is the band width for D, G, D' and 2D bands, as their widths increase with a growing number of defects [31]. Martins Ferreira et al. [31] have reported that the width of G and D' peaks have a less pronounced dependence than the D and 2D bands. According to this assessment, the band widths of the G and D' peaks stay invariable in both the as-exfoliated GNS and the sintered composite (Table 2), while a lower D band width is clearly observed in the GNS after sintering, which supports the decrease of the number of defects mentioned above. Unexpectedly, the 2D band width stays invariable. However, this parameter is not only dependent on the number of defects, but also on other factors such as doping or strain [3]. It has been published that the 2D band width of graphene subjected to strain suffers a broadening and a shift in frequency [3,44,45]. Table 2 shows the positions of the D, G, D' and 2D bands, revealing a shift towards higher frequencies for all of them in the spectra of the GNS sintered composite, in comparison to the spectra of the as-exfoliated GNS. This can be attributed to residual stresses in the GNS imposed by the constraining ceramic matrix [44,45]. Thus, the effect of broadening the 2D band as a consequence of the stresses would counteract the decrease of the band width related to the decrease of defects in the GNs after sintering.

Table 2. Positions and band widths of the D, G, D' and 2D bands obtained for the as-synthesized GNS and the GNS in the sintered composite after fitting the first- and second-order Raman spectra.

Sample	D		G		D'		2D	
	Position (cm^{-1})	Band Width (cm^{-1})	Position (cm^{-1})	Band Width (cm^{-1})	Position (cm^{-1})	Band Width (cm^{-1})	Position * (cm^{-1})	Band Width * (cm^{-1})
As-synthesized GNS	1346.9 ± 0.3	91.5 ± 2.1	1585.34 ± 1.02	52 ± 1	1617.5 ± 0.7	27.2 ± 0.9	2687.6 ± 0.4	103.96 ± 1.6
Sintered 1 vol % GNS/3YTZP	1350.3 ± 0.4	75.3 ± 2.4	1592.1 ± 1.3	53 ± 2	1621.8 ± 0.5	32.7 ± 1.3	2690.9 ± 1.1	105 ± 3

* Values obtained after fitting the 2D band to a pseudo-Voigt function (not shown).

Figure 5 shows the low magnification SEM micrographs acquired on the polished c.s. surface of the nanocomposite using BSE. These images reflect the GNS distribution in the ceramic matrix, as the 3YTZP matrix and the GNS appear in the micrographs as light and dark phases, respectively. A homogeneous distribution of the GNS (marked with thin arrows in the figure) throughout the ceramic matrix is observed, with scarce large GNS agglomerates (marked with a thick arrow). In this c.s. image, most of the observed nanosheets present their side view, which indicates that the ab plane of the graphene layers lies on a plane perpendicular to the compression axis during sintering. This preferential alignment has been previously described for different ceramic composites [19,24,25,41], including composites prepared from powders homogenized using planetary ball milling in wet conditions [20,46]. This structural anisotropy is a consequence of the two-dimensional character of graphene, and the uniaxial pressure applied during the sintering process. When increasing the magnification (inset in Figure 5), very thin GNS with lateral sizes of several microns can be observed throughout the matrix. This could correspond to interconnections of smaller GNS, as previously shown in the HRSEM images of the as-exfoliated GNS (Figure 1b).

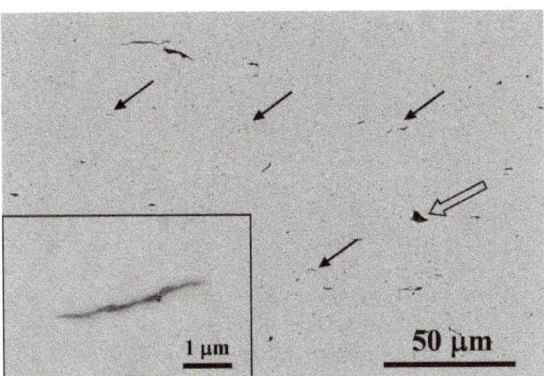

Figure 5. BSE-SEM micrographs of the polished c.s. surface of the sintered composite.

The HRSEM images of the fracture surface of the nanocomposite, shown in Figure 6, give an insight into the morphology of the graphene nanosheets incorporated in the 3YTZP matrix. The GNS (marked with arrows) appear as wavy, semi-transparent tissue covering the ceramic grains, as previously reported in ceramic composites with few-layer graphene [25,47]. Some GNS can be seen from a side view, revealing a very low thickness, in accordance with the HRSEM observations of the as-synthesized nanosheets (Figure 1). The fracture surface presents a mostly intergranular fracture mode, which indicates a strong physical bonding at the interphase between the 3YTZP matrix and the GNS, although some areas with intragranular fracture can also be observed.

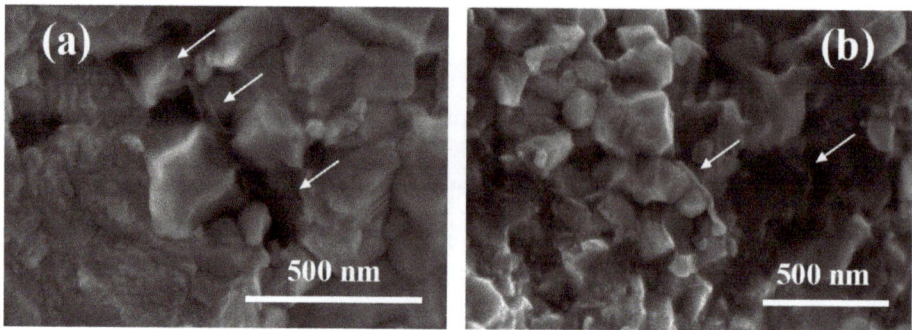

Figure 6. HRSEM micrographs of the fracture surface of the sintered composite. (**a**) Some GNS can be seen as semi-transparent tissue; (**b**) some GNS can be seen from a side view.

3.3. Vickers Hardness of the Nanocomposite

The Vickers hardness of the nanocomposite, evaluated on i.p. and c.s. surfaces, revealed no mechanical anisotropy, as similar hardness values were obtained for both surfaces (Table 3). This may be due to the small lateral size of the GNS. Also, the values were identical to the reported ones for a monolithic 3YTZP ceramic prepared with similar sintering conditions [41], and very similar to the reported ones for a composite with 1 vol % of graphene nanosheets obtained by exfoliation of commercial graphene nanoplatelets by means of high-energy ball milling [23]. These results indicate that ceramic-based composites containing the electrochemically exfoliated GNS may also display good mechanical strength and fracture toughness. Research work to determine this is already underway.

Table 3. Vickers hardness for the composite in this study, compared with a monolithic 3YTZP ceramic and a composite prepared with commercial graphene nanosheets.

Sample	$H_{i.p.}$ (GPa)	$H_{c.s.}$ (GPa)
3YTZP [41]	13.9 ± 0.5	
1 vol % GNS/3YTZP (this work)	14.00 ± 0.13	13.9 ± 0.8
1 vol % GNS/3YTZP [23]	13.6 ± 0.8	12.7 ± 0.7

4. Conclusions

Graphene-like nanosheets (GNS) with a number of graphene layers lower than 10, containing amorphous carbon and vacancy-like defects, and with a lateral size lower than 1 µm were successfully synthesized using a simple, cost-effective and fast electrochemical exfoliation method. The incorporation of 1 vol % GNS to a 3YTZP matrix resulted in a composite material with a homogeneous distribution of GNS into the matrix, despite the high tendency to agglomerate presented by the as-exfoliated GNS. The selected processing method—planetary ball milling of the composite powders and spark plasma sintering (SPS)—produced a fully dense nanocomposite containing GNS with significantly enhanced quality, as revealed by Raman spectroscopy. The GNS were wavy, semi-transparent and with very thin thickness. The microstructural anisotropy—preferential alignment of the GNS in the direction perpendicular to the pressing axis during SPS—revealed by SEM observations was not reflected on the hardness values of the nanocomposite, which were isotropic. The hardness of the composite studied here was very similar to the reported one for a 3YTZP composite with the same content of commercial graphene nanosheets.

Author Contributions: Conceptualization, R.P. and Á.G.-L.; methodology, R.P., R.V., C.M.-F. and Á.G.-L.; validation, R.P. and Á.G.-L.; investigation, R.P., R.V., C.M.-F. and Á.G.-L.; supervision, R.P. and Á.G.-L.; writing—original draft preparation, R.P.; writing—review and editing, Á.G.-L.; project administration, R.P. and Á.G.-L.; funding acquisition, R.P. and Á.G.-L. All authors have read and agreed to the published version of the manuscript.

Funding: This research was funded by project PGC 2018-101377-B-100 (MCIU/AEI/FEDER, UE). Carmen Muñoz-Ferreiro acknowledges the financial support of a VI PPIT-US fellowship through the contract USE-18740-H.

Acknowledgments: Francisco José Gotor and José Manuel Córdoba are gratefully acknowledged for providing access to the planetary ball mill and their helpful advice.

Conflicts of Interest: The authors declare no conflict of interest.

References

1. Novoselov, K.S.; Geim, A.K.; Morozov, S.V.; Jiang, D.; Zhang, Y.; Dubonos, S.V.; Grigorieva, I.V.; Firsov, A.A. Electric field effect in atomically thin carbon films. *Science* **2004**, *306*, 666–669. [CrossRef]
2. Ferrari, A.C.; Bonaccorso, F.; Fal'ko, V.; Novoselov, K.S.; Roche, S.; Bøggild, P.; Borini, S.; Koppens, F.H.L.; Palermo, V.; Pugno, N.; et al. Science and technology roadmap for graphene, related two-dimensional crystals, and hybrid systems. *Nanoscale* **2015**, *7*, 4598–4810. [CrossRef]
3. Milenov, T.I.; Valcheva, E.; Popov, V.N. Ramanspectroscopic study of as-deposited and exfoliated defected graphene grown on (001) Si substrates by CVD. *J. Spectrosc.* **2017**, *2017*. [CrossRef]
4. Milenov, T.I.; Avramova, I.; Valcheva, E.; Avdeev, G.V.; Rusev, S.; Kolev, S.; Balchev, I.; Petrov, I.; Pishinkov, D.; Popov, V.N. Deposition of defected graphene on (001) Si substrates by thermal decomposition of acetone. *Superlattices Microstruct.* **2017**, *111*, 45–56. [CrossRef]
5. Kim, K.S.; Zhao, Y.; Jang, H.; Lee, S.Y.; Kim, J.M.; Kim, K.S.; Ahn, J.H.; Kim, P.; Choi, J.Y.; Hong, B.H. Large-scale pattern growth of graphene films for stretchable transparent electrodes. *Nature* **2009**, *457*, 706–710. [CrossRef] [PubMed]
6. Berger, C.; Song, Z.; Li, T.; Li, X.; Ogbazghi, A.Y.; Feng, R.; Dai, Z.; Alexei, N.; Conrad, M.E.H.; First, P.N.; et al. Ultrathin epitaxial graphite: 2D electron gas properties and a route toward graphene-based nanoelectronics. *J. Phys. Chem. B* **2004**, *108*, 19912–19916. [CrossRef]
7. Buzaglo, M.; Bar, I.P.; Varenik, M.; Shunak, L.; Pevzner, S.; Regev, O. Graphite-to-graphene: Total conversion. *Adv. Mater.* **2017**, *29*, 1–5. [CrossRef]
8. Jeon, I.-Y.; Bae, S.-Y.; Seo, J.-M.; Baek, J.-B. Scalable production of edge-functionalized graphene Nanoplatelets via Mechanochemical Ball-Milling. *Adv. Funct. Mater.* **2015**, *25*, 6961–6975. [CrossRef]
9. Mahdy, S.M.; Gewfiel, E.; Ali, A.A. Production and characterization of three-dimensional graphite nanoplatelets. *J. Mater. Sci.* **2017**, *52*, 5928–5937. [CrossRef]
10. Su, C.Y.; Lu, A.-Y.; Xu, Y.; Chen, F.-R.; Khlobystov, A.N.; Li, L.-J. High-quality thin graphene films from fast elecrochemical exfoliation. *ACS Nano.* **2011**, *5*, 2332–2339. [CrossRef]
11. Roscher, S.; Hoffmann, R.; Prescher, M.; Knittel, P.; Ambacher, O. High voltage electrochemical exfoliation of graphite for high-yield graphene production. *RSC Adv.* **2019**, *9*, 29305–29311. [CrossRef]
12. Posudievsky, O.Y.; Khazieieva, O.A.; Koshechko, V.G.; Pokhodenko, V.D. Preparation of graphene oxide by solvent-free mechanochemical oxidation of graphite. *J. Mater. Chem.* **2012**, *22*, 12465–12467. [CrossRef]
13. Abdelkader, A.M.; Cooper, A.J.; Dryfe, R.A.W.; Kinloch, I.A. How to get between the sheets: A review of recent works on the electrochemical exfoliation of graphene materials from bulk graphite. *Nanoscale* **2015**, *7*, 6944–6956. [CrossRef]
14. Achee, T.C.; Sun, W.; Hope, J.T.; Quitzau, S.G.; Sweeney, C.B.; Shah, S.A.; Habib, T.; Green, M.J. High-yield scalable graphene nanosheet production from compressed graphite using electrochemical exfoliation. *Sci. Rep.* **2018**, *8*, 1–8. [CrossRef] [PubMed]
15. Parvez, K.; Worsley, R.; Alieva, A.; Felten, A.; Casiraghi, C. Water-based and inkjet printable inks made by electrochemically exfoliated graphene. *Carbon* **2019**, *149*, 213–221. [CrossRef]
16. Nieto, A.; Bisht, A.; Lahiri, D.; Zhang, C.; Agarwal, A. Graphene reinforced metal and ceramic matrix composites: A review. *Int. Mater. Rev.* **2017**, *62*, 241–302. [CrossRef]
17. Ahmad, I.; Yazdani, B.; Zhu, Y. Recent advances on carbon nanotubes and graphene reinforced ceramics nanocomposites. *Nanomaterials* **2015**, *5*, 90–114. [CrossRef]

18. Miranzo, P.; Belmonte, M.; Osendi, M.I. From bulk to cellular structures: A review on ceramic/graphene filler composites. *J. Eur. Ceram. Soc.* **2017**, *37*, 3649–3672. [CrossRef]
19. Baskut, S.; Cinar, A.; Seyhan, A.T.; Turan, S. Tailoring the properties of spark plasma sintered SiAlON containing graphene nanoplatelets by using different exfoliation and size reduction techniques: Anisotropic electrical properties. *J. Eur. Ceram. Soc.* **2018**, *38*, 3787–3792. [CrossRef]
20. López-Pernía, C.; Muñoz-Ferreiro, C.; González-Orellana, C.; Morales-Rodríguez, A.; Gallardo-López, Á.; Poyato, R. Optimizing the homogenization technique for graphene nanoplatelet/yttria tetragonal zirconia composites: Influence on the microstructure and the electrical conductivity. *J. Alloys Compd.* **2018**, *767*, 994–1002. [CrossRef]
21. Garvie, R.C.; Hannink, R.H.; Pascoe, R.T. Ceramic steel? *Nature* **1975**, *258*, 703–704. [CrossRef]
22. Chen, F.; Jin, D.; Tyeb, K.; Wang, B.; Han, Y.-H.; Kim, S.; Schoenung, J.M.; Shen, Q.; Zhang, L. Field assisted sintering of graphene reinforced zirconia ceramics. *Ceram. Int.* **2015**, *41*, 6113–6116. [CrossRef]
23. Gallardo-López, Á.; Castillo-Seoane, J.; Muñoz-Ferreiro, C.; López-Pernía, C.; Morales-Rodríguez, A.; Poyato, R. Flexure strength and fracture propagation in zirconia ceramic composites with exfoliated graphene nanoplatelets. *Ceramics* **2020**, *3*, 78–91. [CrossRef]
24. Ramirez, C.; Miranzo, P.; Belmonte, M.; Osendi, M.I.; Poza, P.; Vega-Diaz, S.M.; Terrones, M. Extraordinary toughening enhancement and flexural strength in Si_3N_4 composites using graphene sheets. *J. Eur. Ceram. Soc.* **2014**, *34*, 161–169. [CrossRef]
25. Muñoz-Ferreiro, C.; Morales-Rodríguez, A.; Rojas, T.C.; Jiménez-Piqué, E.; López-Pernía, C.; Poyato, R.; Gallardo-López, A. Microstructure, interfaces and properties of 3YTZP ceramic composites with 10 and 20 vol% different graphene-based nanostructures as fillers. *J. Alloys Compd.* **2019**, *777*, 213–224. [CrossRef]
26. Tapasztó, O.; Puchy, V.; Horváth, Z.E.; Fogarassy, Z.; Bódis, E.; Károly, Z.; Balázsi, K.; Dusza, J.; Tapasztó, L. The effect of graphene nanoplatelet thickness on the fracture toughness of Si3N4 composites. *Ceram. Int.* **2019**, *45*, 6858–6862. [CrossRef]
27. Claramunt, S.; Varea, A.; López-Díaz, D.; Velázquez, M.M.; Cornet, A.; Cirera, A. The importance of interbands on the interpretation of the raman spectrum of graphene oxide. *J. Phys. Chem. C* **2015**, *119*, 10123–10129. [CrossRef]
28. Ferrari, A.C.; Meyer, J.C.; Scardaci, V.; Casiraghi, C.; Lazzeri, M.; Mauri, F.; Piscanec, S.; Jiang, D.; Novoselov, K.S.; Roth, S.; et al. Raman spectrum of graphene and graphene layers. *Phys. Rev. Lett.* **2006**, *97*, 1–4. [CrossRef]
29. Ferrari, A.C.; Basko, D.M. Raman spectroscopy as a versatile tool for studying the properties of graphene. *Nat. Nanotechnol.* **2013**, *8*, 235–246. [CrossRef]
30. Venezuela, P.; Lazzeri, M.; Mauri, F. Theory of double-resonant Raman spectra in graphene: Intensity and line shape of defect-induced and two-phonon bands. *Phys. Rev. B* **2011**, *84*, 1–25. [CrossRef]
31. Ferreira, E.H.M.; Moutinho, M.V.O.; Stavale, F.; Lucchese, M.M.; Capaz, R.B.; Achete, C.A.; Jorio, A. Evolution of the Raman spectra from single-, few-, and many-layer graphene with increasing disorder. *Phys. Rev. B* **2010**, *82*, 125429. [CrossRef]
32. Eckmann, A.; Felten, A.; Mishchenko, A.; Britnell, L.; Krupke, R.; Novoselov, K.S.; Casiraghi, C. Probing the nature of defects in graphene by Raman spectroscopy. *Nano Lett.* **2012**, *12*, 3925–3930. [CrossRef] [PubMed]
33. Vollebregt, S.; Ishihara, R.; Tichelaar, F.D.; Hou, Y.; Beenakker, C.I.M. Influence of the growth temperature on the first and second-order Raman band ratios and widths of carbon nanotubes and fibers. *Carbon* **2012**, *50*, 3542–3554. [CrossRef]
34. Sadezky, A.; Muckenhuber, H.; Grothe, H.; Niessner, R.; Pöschl, U. Raman microspectroscopy of soot and related carbonaceous materials: Spectral analysis and structural information. *Carbon* **2005**, *43*, 1731–1742. [CrossRef]
35. López-Díaz, D.; Holgado, M.L.; García-Fierro, J.L.; Velázquez, M.M. Evolution of the Raman spectrum with the chemical composition of graphene oxide. *J. Phys. Chem. C* **2017**, *121*, 20489–20497. [CrossRef]
36. Herziger, F.; Tyborski, C.; Ochedowski, O.; Schleberger, M.; Maultzsch, J. Double-resonant la phonon scattering in defective graphene and carbon nanotubes. *Phys. Rev. B* **2014**, *90*. [CrossRef]
37. Gupta, A.; Chen, G.; Joshi, P.; Tadigadapa, S.; Eklund, P.C. Raman scattering from high-frequency phonons in supported n-graphene layer films. *Nano Lett.* **2006**, *6*, 2667–2673. [CrossRef]

38. King, A.A.K.; Davies, B.R.; Noorbehesht, N.; Newman, P.; Church, T.L.; Harris, A.T.; Razal, J.M.; Minett, A.I. A new raman metric for the characterisation of graphene oxide and its derivatives. *Sci. Rep.* **2016**, *6*, 1–6. [CrossRef]
39. Díez-Betriu, X.; Álvarez-García, S.; Botas, C.; Álvarez, P.; Sánchez-Marcos, J.; Prieto, C.; Menéndez, R.; de Andrés, A. Raman spectroscopy for the study of reduction mechanisms and optimization of conductivity in graphene oxide thin films. *J. Mater. Chem. C* **2013**, *1*, 6905–6912. [CrossRef]
40. Malard, L.M.; Pimenta, M.A.; Dresselhaus, G.; Dresselhaus, M.S. Raman spectroscopy in graphene. *Phys. Rep.* **2009**, *473*, 51–87. [CrossRef]
41. Gallardo-López, A.; Márquez-Abril, I.; Morales-Rodríguez, A.; Muñoz, A.; Poyato, R. Dense graphene nanoplatelet/yttria tetragonal zirconia composites: Processing, hardness and electrical conductivity. *Ceram. Int.* **2017**, *43*, 11743–11752. [CrossRef]
42. Jayakumar, S.; Ananthapadmanabhan, P.V.; Perumal, K.; Thiyagarajan, T.K.; Mishra, S.C.; Su, L.T.; Tok, A.I.Y.; Guo, J. Characterization of nano-crystalline ZrO$_2$ synthesized via reactive plasma processing. *Mater. Sci. Eng. B* **2011**, *176*, 894–899. [CrossRef]
43. Petrescu, S.; Constantinescu, M.; Anghel, E.M.; Atkinson, I.; Olteanu, M.; Zaharescu, M. Structural and physico-chemical characterization of some soda lime zinc alumino-silicate glasses. *J. Non. Cryst. Solids.* **2012**, *358*, 3280–3288. [CrossRef]
44. Tsoukleri, G.; Parthenios, J.; Papagelis, K.; Jalil, R.; Ferrari, A.C.; Geim, A.K.; Novoselov, K.S.; Galiotis, C. Subjecting a graphene monolayer to tension and compression. *Small* **2009**, *5*, 2397–2402. [CrossRef] [PubMed]
45. Androulidakis, C.; Koukaras, E.N.; Parthenios, J.; Kalosakas, G.; Papagelis, K.; Galiotis, C. Graphene flakes under controlled biaxial deformation. *Sci. Rep.* **2015**, *5*, 1–11. [CrossRef] [PubMed]
46. Michálková, M.; Kašiarová, M.; Tatarko, P.; Dusza, J.; Šajgalík, P. Effectof homogenization treatment on the fracture behaviour of silicon nitride/graphene nanoplatelets composites. *J. Eur. Ceram. Soc.* **2014**, *34*, 3291–3299. [CrossRef]
47. Fan, Y.; Estili, M.; Igarashi, G.; Jiang, W.; Kawasaki, A. The effect of homogeneously dispersed few-layer graphene on microstructure and mechanical properties of Al2O3 nanocomposites. *J. Eur. Ceram. Soc.* **2014**, *34*, 443–451. [CrossRef]

© 2020 by the authors. Licensee MDPI, Basel, Switzerland. This article is an open access article distributed under the terms and conditions of the Creative Commons Attribution (CC BY) license (http://creativecommons.org/licenses/by/4.0/).

Article

Controlling the Electronic Properties of a Nanoporous Carbon Surface by Modifying the Pores with Alkali Metal Atoms

Michael M. Slepchenkov [1], Igor S. Nefedov [2,3] and Olga E. Glukhova [1,4,*]

[1] Department of Physics, Saratov State University, Astrakhanskaya street 83, 410012 Saratov, Russia; slepchenkovm@mail.ru
[2] School of Electrical Engineering, Aalto University, P.O. Box 13000, 00076 Aalto, Finland; igor.nefedov@aalto.fi
[3] Faculty of Science, People's Friendship University of Russia (RUDN University) 6 Miklukho-Maklaya St, 117198 Moscow, Russia
[4] Laboratory of Biomedical Nanotechnology, I.M. Sechenov First Moscow State Medical University, Bolshaya Pirogovskaya street 2-4, 119991 Moscow, Russia
* Correspondence: glukhovaoe@info.sgu.ru; Tel.: +7-8452-514562

Received: 30 December 2019; Accepted: 27 January 2020; Published: 30 January 2020

Abstract: We investigate a process of controlling the electronic properties of a surface of nanoporous carbon glass-like thin films when the surface pores are filled with potassium atoms. The presence of impurities on the surface in the form of chemically adsorbed hydrogen and oxygen atoms, and also in the form of hydroxyl (OH) groups, is taken into account. It is found that even in the presence of impurities, the work function of a carbon nanoporous glass-like film can be reduced by several tenths of an electron volt when the nanopores are filled with potassium atoms. At the same time, almost all potassium atoms are ionized, losing one electron, which passes to the carbon framework of the film. This is due to the nanosizes of the pores in which the electron clouds of the potassium atom interact maximally with the electrons of the carbon framework. As a result, this leads to an improvement in the electrical conductivity and an increase in the electron density at the Fermi level. Thus, we conclude that an increase in the number of nanosized pores on the film surface makes it possible to effectively modify it, providing an effective control of the electronic structure and emission properties.

Keywords: nanoporous carbon surface; electronic structure; emission properties; work function; potassium atoms; charge transfer

1. Introduction

It is well known that carbon nanoporous glass-like materials are actively used in modern emission electronics [1–6]. In particular, they are very promising for the manufacture of field emission cathodes based on them. Such cathodes exhibit promising field emission parameters [7–9]. The edges of pores in these carbon glass-like materials are sharp blade structures that make them emission centers with a field enhancement factor, β. As already known, the emission tips of matrix cathodes with a high emission current density are traditionally produced on the basis of this mechanically strong material. Carbon glass-like materials have high strength, high heat resistance, abrasion resistance, and chemical inertness, as well as isotropic electrical conductivity [10,11]. It should also be noted that modern technology allows us to control the size of nanopores. For example, by forming nanoporous carbon on a conductive substrate, the nanopore size can be varied from 0.6 to 4.0 nm [12,13]. The samples of such nanoporous carbon were produced as a result of thermochemical treatment of different carbides with chlorine at different temperatures. It should also be noted that interest in the glass-like carbon nanomaterial increases in connection with the increasing possibilities of modifying its pores with atoms of various chemical elements [14–16]. With the increased capabilities of computer modeling, it has become possible to study all the features of carbon surface modification at atomic level [17]. The atomic structure of the glass-like carbon nanomaterial was reproduced due to modern microscopy. The glass-like carbon can be divided into two types: anisotropic material of multi-layer graphene flakes forming a likeness of layers (graphitized glass-like carbon) and randomly arranged graphite flakes (non-graphitized glass-like carbon) [18,19]. Non-graphitized glass-like carbon, in turn, can be divided into two classes: the first class has mainly randomly distributed curved layers of graphene fragments [20], the second class has self-organizing nanometer fullerene-like spheroids and 3D disordered multilayer graphene flakes [21]. In some recent papers, the carbon material between few-layer graphene and fullerenes are also called porous glass-like carbon [22].

At the same time, the question of the effect of the emitting tip surface on the emission properties remains topical. At present, it is possible to modify and nanostructure the surface of porous carbon glass-like materials [2], which makes it possible to sharply increase the field enhancement factor β by 4–5, because a thin surface layer of the emitting material plays a decisive role in autocathodes. This layer may be compared with a film of a thickness of several tens of nanometers, as shown by Gay et al. [2]. A typical scanning electron microscope (SEM) image of the emitting surface of a nanoporous carbon glass-like emitter is shown in Figure 1. This image was taken using a JEOL JEM-2100 Plus transmission electron microscope at an accelerating voltage of 10 kV. The diameter of the tip at the base is 32 microns; the diameter of the apex is 16 microns. As can be seen in the figure, the working surface of the emitting tip is a rough structure which ensures a high density of field emission current. The technology for producing such films using laser radiation is described in detail by Bessonov et al. [23].

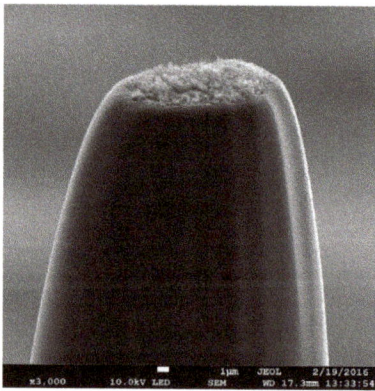

Figure 1. SEM image of the emitting tip surface of the porous glass-like nanomaterial.

However, the complexity of the surface modification problem is that the surface always contains at least a small number of impurity atoms [24]. First of all, these are atoms of oxygen, hydrogen, and OH groups. Even despite the technical vacuum conditions (10^{-6}–10^{-7} Torr) under which autocathodes work in most cases, a certain proportion of adsorbates remain. The presence of impurities inevitably leads to a deterioration of the emission properties [25,26]. However, there is a method for an efficient surface modification, for example, by doping with alkali metal atoms, which contribute to a decrease in the work function [27,28]. Thus, the emitter surface plays the most important role in all emission processes [29,30]. Depending on the type of impurities adsorbed on the emitter surface, as well as the surface topology, the emission properties of the material may deteriorate or improve [31–36]. Moreover, the impurities and topology determine the energy of the surface and the electron charge density distribution. The combination of these factors determines the work function and the electrical conductivity of the emitting surface.

This work is devoted to the in silico study of the patterns of change in the electronic properties of thin films of carbon nanoporous glass-like nanomaterial upon oxidation and modification by hydroxyl groups and potassium atoms, from the standpoint of the effect on the emission properties and the expansion of the applications of this nanomaterial in electronics. As the carbon nanoporous film, we considered a film with a thickness of 4–5 nm, whose surface had a nanopore system formed by the fragments of single- and few-layer graphene flakes and fullerenes.

2. Atomistic Model of the Nanoporous Carbon Surface

A super-cell of porous glass-like carbon nanomaterial film was constructed for this investigation. Earlier, we had created an atomistic model of a porous glass-like carbon nanomaterial [37], which was a combination of interacting graphene flakes and fullerene fragments (see Figure 2a). The atomic network of the super-cell was obtained as a result of the optimization of the atomic cell by the self-consistent charge density functional tight-binding (SCC-DFTB) method [38]. The calculations were performed using the DFTB+ package [39]. This program uses one of the best parameterizations of the SCC-DFTB method. This parameterization is very popular in academia. According to Google, it is mentioned in more than 20,000 scientific publications as a calculation method. A comparative analysis of the results of using density functional theory (DFT) and DFTB, using the example of defective graphene, showed that the DFTB method reproduced both the structural and energy parameters of nanostructures with high quantitative accuracy [40]. The error of the SCC-DFTB method depended on the class of problems being solved. In particular, the error in calculating the atomic structure and electronic parameters (energy gap of the band structure, Fermi level, and distribution of the electron charge over atoms) was several percent. A super-cell of 2D porous carbon material was a thin layer of glass-like nanomaterial, the surface of which corresponded to the equilibrium state. The approach of super-cell formation had previously been used by the authors to construct a bulk sample of a porous glass-like material. As a result, the structural density of the created film model was 1.21 g/cm^3 with the dimensions of the super-cell 4.2 nm × 4.2 nm × 4.1 nm. As the film was a two-dimensional structure, the super-cell of the film was only translated in the X and Y directions, as shown in Figure 2a. The number of atoms in the super-cell was 3874 (not taking into account the impurity). The pore size in our model was 0.25–1 nm, which corresponded completely to the glass-like porous nanomaterial [19,21]. The fraction ratio of non-hexagonal elements was 9.2%, which also fully corresponded to the structure of the glass-like porous nanomaterial [21], which was characterized by 10%–15% of non-hexagonal elements. The super-cell translated in the X and Y directions reproduced the surface of the emitting tip with some approximation. In the Z direction, the structure had freely oriented fullerene fragments and graphene flakes. The atomic networks of these fragments were obtained as a result of optimization by the SCC-DFTB method. The surface of the porous film obtained on the basis of the created super-cell is shown in Figure 2b. If we compare the obtained surface with the SEM image shown in Figure 1, one can see the similarity of topology, which is represented by a large number of individual fragments of nanostructures. The developed surface of the atomistic model of a nanoporous carbon material had numerous fragments of graphene and fullerene flakes. Nanopores were located on the surface, no deeper than 2.6 nm.

The modification of the surface with potassium atoms was carried out as a result of bombarding the surface of the glass-like carbon with potassium atoms using the molecular dynamics (MD) SCC-DFTB method under normal external conditions [41]. The calculations were performed using open-source Kvazar [42]. The initial conditions were taken from experimental data [43], in which the beam energy was from 50 to 150 keV at a beam density of 10^{15}–10^{17} atom/cm^2. We considered the case when the beam density was 10^{16} atom/cm^2, and the energy was equal to 100 keV. The velocity of the atoms was 0.70 m/s under such conditions. The main regularities in filling the pores and reaching the maximum potassium concentration were found as a result of a series of 20 numerical experiments. This process was accompanied by temperature fluctuations within 300–800 K. It was established that the maximum possible potassium concentration by mass was 4.65% for this structure of glass-like carbon surface.

Figure 2c shows the course of filling the nanopores with potassium atoms. The ordinate represents the depth at which the nanopores were located (negative direction of the Z axis) and the abscissa—the number of potassium atoms filling these nanopores. One can see a clear regularity in the filling of the nanocavities of the glass-like carbon surface. Even at a low concentration of 0.59%, two potassium atoms reached a depth of 2.5 nm with respect to the surface layer. During the process of bombarding the surface of glass-like carbon, all the nanocavities were uniformly filled with potassium atoms. This is well demonstrated by the inset of Figure 2c, which corresponded to the maximum concentration, where the potassium atoms uniformly filled the surface, deep, and middle cavities (potassium atoms are marked with ocher-colored balls). Thus, all pores were filled at a depth of 2.5–2.6 nm.

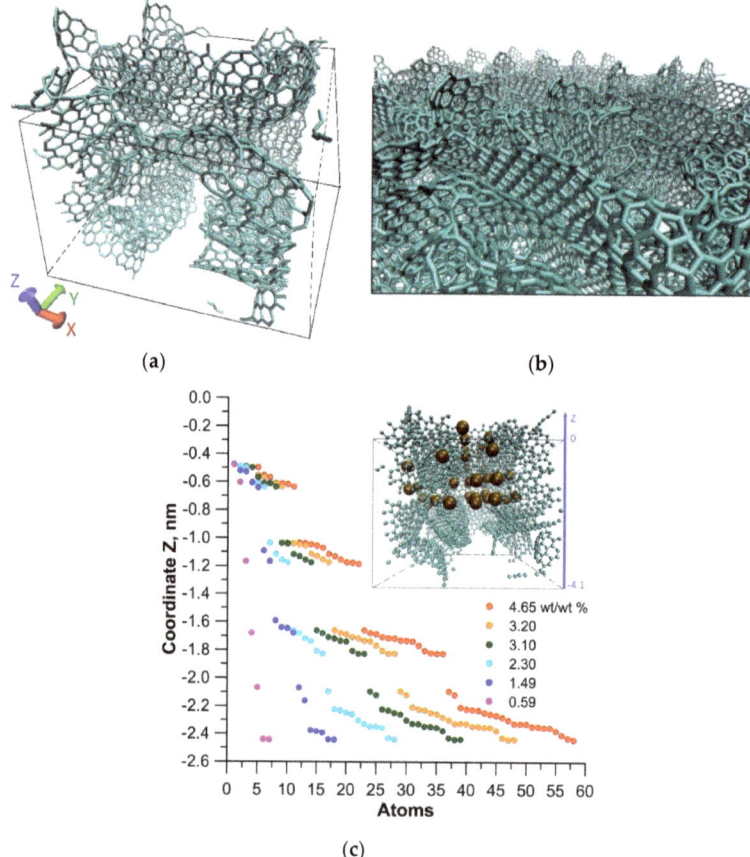

Figure 2. A surface of porous glass-like nanomaterial: (**a**) a super-cell of the film, which is the glass-like nanomaterial surface; (**b**) a topology of the atomic model surface; (**c**) a distribution of potassium atoms (balls of ocher color) by nanopores at different concentrations (mass fraction).

3. Results and Discussion

As already known, the working nanostructured surface of the emitting tip is not absolutely clean. It always contains adsorbed oxygen atoms, hydrogen, OH groups, or a combination of these. These impurity atoms/atomic groups always increase the work function of the material, especially oxygen. In connection with this, we conducted an in silico investigation of various combinations of impurity elements with different concentrations.

One series of numerical experiments was carried out by filling the nanopores with potassium atoms during the hydrogenation of fragmentary surface elements. Figure 3a shows a super-cell of a nanocarbon film whose surface contained 3.56% atomic hydrogen (hydrogen atoms are shown in white and gray color). Hydrogen atoms were located on the edges of the atomic network fragments, because these atoms of the carbon framework were most chemically active. Balls of ocher color represent potassium atoms. It should be noted that the impurity atoms of hydrogen did not change the process of filling the nanopores with alkali metal atoms. The directions of the X, Y, and Z axes remained the same, as shown in Figure 2c. We constructed super-cells and calculated the band structures and densities of states (DOS) with determination of the Fermi level for various values of the atomic hydrogen concentration. Numerical simulations of the process of filling the nanopores with potassium atoms were performed for each case until the maximum concentration was achieved. Noticeable changes in the energy of the film and, in particular, in the position of the Fermi level, which determines the work function, were detected during the filling of the nanopores. We found that the work function of the film decreased as the nanopore was filled with potassium atoms regardless of the amount of adsorbed atomic oxygen. When the initial value of the work function was 5.1–5.3 eV, then it amounted to 4.6 eV with the maximum filling of the nanopores. This result was extremely important for the process of electron emission by the application of an electric field. It is characteristic that a lower value of the work function was achieved with maximum hydrogenation of the film surface in this case. It should also be noted that the work function of the film with the surface adsorbing the impurity atoms (4.96 eV) was always greater than the work function of the pure carbon framework. We noted that a decrease in the work function during hydrogenation was provided by an increase in the potassium concentration at a given fixed fraction of hydrogen atoms on the surface. We can say that the hydrogenation did not prevent a reduction in the work function during the filling of pores by potassium atoms.

Another series of experiments was devoted to the study of the band structure, the position of the Fermi level, the oxidation of the surface, the adsorption of OH groups, and also the adsorption of oxygen atoms, hydrogen, and OH groups simultaneously. Figure 3b shows the super-cell at the oxidation (oxygen atoms are shown in red) of the surface fragments of the atomic cell and at the maximum filling of the nanopores with potassium. The concentration of oxygen atoms was 0.81%. Figure 3c shows the super-cell of the film upon adsorption of OH groups on its surface (0.58%). The band structures and Fermi levels were calculated for all cases of different concentrations of impurity atoms. The analysis showed that the type of impurity directly determined the nature of the change in the energy structure and the position of the Fermi level. Figure 3d shows a graph of the work function as a function of the mass fraction of potassium for various variants of impurity atoms and their concentration. Characteristic groups of curves were distinguished. Firstly, all cases with hydrogenation were characterized by a decrease in the work function when filling the nanopores with potassium.

As can be seen in Figure 3d, with an increase in the concentration of hydrogen atoms on the surface, the work function decreased against the background of increasing potassium mass fraction. Secondly, oxidation impaired the emission properties of the surface regardless of the concentration of oxygen atoms. The work function only increased, in particular by 0.5–0.6 eV, regardless of the number of oxygen atoms and the degree of presence of potassium atoms. Other variants of impurity atoms showed a noticeable deterioration of the emission properties, regardless of the concentration. The results of the calculations showed that the presence of OH groups increased the work function by 0.5 eV, in comparison with hydrogen. However, the modification of pores with potassium made it possible to compensate increasing work function, as can be seen in Figure 3d (curve with triangles). It should also be noted that the presence of both O atoms and H atoms on the surface simultaneously, as with oxidation, only contributed to an increase in the work function.

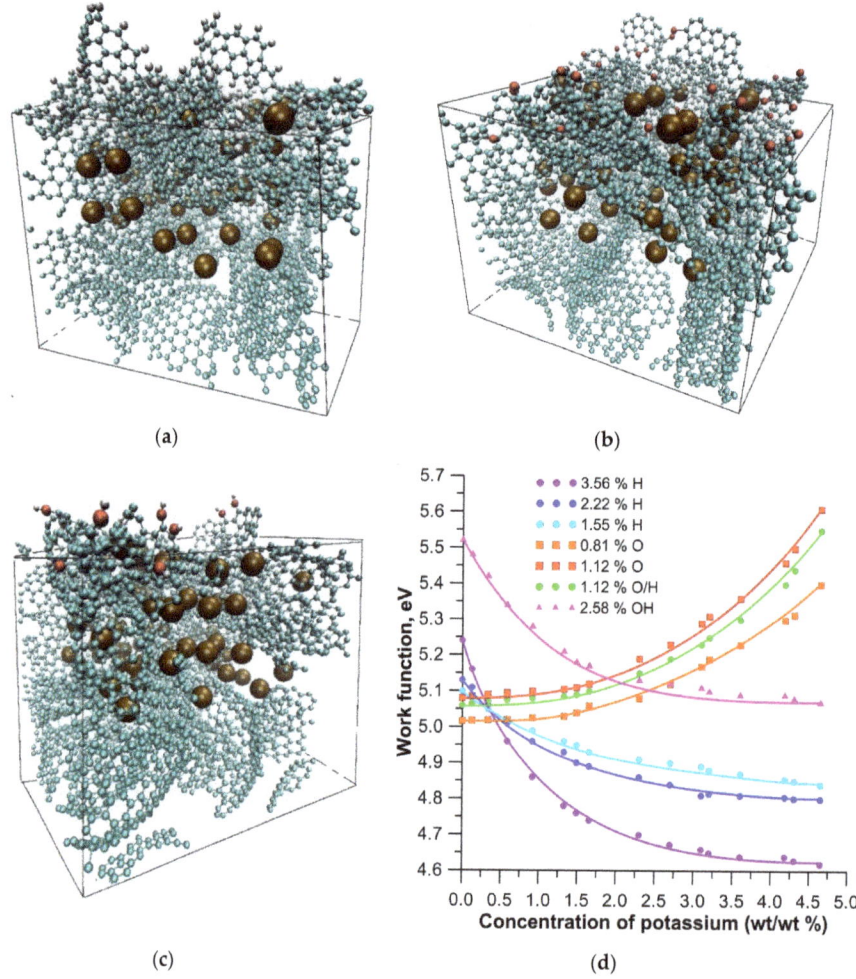

Figure 3. The atomic structure and the work function of a nanocarbon porous film with different types of impurity atoms: (**a**) modification by hydrogen atoms (3.56%); (**b**) modification by oxygen atoms (0.81%); (**c**) modification by OH groups (2.58%); (**d**) a graph of the change in the work function with an increase in the mass fraction of potassium at different concentrations of impurity atoms.

In order to find the physical phenomena that occurred during such changes in the energy structure of the film and in the displacement of the Fermi level, we investigated the charge transfer between atoms of potassium and the carbon framework with adsorbed atoms/atomic groups. Table 1 shows the calculated values of Mulliken charges—charges localized on the carbon framework, on potassium (K) atoms, and on other atoms of the surface. The same cases of impurity atoms on the surface are given as in Figure 3d. Potassium atoms always gave up their charge. The dependence of the amount of overflowing charge from potassium atoms on the carbon framework was non-linear. An analysis of the pattern of the electron density distribution showed that the potassium atoms lost almost an entire electron (0.8–0.9 e) directly near the framework and that the atoms inside the group of potassium atoms practically did not lose anything (they had a charge of 0.26 e). Therefore, the ratio of the charge to the number of potassium atoms in the cell varied from 0.86 to 0.73 e/atom.

Table 1. Charge distribution at the maximum mass fraction of potassium (4.65%).

N/N atoms, %	Charge, |e|				
	K	H	O	OH	Carbon
3.56% H	+39.95	+10.39	-	-	−50.34
2.22% H	+40.20	+7.10	-	-	−47.30
1.55% H	+40.46	+5.14	-	-	−45.60
0.81% O	+41.28	-	−10.38	-	−30.90
1.12% O	+41.45	-	−15.04	-	−26.41
2.58% OH	+40.10	-	-	−1.20	−38.90
1.12% O, 1.12% H	+40.75	+1.60	−8.42	-	−33.93

Figure 4a shows a super-cell with impurity hydrogen atoms (3.56% H) and a maximum mass fraction of 4.65%. The charge is displayed in color—the potassium atoms change color from red (charge +0.96 e) to yellow green (+0.50 e) and green (+0.26 e). The carbon framework had a practically uniformly distributed charge −0.1 e to −0.5 e, therefore it is displayed in blue. Hydrogen atoms, like potassium, gave up a partial charge, so they are shown in green. Their charge varied from +0.1 to +0.23 e. Thus, analysis of the electron charge density distribution data (data in Table 1) and comparison with the graphs in Figure 3d shows that the minimum work function was observed for the case of the maximum transfer of electron charge to the carbon framework. This was due to an increase in DOS at the Fermi level. Figure 4b shows DOS for the case of a maximum displacement of the Fermi level upward and a corresponding decrease in the work function (lilac curve). It can be clearly seen that in this case the DOS peak fell on the Fermi level, which improved electrical conductivity and reduced the work function. The gray color represents DOS surfaces with 3.56% H, but without potassium atoms. When oxidizing the surface, even pouring potassium into the pores did not improve the emission properties. Figure 4c shows DOS for the oxidized surface with a concentration of 1.12% O (gray curve) and for the same surface at the maximum mass fraction of potassium (red curve). It can be seen that at the Fermi level the DOS of the surfaces with potassium were smaller in comparison to the case without potassium. For comparison, we calculated the DOS of a clean surface (without impurity atoms) and a surface with the maximum mass fraction of potassium. These calculation results are presented in Figure 4d. The DOS increased at the Fermi level.

Calculations of the charge of the carbon framework in all considered cases of surface modification showed that there was a clear regularity. As the electron charge flowed over the carbon framework, the work function decreased. Figure 5 shows the graphs of the change in the work function of the film with an increase in charge on the carbon framework for various cases of filling the nanopores with potassium.

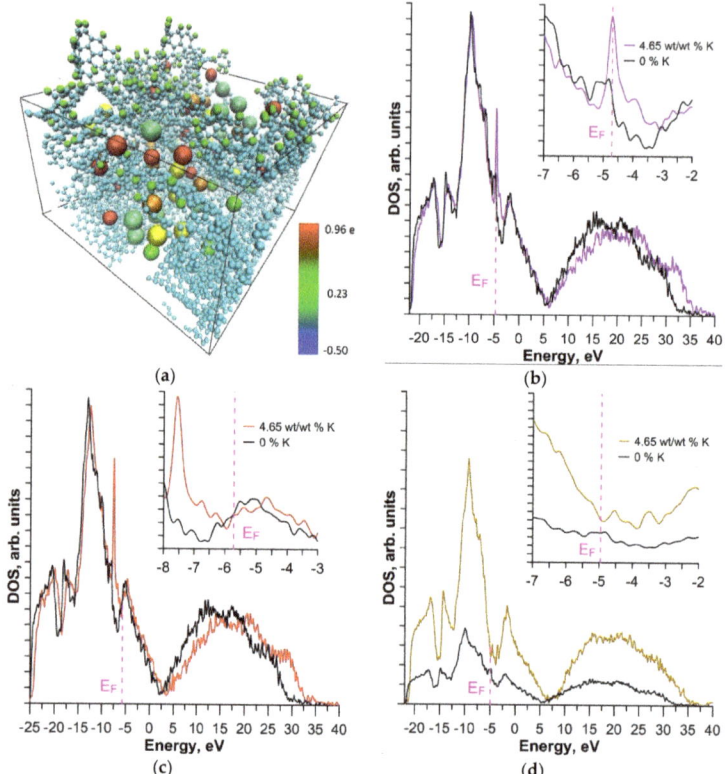

Figure 4. Distribution of the electron charge: (**a**) on a surface modified by hydrogen atoms with a concentration of 3.56%, with a maximum mass fraction of potassium; (**b**) Density of states (DOS) of hydrogenated (3.56%) surface without potassium (gray curve) and with potassium (lilac); (**c**) DOS of oxidized (1.12%) surface without potassium (gray curve) and with potassium (red); (**d**) DOS of a clean surface without potassium (gray curve) and with potassium (ocher color).

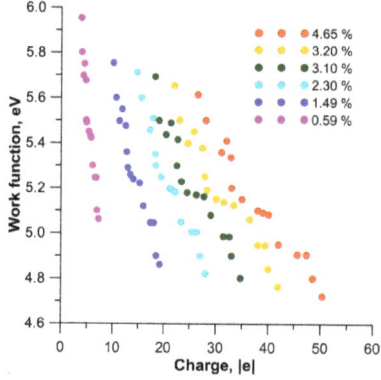

Figure 5. The change in the work function of the film with an increase in charge on the carbon framework for various cases of filling the nanopores with potassium.

4. Conclusions

Regularities in the electronic structure and charge transfer in the nanoporous surface of glass-like thin films when the surface pores were filled with potassium atoms were investigated using MD DFTB calculations. We constructed an atomistic model of the porous carbon film surface, completely reproducing the features of the glass-like porous nanomaterial structure. It was found that the maximum possible concentration of potassium by weight was 4.65% for the examined glass-like carbon surface. In this case, the potassium atoms uniformly filled the surface, deep, and middle cavities at a depth of 2.5–2.6 nm. The presence of impurities in the form of chemically adsorbed hydrogen atoms, oxygen, and OH groups was taken into account during the calculation of the electronic structure of the glass-like carbon surface doped with potassium. The calculation results showed that the type of impurity determines directly the nature of the change in the band structure and the position of the Fermi level of the glass-like carbon surface doped with potassium. Hydrogen saturation reduces the work function, oxidation increases the work function by 0.5–0.6 eV regardless of the oxygen concentration, and, finally, the presence of OH groups increases the work function by 0.5 eV, compared with hydrogen. When the surface nanopores filled with potassium atoms, all the potassium atoms were ionized, losing one electron, which transferred over to the carbon framework of the film. It was found that changing the mass fraction of potassium atoms on the surface can reduce the work function even in the presence of impurity atoms, which impair the emission properties.

Author Contributions: Conceptualization, O.E.G.; methodology, O.E.G. and I.S.N.; funding acquisition, I.S.N., M.M.S.; investigation, O.E.G., I.S.N., and M.M.S.; writing—original draft preparation, O.E.G. and M.M.S.; writing—review and editing, I.S.N. and O.E.G.; supervision, O.E.G. All authors have read and agreed to the published version of the manuscript.

Funding: This research was funded by the Council of the President of the Russian Federation (project no. MK-2373.2019.2) and with the support of the «RUDN University Program 5-100» (Igor Nefedov).

Acknowledgments: The authors thank V.I. Shesterkin (Joint-Stock Company (JSC) "Research & Production Enterprise "Almaz", Russia") for providing the SEM image of the tip of the emitting surface.

Conflicts of Interest: The authors declare no conflict of interest.

References

1. Sharma, S. Glassy Carbon: A Promising Material for Micro- and Nanomanufacturing. *Materials* **2018**, *11*, 1857. [CrossRef]
2. Gay, S.; Orlanducci, S.; Passeri, D.; Rossic, M.; Terranova, M.L. Nanoshaping field emitters from glassy carbon sheets: A new functionality induced by H-plasma etching. *Phys. Chem. Chem. Phys.* **2016**, *18*, 25364–25372. [CrossRef] [PubMed]
3. Smith, B.C.; Hunt, C.E.; Brodiec, I.; Carpenter, A.C. High-performance field-emission electron gun using a reticulated vitreous carbon cathode. *J. Vac. Sci. Technol. B* **2011**, *29*, 02B108. [CrossRef]
4. Cao, M.M.; Chacon, R.J.; Hunt, C.E. A Field Emission Light Source Using a Reticulated Vitreous Carbon (RVC) Cathode and Cathodoluminescent Phosphors. *J. Disp. Technol.* **2011**, *7*, 467–472. [CrossRef]
5. Matsubara, E.Y.; Rosolen, J.M.; Silva, S.R.P. Composite electrode of carbon nanotubes and vitreous carbon for electron field emission. *J. Appl. Phys.* **2008**, *104*, 054303. [CrossRef]
6. Carpenter, A.C.; Hunt, C.E. High-current, low-cost field emission triode using a reticulated vitreous carbon cathode. *J. Vac. Sci. Technol. B* **2010**, *28*, C2C37–C2C40. [CrossRef]
7. Chakhovskoi, A.G.; Hunt, C.E. Reticulated vitreous carbon field emission cathodes for light source applications. *J. Vac. Sci. Technol. B* **2003**, *21*, 571–575. [CrossRef]
8. Egorov, N.; Sheshin, E. Carbon-Based Field-Emission Cathodes. In *Field Emission Electronics*, 1st ed.; Springer: Berlin, Germany, 2017; Volume 60, pp. 295–367.
9. Hunt, C.E.; Chakhovskoi, A.G.; Wang, Y. Ion-beam morphological conditioning of carbon field emission cathode surfaces. *J. Vac. Sci. Tech. B* **2005**, *23*, 731–734. [CrossRef]
10. Jovanovic, Z.; Kalijadis, A.; Vasiljevic-Radovic, D.; Eric, M.; Lausevic, M.; Mentus, S.; Lausevic, Z. Modification of glassy carbon properties under low energy proton irradiation. *Carbon* **2011**, *49*, 3737–3746. [CrossRef]

11. Serp, P.; Machado, B. *Nanostructured Carbon Materials for Catalysis*, 1st ed.; Royal Society of Chemistry: London, UK, 2015; pp. 1–45.
12. Arkhipov, A.; Davydov, S.; Gabdullin, P.; Gnuchev, N.; Kravchik, A.; Krel, S. Field-induced electron emission from nanoporous carbons. *J. Nanomater.* **2014**, *2014*, 190232. [CrossRef]
13. Kravchik, A.E.; Kukushkina, J.A.; Sokolov, V.V.; Tereshchenko, G.F. Structure of nanoporous carbon produced from boron carbide. *Carbon* **2006**, *44*, 3263–3268. [CrossRef]
14. Benzigar, M.R.; Talapaneni, S.N.; Joseph, S.; Ramadass, K.; Singh, G.; Scaranto, J.; Ravon, U.; Al-Bahily, K.; Vinu, A. Recent advances in functionalized micro and mesoporous carbon materials: Synthesis and applications. *Chem. Soc. Rev.* **2018**, *47*, 2680–2721. [CrossRef]
15. Figueiredo, J.L. Functionalization of porous carbons for catalytic applications. *J. Mater. Chem. A* **2013**, *1*, 9351–9364. [CrossRef]
16. Long, C.; Jiang, L.; Wu, X.; Jiang, Y.; Yang, D.; Wang, C.; Wei, T.; Fan, Z. Facile synthesis of functionalized porous carbon with three-dimensional interconnected pore structure for high volumetric performance supercapacitors. *Carbon* **2015**, *93*, 412–420. [CrossRef]
17. Pykal, M.; Jurečka, P.; Karlický, F.; Otyepka, M. Modelling of graphene functionalization. *Phys. Chem. Chem. Phys.* **2016**, *18*, 6351–6372. [CrossRef]
18. Jurkiewicz, K.; Duber, S.; Fischer, H.E.; Burian, A. Modelling of glass-like carbon structure and its experimental verification by neutron and X-ray diffraction. *J. Appl. Cryst.* **2017**, *50*, 36–48. [CrossRef]
19. Harris, P.J.F. Fullerene-related structure of commercial glassy carbons. *Philos. Mag.* **2004**, *84*, 3159–3167. [CrossRef]
20. Zhao, Z.; Wang, E.F.; Yan, H.; Kono, Y.; Wen, B.; Bai, L.; Shi, F.; Zhang, J.; Kenney-Benson, C.; Park, C.; et al. Nanoarchitectured materials composed of fullerene-like spheroids and disordered grapheme layers with tunable mechanical properties. *Nat. Commun.* **2015**, *6*, 6212. [CrossRef]
21. Harris, P.J.F. Fullerene-like models for microporous carbon. *J. Mater Sci.* **2013**, *48*, 565–577. [CrossRef]
22. Liang, H.; Ma, X.; Yang, Z.; Wang, P.; Zhang, X.; Ren, Z.; Xue, M.; Chen, G. Emergence of superconductivity in doped glassy-carbon. *Carbon* **2016**, *99*, 585–590. [CrossRef]
23. Bessonov, D.A.; Sokolova, T.N.; Shesterkin, V.I.; Surmenko, E.L.; Popov, I.A.; Chebotarevsky, Y.V. Laser formation of tip emitting structures with high aspect ratio on glass-carbon field-emission cathodes. *J. Phys. Conf. Ser.* **2016**, *741*, 012166. [CrossRef]
24. Araujo, P.T.; Terrones, M.; Dresselhaus, M.S. Defects and impurities in graphene-like materials. *Mater. Today* **2012**, *15*, 98–109. [CrossRef]
25. Koh, A.L.; Gidcumb, E.; Zhoubc, O.; Sinclair, R. The dissipation of field emitting carbon nanotubes in an oxygen environment as revealed by in situ transmission electron microscopy. *Nanoscale* **2016**, *8*, 16405–16415. [CrossRef]
26. Giubileoa, F.; Di Bartolomeoa, A.; Scarfatoa, A.; Iemmoa, L.; Bobbaa, F.; Passacantandob, M.; Santuccib, S.; Cucoloa, A.M. Local probing of the field emission stability of vertically aligned multi-walled carbon nanotubes. *Carbon* **2009**, *47*, 1074–1080. [CrossRef]
27. Wang, Y.; Yao, X.H.; Huang, G.; Shao, Q.Y. The enhanced field emission properties of K and Rb doped (5,5) capped single-walled carbon nanotubes. *RSC Adv.* **2015**, *5*, 16718–16722. [CrossRef]
28. Izrael'yants, K.R.; Orlov, A.P.; Ormont, A.B.; Chirkov, E.G. Effect of the cesium and potassium doping of multiwalled carbon nanotubes grown in an electrical arc on their emission characteristics. *Phys. Solid State* **2017**, *59*, 838–844. [CrossRef]
29. Ye, D.; Moussa, S.; Ferguson, J.D.; Baski, A.A.; El-Shall, M.S. Highly Efficient Electron Field Emission from Graphene Oxide Sheets Supported by Nickel Nanotip Arrays. *Nano Lett.* **2012**, *12*, 1265–1268. [CrossRef]
30. Iemmo, L.; Di Bartolomeo, A.; Giubileo, F.; Luongo, G.; Passacantando, M.; Niu, G.; Hatami, F.; Skibitzki, O.; Schroeder, T. Graphene enhanced field emission from InP nanocrystals. *Nanotechnology* **2017**, *28*, 495705. [CrossRef]
31. Kim, J.P.; Chang, H.B.; Kim, B.J.; Park, J.S. Enhancement of electron emission and long-term stability of tip-type carbon nanotube field emitters via lithium coating. *Thin Solid Film.* **2013**, *528*, 242–246. [CrossRef]
32. Giubileo, F.; Di Bartolomeo, A.; Iemmo, L.; Luongo, G.; Urban, F. Field Emission from Carbon Nanostructures. *Appl. Sci.* **2018**, *8*, 526. [CrossRef]

33. Gupta, B.K.; Kedawat, G.; Kumar, P.; Singh, S.; Suryawanshi, S.R.; Agrawal (Garg), N.; Gupta, G.; Kim, A.R.; Gupta, R.K.; More, M.A.; et al. Field emission properties of highly ordered low-aspect ratio carbon nanocup arrays. *RSC Adv.* **2016**, *6*, 9932–9939. [CrossRef]
34. Parveen, S.; Husain, S.; Kumar, A.; Ali, J.; Harsh; Husain, H.M. Improved field emission properties of carbon nanotubes by dual layer deposition. *J. Exp. Nanosci.* **2015**, *10*, 499–510. [CrossRef]
35. Yu, J.; Chua, D.H.C. Enhanced field emission properties of hydrogenated tetrahedral amorphous carbon/carbon nanotubes nanostructures electrochem. *Solid-State Lett.* **2010**, *13*, K80–K82. [CrossRef]
36. Varshney, D.; Makarov, V.I.; Saxena, P.; González-Berríos, A.; Scott, J.F.; Weiner, B.R.; Morell, G. Fabrication and field emission study of novel rod-shaped diamond-like carbon nanostructures. *Nanotechnology* **2010**, *21*, 285301. [CrossRef]
37. Glukhova, O.E.; Slepchenkov, M.M. Electronic Properties of the Functionalized Porous Glass-like Carbon. *J. Phys. Chem. C* **2016**, *120*, 17753–17758. [CrossRef]
38. Elstner, M.; Seifert, G. Density functional tight binding. *Phil. Trans. R. Soc. A* **2014**, *372*, 20120483. [CrossRef]
39. Aradi, B.; Hourahine, B.; Frauenheim, T. DFTB+, a sparse matrix-based implementation of the DFTB method. *J. Phys. Chem. A* **2007**, *111*, 5678–5684. [CrossRef]
40. Zobelli, A.; Ivanovskaya, V.; Wagner, P.; Suarez-Martinez, I.; Yaya, A.; Ewels, C.P. A comparative study of density functional and density functional tight binding calculations of defects in graphene. *Phys. Status Solidi B* **2012**, *249*, 276–282. [CrossRef]
41. Shunaev, V.V.; Savostyanov, G.V.; Slepchenkov, M.M.; Glukhova, O.E. Phenomenon of current occurrence during the motion of a C_{60} fullerene on substrate-supported graphene. *RSC Adv.* **2015**, *5*, 86337–86346. [CrossRef]
42. Glukhova, O.E. Molecular Dynamics as the Tool for Investigation of Carbon Nanostructures Properties. In *Thermal Transport in Carbon-Based Nanomaterials*, 1st ed.; Zhang, G., Ed.; Elsevier: Oxford, UK, 2017; pp. 267–289.
43. Nakao, A.; Iwaki, M.; Yokoyama, Y. Potassium ion implantation into glassy carbon. *Nucl. Instrum. Methods Phys. Res. Sect. B* **2003**, *206*, 211–214. [CrossRef]

© 2020 by the authors. Licensee MDPI, Basel, Switzerland. This article is an open access article distributed under the terms and conditions of the Creative Commons Attribution (CC BY) license (http://creativecommons.org/licenses/by/4.0/).

Article

ZnO-Controlled Growth of Monolayer WS$_2$ through Chemical Vapor Deposition

Zhuhua Xu [1], Yanfei Lv [1,*], Feng Huang [2], Cong Zhao [1], Shichao Zhao [1] and Guodan Wei [2,*]

[1] College of Materials & Environmental Engineering, Hangzhou Dianzi University, Hangzhou 310018, China; zhuhuaxu@hdu.edu.cn (Z.X.); grafengh@hdu.edu.cn (F.H.); zhaoshichao@hdu.edu.cn (S.Z.)
[2] Tsinghua-Berkeley Shenzhen Institute (TBSI), Tsinghua University, Shenzhen 518055, China; zhao-c18@mails.tsinghua.edu.cn
* Correspondence: lvyanfei@hdu.edu.cn (Y.L.); weiguodan@sz.tsinghua.edu.cn (G.W.)

Received: 16 May 2019; Accepted: 6 June 2019; Published: 12 June 2019

Abstract: Monolayer tungsten disulfide (2D WS$_2$) films have attracted tremendous interest due to their unique electronic and optoelectronic properties. However, the controlled growth of monolayer WS$_2$ is still challenging. In this paper, we report a novel method to grow WS$_2$ through chemical vapor deposition (CVD) with ZnO crystalline whisker as a growth promoter, where partially evaporated WS$_2$ reacts with ZnO to form ZnWO$_4$ by-product. As a result, a depletion region of W atoms and S-rich region is formed which is favorable for subsequent monolayer growth of WS$_2$, selectively positioned on the silicon oxide substrate after the CVD growth.

Keywords: monolayer WS$_2$; ZnO; CVD; controlled growth

1. Introduction

Two dimensional (2D) materials such as graphene, hexagonal boron nitride, and transition metal dichalcogenides have attracted tremendous attention from scientists in materials science, physics, and chemistry for their monolayer structure and properties [1–3]. Monolayer WS$_2$ is one of the typical 2D materials with a suitable direct band gap ca 2.0 eV showing potential applications in sensors, electronics, and optoelectronics [4–8]. Recently, many methods have been used to prepare monolayer WS$_2$, such as mechanical exfoliation, as well as wet chemical and hydrothermal synthesis [9–12]. Chemical vapor deposition (CVD) is considered the most suitable method for the preparation of monolayer WS$_2$ used for thin film devices [13]. In the CVD method, tungsten oxide and sulfide powders are commonly used as precursors, which evaporate at high temperature and react to form WS$_2$. Large-scale monolayer WS$_2$ has been reported to be grown on an Au substrate [14]. To integrate with silicon integrated circuit technology, the growth of WS$_2$ on a silicon dioxide (SiO$_2$) substrate is preferred [15,16]. Many works have focused on the deposition of the monolayer WS$_2$ on the SiO$_2$ substrate by CVD [17–19]. However, the deposition has poor reproducibility caused by the growth conditions [20]. In addition, the 2D WS$_2$ obtained is combined with monolayer and multilayers [19,21,22]. The selective growth and the positioning of monolayer WS$_2$ are not under control. Besides the controlled growth, the mechanism of the monolayer WS$_2$ formation is still under discussion. Fan et al. provided an understanding of the dislocation-driven growth mechanism of 2D nanostructures in their work [23]. Cain et al. found that transition metal dichalcogenide monolayer growth proceeds from nominal lyoxi-chalcogenide nanoparticles which act as heterogeneous nucleation sites for monolayer growth [24]. The reaction of tungsten oxide in sulfur vapor suggests that the growth of WS$_2$ is thermodynamically correlated with the sulfur concentration. We hypothesized that if the sulfur concentration could be controlled, it would be possible to deposit WS$_2$ with the desired thickness.

Herein, we report a novel method to selectively grow monolayer WS$_2$ on SiO$_2$ substrate. ZnO crystal whisker is used to mediate the spacial distribution of sulfur concentration. We show that monolayer WS$_2$ symmetrically distributes on both sides of the ZnO crystal whisker. By constructing a concentration distribution model, the monolayer growth mechanism can be discussed.

2. Materials and Methods

2.1. WS$_2$ Monolayer Preparation

Figure 1a shows a schematic diagram of the home-built CVD system. WS$_2$ powders (2.0 g, purity 99.5%, Aladdin, Shanghai, China) used as precursor were loaded into a quartz boat and put at the center of a quartz tube (1 inch in diameter). Silicon wafer with a 300 nm thick oxide layer used as substrate (SiO$_2$/Si) was placed at the low temperature region of the quartz tube downstream of the carrier gas. Ar/H$_2$ (5% H$_2$) was used as carrier gas. ZnO crystal whiskers (1.5 cm in length) were obtained by thermal evaporation of ZnO powders according to the reference and transferred onto the substrate [25]. Figure 1b shows a typical ZnO crystal whisker with a size of 67.0 μm × 45.9 μm × 1500 μm.

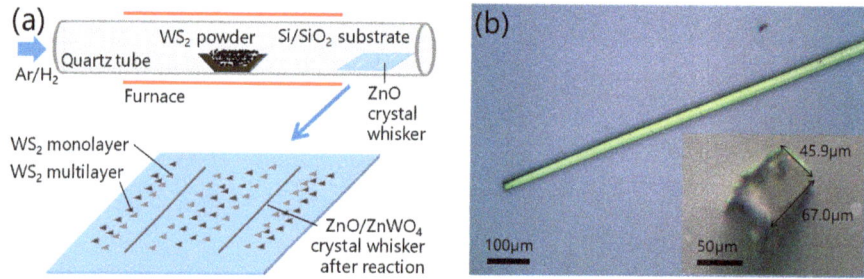

Figure 1. (a) Schematic diagram of the home-built chemical vapor deposition (CVD) system and tungsten disulfide (WS$_2$) after CVD growth. (b) Optical microscopy image of ZnO crystal whisker transferred onto the surface of SiO$_2$/Si substrate before WS$_2$ growth. Insert illustration: cross section of ZnO crystal whisker with a length and width of 67.0 μm and 45.9 μm, respectively.

For the growth of the WS$_2$, the carrier gas with 35 sccm was introduced into the CVD system, which was evacuated to 70 torr. Then the furnace was heated from room-temperature to 1000 °C in 40 min and kept for 1 h. After that, the furnace was cooled down from 1000 °C to room-temperature under carrier gas flow.

2.2. Characterizations

Optical and photoluminescence (PL) imagings were carried out on a Jiangnan MV3000 digital microscope (Nanjing Jiangnan Novel Optics Co., Ltd.; Nanjing, China). Scanning electron microscopy (SEM) was conducted on a field emission scanning electron microscope (FESEM, ULTRA 55, Zeiss, Heidenheim, Germany). Photoluminescence (PL) and Raman spectra were acquired on a home-built Raman system, consisting of an inverted microscope (Ti eclipse, Nikon, Tokyo, Japan), a Raman spectrometer (iHR320, Horiba, Kyoto, Japan) with CCD detector (Syncerity, Horiba, Kyoyo, Japan) and a semiconductor laser at 532 nm (Uniklasers, Glasgow, UK). All measurements were performed at room temperature.

3. Results and Discussion

Figure 2 shows the SEM image of the WS$_2$ grown on the SiO$_2$/Si substrate. We can see the separated WS$_2$ domains of a triangular and hexagonal shape. The maximum size of the domain is up to 28.3 μm.

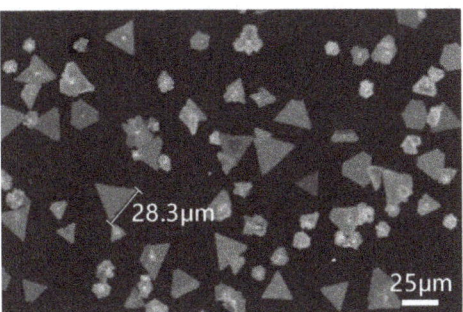

Figure 2. Scanning electron microscopy (SEM) image of WS$_2$ grown on SiO$_2$/Si substrate.

Figure 3a shows the optical microscopy image of the sample. The dark region marked with a blue triangle is monolayer WS$_2$ while the bright region marked with a red triangle is due to multilayer WS$_2$. The uniform color contrast of the monolayer indicates the thickness uniformity of the WS$_2$ monolayer. Figure 3b shows the PL image corresponding to the sample in Figure 3a taken at the same location. The excitation wavelength was 485 nm. The monolayer WS$_2$ displays super-bright red light emitting under irradiation. The patterns with the red color in Figure 3b remain as features of the monolayer WS$_2$ domains in Figure 3a. However, the PL emission of multilayer WS$_2$ is not strong enough to be detected by photoluminescence microscopy. The difference of the PL behavior between the monolayer and multilayer lies in the different electrical structures. Monolayer WS$_2$ is a direct band gap, while multilayer is an indirect band gap [26]. The fluorescence quantum efficiency of the direct band gap semiconductor is much higher than that of the indirect [27,28]. Therefore, we only observe the PL image in the monolayer WS$_2$.

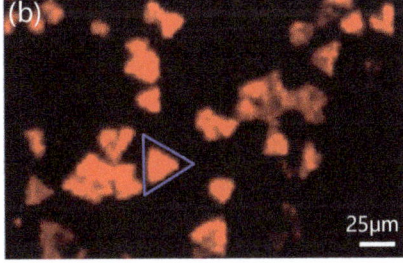

Figure 3. (a) Optical microscopy image of WS$_2$ monolayer (marked with a blue triangle) and multilayer (marked with a red triangle). (b) The photoluminescence (PL) image corresponding to the sample (a). (a,b) are taken at the same location. The pattern with red color in (b) is due to the PL emission of monolayer WS$_2$.

Figure 4 shows the Raman and PL spectra of the WS$_2$ monolayer (red line) and multilayer (black line). Raman spectroscopy was used to identify the number of two dimensional material layers. Raman peaks at 351.1 cm^{-1}, 417.2 cm^{-1} are the fingerprint peaks of monolayer WS$_2$, which are due to the second-order longitudinal acoustic mode (2LA(M)), and out-of-plane vibration mode (A$_{1g}$), respectively [29]. With the increase of the number of layers, the 2LA(M) peak red shifts and the A$_{1g}$ peak blueshifts. The 2LA(M) and A$_{1g}$ peaks are observed in Figure 4a. For the domain marked with a blue triangle in Figure 3a, Raman peaks located at 351.0 cm^{-1} (2LA(M)) and 417.8 cm^{-1} (A$_{1g}$) were observed, indicating the thickness of the domain is monolayer. For the domain marked with a red triangle in Figure 3a, the 2LA(M) peak redshifted to 350 cm^{-1} and A$_{1g}$ peak blueshifted to 419.6 cm^{-1}, indicating the thickness of the domain is multilayer [27,29,30]. An intense PL emission peak at 625.7 nm was observed in Figure 3b, which is related to the direct band gap. Compared to the monolayer, the multilayer undergoes a transition from direct to indirect band gap resulting in a redshift of the PL peak and a sharp decrease of the PL intensity at 631.9 nm [27].

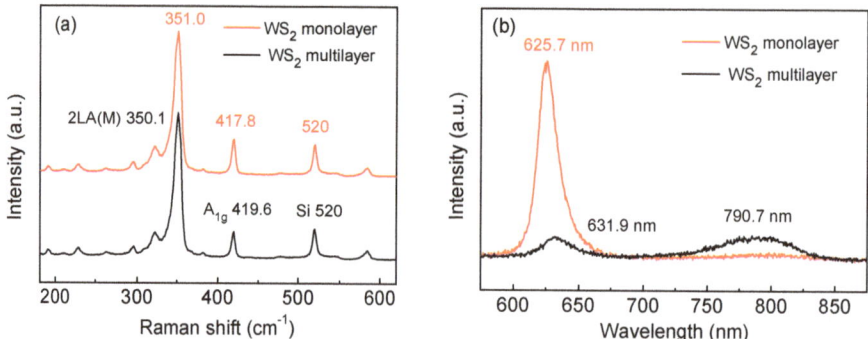

Figure 4. (a) Raman and (b) PL spectra of WS_2 monolayer (red line) and multilayer (black line) corresponding to the WS_2 domains in Figure 3a.

Figure 5a,b shows the optical and PL image of WS_2 grown near the region of ZnO whisker, respectively. To investigate the influence of ZnO crystal whisker on the growth of WS_2 monolayer, we made distribution statistics of the WS_2 monolayer and multilayer at one side of the ZnO crystal whisker. Figure 5c shows the distribution of statistical estimates of the WS_2 domain density on the substrate. The direction of the horizontal axis is perpendicular to the growth axis of ZnO whisker. The coordinate of the ZnO whisker on the horizontal axis is zero.

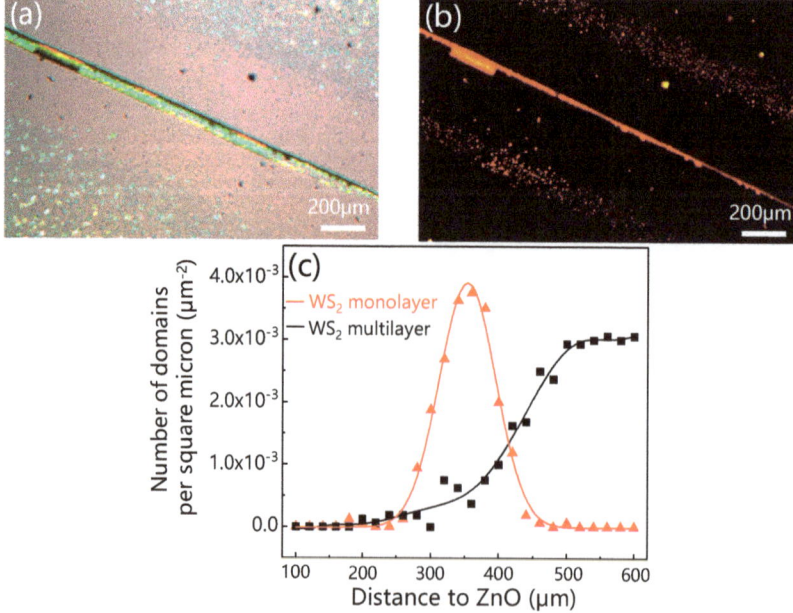

Figure 5. (a) The optical image of WS_2 domains prepared with the assistance of ZnO whisker. (b) PL image corresponding to the image of (a). (c) WS_2 domain distribution is a function of distance from the ZnO whisker. The origin is the ZnO crystal whisker. (a) and (b) were taken at the same location. The straight lines are related to ZnO whisker and the scale bars represent 200 μm.

WS$_2$ domains hardly grew in the region (0–250 µm) close to the ZnO whisker. In the region (250–450 µm), a little farther away from the WS$_2$ whisker, monolayer WS$_2$ domains symmetrically grew on both sides of the ZnO crystal whisker, suggesting the ZnO crystal whisker played a crucial role in the growth of the WS$_2$ monolayer. In the region (larger than 450 µm) far away from the ZnO whisker, where ZnO has hardly any impact on the growth of WS$_2$, multilayer WS$_2$ domains were observed. Therefore, monolayer and multilayers were grown separately in different zones of the substrate. To date, the position control of the WS$_2$ monolayer has not been reported. The merit of our growth method lies in the accurate positioning of the WS$_2$ monolayer using ZnO whisker.

Before analyzing the mechanism of monolayer WS$_2$ growth in the presence of ZnO, we characterized the product of ZnO after WS$_2$ growth by SEM, EDS, and Raman spectra. Besides oxygen and zinc, we found tungsten (W) element in the sample. The EDS spectrum intensity of W is even greater than that of Zn. In addition, the morphology transformed from the crystal whisker of ZnO to capsules as shown in the insert of Figure 6a. The results of EDS and SEM indicate that the chemical composition of the ZnO whisker may have been changed after the growth of WS$_2$. Raman results further verified our assumption. Figure 6b is the Raman spectrum of ZnO crystal whisker after WS$_2$ deposition. We find all the Raman peaks are due to ZnWO$_4$. The slight difference of Raman peak position between our experiment and the reference may lie in the test conditions and/or strain in the sample. Therefore, ZnO transformed into ZnWO$_4$ during the growth of WS$_2$ domains.

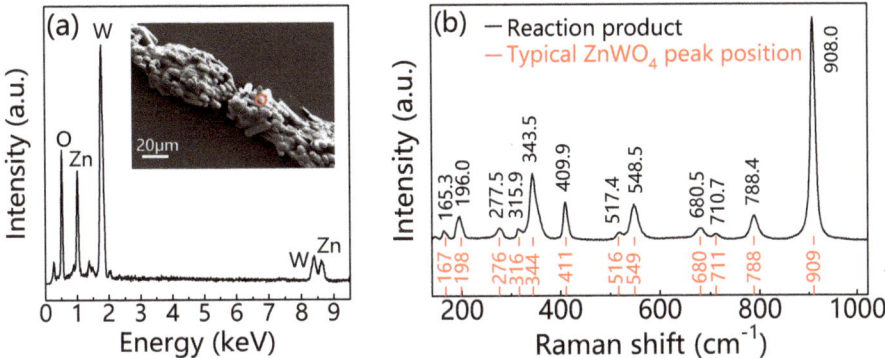

Figure 6. (a) EDS and (b) Raman spectra of ZnO crystal whisker after WS$_2$ growth. Insert in (a) is the SEM image with a red circle where the EDS spectrum is taken from. The data in black and red color in (b) correspond to our experiment and reference [31,32], respectively.

At the high temperature of 1000 °C and low pressure of 70 torr, WS$_2$ powder evaporated and decomposed into W and S atoms. According to the elementary composition of reaction product ZnWO$_4$, W atoms reacted with ZnO and were consumed resulting in the depletion region of W atoms and the S-rich region around ZnO.

To discuss the growth mechanism, we built a distance dependent model of W and S distribution and WS$_2$ domain growth. Figure 7a shows a schematic diagram of W and S distribution around ZnO crystal whisker. Figure 7b shows the distribution of estimates of W and S atoms concentration around the ZnO whisker. The direction of the horizontal axis is perpendicular to the growth axis of ZnO whisker. The coordinate of the ZnO whisker on the horizontal axis is zero.

In the area far from the ZnO (larger than 450 µm), the growth of WS$_2$ is barely influenced by ZnO. In this area, multilayer WS$_2$ domains are grown. We consider the deposition of WS$_2$ domains achieves chemical equilibrium. The equilibrium constant is denoted as K$_C$.

In the area (0–450 µm) close to ZnO crystal whisker, W atoms reacted with ZnO to form ZnWO$_4$ resulting in a decrease of the W atom concentration. WS$_2$ precursors far from this area will diffuse to this region to keep the reaction running. With the reaction and diffusion of the WS$_2$ precursor,

W atoms are consumed and S atoms are accumulated in this area. With the distance increase, the W atom concentration increases and the S atom concentration decreases. They all reach equilibrium concentration when the distance is larger than 450 µm. Reaction quotient (Q_P) increases first, and reaches the maximum value (250–450 µm), then decreases to the equilibrium constant K_C. In the area of 0–250 µm to ZnO crystal whisker, the W atom concentration is low and the reaction quotient Q_P is no more than K_C. Therefore, the WS_2 domains hardly grow. In the area of 250–450 µm to ZnO crystal whisker, the larger Q_P ($>K_C$) and S atom concentration promote the growth of monolayer WS_2. The excess S atom is the key parameter for the monolayer growth.

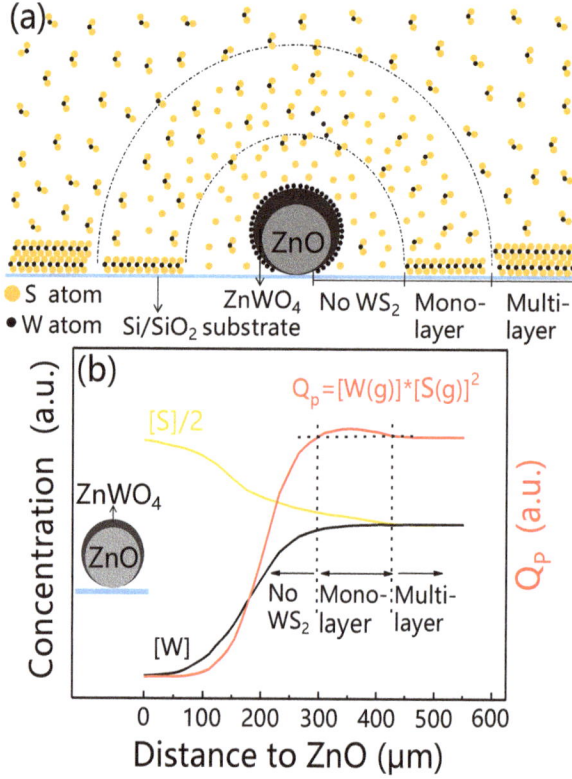

Figure 7. (a) Schematic diagram of W and S distribution around ZnO crystal whisker. (b) Distance dependence of W atom concentration, S atom concentration, and the reaction quotient (Q_P). The direction of the horizontal axis is perpendicular to the growth axis of ZnO whisker. The coordinate of the ZnO whisker on the horizontal axis is zero. [W] and [S] represent W and S atom concentration in (b), respectively.

4. Conclusions

In summary, we successfully prepared monolayer WS_2 by a novel method. ZnO crystal whisker was used to position and promote the growth of WS_2. The distribution statistics show monolayer WS_2 was grown on both sides of the ZnO crystal whisker. By constructing a concentration distribution model, we were able to discuss the monolayer growth mechanism. The results reveal that gaseous sulfur and tungsten concentration are crucial for the thickness control of WS_2. Higher concentration of sulfur and lower concentration of tungsten are of tremendous benefit for monolayer WS_2 growth. This method would provide a way to grow and pattern monolayer WS_2 and other two dimensional transition metal disulfides on silicon substrate for the fabrication of nano-optoelectronic devices.

Author Contributions: Y.L. and G.W. conceived and designed the experiments and wrote the paper; Z.X. and F.H. performed the experiments and wrote the paper; C.Z. and S.Z. analyzed the data and wrote the paper.

Funding: This work was supported by the Natural Science Foundation of Zhejiang Province, China Projects (LY16E020008) and Chinese NSF Projects (61106100).

Conflicts of Interest: The authors declare no conflict of interest.

References

1. Song, L.; Ci, L.; Lu, H.; Sorokin, P.B.; Jin, C.; Ni, J.; Kvashnin, A.G.; Kvashnin, D.G.; Lou, J.; Yakobson, B.I.; et al. Large Scale Growth and Characterization of Atomic Hexagonal Boron Nitride Layers. *Nano Lett.* **2010**, *10*, 3209–3215. [CrossRef]
2. Geim, A.K.; Novoselov, K.S. The rise of graphene. *Nat. Mater.* **2007**, *6*, 183–191. [CrossRef] [PubMed]
3. Wang, Q.H.; Kalantar-Zadeh, K.; Kis, A.; Coleman, J.N.; Strano, M.S. Electronics and optoelectronics of two-dimensional transition metal dichalcogenides. *Nat. Nanotechnol.* **2012**, *7*, 699–712. [CrossRef] [PubMed]
4. Xue, Y.; Zhang, Y.; Liu, Y.; Liu, H.; Song, J.; Sophia, J.; Liu, J.; Xu, Z.; Xu, Q.; Wang, Z.; et al. Scalable Production of a Few-Layer MoS_2/WS_2 Vertical Heterojunction Array and Its Application for Photodetectors. *ACS Nano* **2016**, *10*, 573–580. [CrossRef] [PubMed]
5. Liu, Y.; Weiss, N.O.; Duan, X.; Cheng, H.; Huang, Y.; Duan, X. Van der Waals heterostructures and devices. *Nat. Rev. Mater.* **2016**, *1*, 160429. [CrossRef]
6. Perea-Lopez, N.; Elias, A.L.; Berkdemir, A.; Castro-Beltran, A.; Gutierrez, H.R.; Feng, S.; Lv, R.; Hayashi, T.; Lopez-Urias, F.; Ghosh, S.; et al. Photosensor Device Based on Few-Layered WS_2 Films. *Adv. Funct. Mater.* **2013**, *23*, 5511–5517. [CrossRef]
7. Ma, X.L.; Zhang, R.J.; An, C.H.; Wu, S.; Hu, X.D.; Liu, J. Efficient doping modulation of monolayer WS2 for optoelectronic applications. *Chin. Phys. B* **2019**, *28*, 037803. [CrossRef]
8. Fiori, G.; Bonaccorso, F.; Iannaccone, G.; Palacios, T.; Neumaier, D.; Seabaugh, A.; Banerjee, S.K.; Colombo, L. Electronics based on two-dimensional materials. *Nat. Nanotechnol.* **2014**, *9*, 768–779. [CrossRef]
9. Therese, H.A.; Li, J.X.; Kolb, U.; Tremel, W. Facile large scale synthesis of WS_2 nanotubes from WO_3 nanorods prepared by a hydrothermal route. *Solid State Sci.* **2005**, *7*, 67–72. [CrossRef]
10. Coleman, J.N.; Lotya, M.; O'Neill, A.; Bergin, S.D.; King, P.J.; Khan, U.; Young, K.; Gaucher, A.; De, S.; Smith, R.J.; et al. Two-Dimensional Nanosheets Produced by Liquid Exfoliation of Layered Materials. *Science* **2011**, *331*, 568–571. [CrossRef]
11. Biccai, S.; Barwich, S.; Boland, D.; Harvey, A.; Hanlon, D.; McEvoy, N.; Coleman, J.N. Exfoliation of 2D materials by high shear mixing. *2D Mater.* **2019**, *6*, 0150081. [CrossRef]
12. Lv, R.; Robinson, J.A.; Schaak, R.E.; Sun, D.; Sun, Y.; Mallouk, T.E.; Terrones, M. Transition Metal Dichalcogenides and Beyond: Synthesis, Properties, and Applications of Single- and Few-Layer Nanosheets. *Acc. Chem. Res.* **2015**, *48*, 56–64. [CrossRef] [PubMed]
13. Lan, C.; Li, C.; Yin, Y.; Liu, Y. Large-area synthesis of monolayer WS_2 and its ambient-sensitive photo-detecting performance. *Nanoscale* **2015**, *7*, 5974–5980. [CrossRef] [PubMed]
14. Yun, S.J.; Chae, S.H.; Kim, H.; Park, J.C.; Park, J.; Han, G.H.; Lee, J.S.; Kim, S.M.; Oh, H.M.; Seok, J.; et al. Synthesis of Centimeter-Scale Monolayer Tungsten Disulfide Film on Gold Foils. *ACS Nano* **2015**, *9*, 5510–5519. [CrossRef] [PubMed]
15. Joo, M.; Yun, Y.; Yun, S.; Lee, Y.H.; Suh, D. Strong Coulomb scattering effects on low frequency noise in monolayer WS_2 field-effect transistors. *Appl. Phys. Lett.* **2016**, *109*, 153102. [CrossRef]
16. Forcherio, G.T.; Bonacina, L.; Riporto, J.; Mugnier, Y.; Le Dantec, R.; Dunklin, J.R.; Benamara, M.; Roper, D.K. Integrating plasmonic metals and 2D transition metal dichalcogenides for enhanced nonlinear frequency conversion. In *Physical Chemistry of Semiconductor Materials and Interfaces XVII, Proceedings of the SPIE 2018, San Diego, CA, USA, 20–23 August 2018*; Bronstein, H.A., Deschler, F., Kirchartz, T., Eds.; SPIE: Bellingham, WA, USA, 2018.
17. Liu, P.; Luo, T.; Xing, J.; Xu, H.; Hao, H.; Liu, H.; Dong, J. Large-Area WS_2 Film with Big Single Domains Grown by Chemical Vapor Deposition. *Nanoscale Res. Lett.* **2017**, *12*, 558. [CrossRef] [PubMed]
18. O'Brien, M.; McEvoy, N.; Hanlon, D.; Hallam, T.; Coleman, J.N.; Duesberg, G.S. Mapping of Low-Frequency Raman Modes in CVD-Grown Transition Metal Dichalcogenides: Layer Number, Stacking Orientation and Resonant Effects. *Sci. Rep.* **2016**, *6*, 19476. [CrossRef]

19. Cong, C.; Shang, J.; Wu, X.; Cao, B.; Peimyoo, N.; Qiu, C.; Sun, L.; Yu, T. Synthesis and Optical Properties of Large-Area Single-Crystalline 2D Semiconductor WS$_2$ Monolayer from Chemical Vapor Deposition. *Adv. Opt. Mater.* **2014**, *2*, 131–136. [CrossRef]
20. Rajan, A.G.; Warner, J.H.; Blankschtein, D.; Strano, M.S. Generalized mechanistic model for the chemical vapor deposition of 2D transition metal dichalcogenide monolayers. *ACS Nano* **2016**, *10*, 4330–4344. [CrossRef]
21. Shi, Y.; Li, H.; Li, L.-J. Recent Advances in Controlled Synthesis of Two-Dimensional Transition Metal Dichacogenides via Vapour Deposition Techniques. *Chem. Soc. Rev.* **2015**, *44*, 2744–2756. [CrossRef]
22. Zhang, Y.; Zhang, Y.; Ji, Q.; Ju, J.; Yuan, H.; Shi, J.; Gao, T.; Ma, D.; Liu, M.; Chen, Y.; et al. Controlled Growth of High-Quality Monolayer WS$_2$ Layers on Sapphire and Imaging Its Grain Boundary. *ACS Nano* **2013**, *7*, 8963–8971. [CrossRef] [PubMed]
23. Fan, X.; Zhao, Y.; Zheng, W.; Li, H.; Wu, X.; Hu, X.; Zhang, X.; Zhu, X.; Zhang, Q.; Wang, X.; et al. Controllable Growth and Formation Mechanisms of Dislocated WS$_2$ Spirals. *Nano Lett.* **2018**, *18*, 3885–3892. [CrossRef] [PubMed]
24. Cain, J.D.; Shi, F.; Wu, J.; Dravid, V.P. Growth Mechanism of Transition Metal Dichalcogenide Monolayers: The Role of Self-Seeding Fullerene Nuclei. *ACS Nano* **2016**, *10*, 5440–5445. [CrossRef] [PubMed]
25. Ji, Z.; Hao, F.; Wang, C.; Xi, J. Centimetre-long single crystalline ZnO fibres prepared by vapour transportation. *Chin. Phys. Lett.* **2008**, *25*, 3467–3469.
26. Heo, H.; Sung, J.H.; Cha, S.; Jang, B.; Kim, J.; Jin, G.; Lee, D.; Ahn, J.; Lee, M.; Shim, J.H.; et al. Interlayer orientation-dependent light absorption and emission in monolayer semiconductor stacks. *Nat. Commun.* **2015**, *6*, 7372. [CrossRef]
27. Gutierrez, H.R.; Perea-Lopez, N.; Elias, A.L.; Berkdemir, A.; Wang, B.; Lv, R.; Lopez-Urias, F.; Crespi, V.H.; Terrones, H.; Terrones, M. Extraordinary Room-Temperature Photoluminescence in Triangular WS$_2$ Monolayers. *Nano Lett.* **2013**, *13*, 3447–3454. [CrossRef]
28. Wang, X.H.; Ning, J.Q.; Zheng, C.C.; Zhu, B.R.; Xie, L.; Wu, H.S.; Xu, S.J. Photoluminescence and Raman mapping characterization of WS$_2$ monolayers prepared using top-down and bottom-up methods. *J. Mater. Chem. C* **2015**, *3*, 2589–2592. [CrossRef]
29. Berkdemir, A.; Gutierrez, H.R.; Botello-Mendez, A.R.; Perea-Lopez, N.; Elias, A.L.; Chia, C.; Wang, B.; Crespi, V.H.; Lopez-Urias, F.; Charlier, J.; et al. Identification of individual and few layers of WS$_2$ using Raman Spectroscopy. *Sci. Rep.* **2013**, *3*, 1755. [CrossRef]
30. Zhang, X.; Qiao, X.; Shi, W.; Wu, J.; Jiang, D.; Tan, P. Phonon and Raman scattering of two-dimensional transition metal dichalcogenides from monolayer, multilayer to bulk material. *Chem. Soc. Rev.* **2015**, *44*, 2757–2785. [CrossRef]
31. Basiev, T.T.; Karasik, A.Y.; Sobol, A.A.; Chunaev, D.S.; Shukshin, V.E. Spontaneous and stimulated Raman scattering in ZnWO$_4$ crystals. *Quantum Electron.* **2011**, *41*, 370–372. [CrossRef]
32. Huang, G.; Zhu, Y. Synthesis and photocatalytic performance of ZnWO$_4$ catalyst. *Mater. Sci. Eng. B-Solid State Mater. Adv. Technol.* **2007**, *139*, 201–208. [CrossRef]

© 2019 by the authors. Licensee MDPI, Basel, Switzerland. This article is an open access article distributed under the terms and conditions of the Creative Commons Attribution (CC BY) license (http://creativecommons.org/licenses/by/4.0/).

Article

Borocarbonitride Layers on Titanium Dioxide Nanoribbons for Efficient Photoelectrocatalytic Water Splitting

Nuria Jiménez-Arévalo [1,*], Eduardo Flores [2], Alessio Giampietri [3], Marco Sbroscia [3], Maria Grazia Betti [3], Carlo Mariani [3], José R. Ares [1], Isabel J. Ferrer [1,4] and Fabrice Leardini [1,4]

1. Departamento de Física de Materiales, Campus de Cantoblanco, Universidad Autónoma de Madrid, E-28049 Madrid, Spain; joser.ares@uam.es (J.R.A.); isabel.j.ferrer@uam.es (I.J.F.); fabrice.leardini@uam.es (F.L.)
2. Centro de Nanociencias y Nanotecnología (CNyN), Universidad Nacional Autónoma de México (UNAM), Ensenada 22860, BC, Mexico; eduardoe.floresc@gmail.com
3. Dipartimento di Fisica, Università di Roma 'La Sapienza', I-00185 Rome, Italy; alessio.giampietri@uniroma1.it (A.G.); marco.sbroscia@uniroma1.it (M.S.); maria.grazia.betti@roma1.infn.it (M.G.B.); carlo.mariani@uniroma1.it (C.M.)
4. Instituto Nicolás Cabrera, Campus de Cantoblanco, Universidad Autónoma de Madrid, E-28049 Madrid, Spain
* Correspondence: nuria.jimeneza@uam.es

Abstract: Heterostructures formed by ultrathin borocarbonitride (BCN) layers grown on TiO_2 nanoribbons were investigated as photoanodes for photoelectrochemical water splitting. TiO_2 nanoribbons were obtained by thermal oxidation of TiS_3 samples. Then, BCN layers were successfully grown by plasma enhanced chemical vapour deposition. The structure and the chemical composition of the starting TiS_3, the TiO_2 nanoribbons and the TiO_2-BCN heterostructures were investigated by Raman spectroscopy, X-ray diffraction and X-ray photoelectron spectroscopy. Diffuse reflectance measurements showed a change in the gap from 0.94 eV (TiS_3) to 3.3 eV (TiO_2) after the thermal annealing of the starting material. Morphological characterizations, such as scanning electron microscopy and optical microscopy, show that the morphology of the samples was not affected by the change in the structure and composition. The obtained TiO_2-BCN heterostructures were measured in a photoelectrochemical cell, showing an enhanced density of current under dark conditions and higher photocurrents when compared with TiO_2. Finally, using electrochemical impedance spectroscopy, the flat band potential was determined to be equal in both TiO_2 and TiO_2-BCN samples, whereas the product of the dielectric constant and the density of donors was higher for TiO_2-BCN.

Keywords: borocarbonitride; TiO_2-BCN heterostructures; water splitting; photoelectrocatalysis; X-ray photoelectron spectroscopy; graphene analogues; hybrid structures

1. Introduction

The current energetic model based on fossil fuels is unsustainable from an environmental perspective, as it is one of the leading causes of global warming and climate change [1–3]. The focus is now placed on solar and wind energy, which have the problem of being intermittent, which points to the necessity of developing new ways of storing energy.

Among all the energy storage methods, energy storage using molecular bonding stands out, such as the one in the hydrogen molecule. Hydrogen has been reported to be a suitable energy vector and a clean energy fuel if its production comes from renewable sources [2,4].

In 1972, Honda and Fujishima reported a way to obtain hydrogen by carrying out a photoassisted water splitting reaction using TiO_2 as the photoanode [5]. Since then, this effect has been considered one of the cleanest methods to obtain green hydrogen and a promising strategy to overcome the environmental and energy crises [6,7]. The water-splitting reaction consists of two partial reactions, the oxygen evolution reaction (OER) and the hydrogen evolution reaction (HER). The OER is the rate-determining step as it involves the transfer of four electrons [8], and this is the reaction we will tackle in this paper.

The low cost, stability and non-toxicity of TiO_2, as well as its adequacy to carry out the water splitting reaction, has drawn the attention of many groups who have reported the good properties of this material by synthesizing it in different structures and nanostructures [9,10].

To increase the charge transfer between the electrode and the electrolyte, metals nanoparticles, such as Pt and Ni, are commonly used as active electrocatalytic sites for water splitting [11]. The main focus is now placed on developing new metal-free compounds to be used as catalysts for the oxygen and hydrogen evolution reactions.

Graphene analogues and other 2D materials have demonstrated to be highly interesting metal-free compounds with a wide range of applications in electrocatalysis [12]. These layers have the advantage of being distributed along all the electrode surface, increasing the reaction area in comparison to the metal nanoparticles. Among these 2D materials borocarbonitride compounds (BCNs hereafter) stand out, which are low cost and highly stable materials formed by h-BN and graphite domains. BCNs have been reported to be efficient electrocatalysts for the HER [13,14] and, most interestingly, have recently been proved to be an efficient electrocatalysts for the OER, improving the properties of TiO_x substrates for this reaction [15].

In this article we have first confirmed the utility and versatility of plasma enhanced chemical vapor deposition to grow BCN on samples with different morphologies. This technique has been previously used to grow BCN on TiO_x, Cu, and other flat substrates [15,16]. In this work, BCN was grown, for the first time, on nanostructured TiO_2 samples. In particular, BCN was grown on TiO_2 nanoribbons, obtained by the thermal annealing of TiS_3 [17]. A deep characterization of TiO_2 nanoribbons, with and without BCN, as well as the starting material, TiS_3, was made with scanning electron microscopy (SEM), X-ray diffraction (XRD), Raman spectroscopy and diffuse reflectance measurements. X-ray photoelectron spectroscopy (XPS) characterization of bare and BCN-covered TiO_2 nanoribbons was performed.

Finally, we demonstrated the good properties of the BCN as an electrocatalyst of the OER by improving the charge transfer between the TiO_2 nanoribbons electrode and the KOH aqueous electrolyte, under dark and light conditions.

2. Materials and Methods

2.1. Synthesis

The starting TiS_3 material has been obtained by the sulfuration of Ti disks (15 mm diameter, Good Fellow 99.5%) in sealed Pyrex ampoules for 20 h at 550 °C, using sulfur powder as sulfur source [18,19].

TiS_3 nanoribbons were squashed in one direction and oxidized on a hot plate at 300 °C in air (the decomposition temperature of TiS_3 [17]) for 20–30 s, which allowed us to obtain the desired TiO_2 nanoribbons. Figure S1 shows the change in the color of the samples, from black to white, at a glance.

BCN was grown on TiO_2 using PE-CVD to get the TiO_2-BCN heterostructures. The TiO_2 sample and a single-source molecular precursor (methylamine-borane, $BH_3NH_2CH_3$) were placed inside a Pyrex ampoule immersed in liquid nitrogen and then sealed under vacuum at a pressure around 10^{-5} mbar. When the sealed ampoule reaches room temperature, the molecular precursor is in equilibrium with its vapor pressure in the 10^{-2} mbar range. Then a plasma was activated inside the ampoule using the radiation of a conventional microwave oven. More details about this technique can be found elsewhere [15,16].

2.2. Characterization

The starting material, TiS_3, as well as TiO_2 and TiO_2-BCN heterostructures, were characterized by using different techniques.

The morphology of the samples was investigated by scanning electron microscopy (SEM) using a Hitachi S3000 instrument. Additionally, they were characterized using Raman spectroscopy with a WITec ALPHA 300AR instrument using a confocal microscope with lenses of 20× and 100×. The used laser had a power of 0.2 mW and an excitation wavelength of 532.3 nm.

The structural properties were measured using a Panalytical X'Pert Pro X-ray diffractometer at glancing angle configuration (incident angle of 1.7°, CuKα radiation).

The optical reflectance spectra were recorded in a UV/VIS/NIR Perkin-Elmer LAMBDA 950 spectrophotometer equipped with an integrating sphere to collect the reflecting flux, using a spot size of 21 mm^2 in the 300–2000 nm spectral range.

The composition on the surface of TiO_2 and TiO_2-BCN samples was investigated by X-ray photoelectron spectroscopy (XPS). These measurements have been carried out in an ultrahigh-vacuum (UHV) chamber, with a base pressure in the low 10^{-10} mbar range. Photoelectrons excited by an Al Kα photon source ($hv = 1486.7$ eV), were measured by a hemispherical electron analyzer (VG Microtech Clam-2) in a pass energy mode set at 50 eV for Ti and C, and 100 eV for B and N. Further details about the procedure are available in [15,16,20,21]. The binding energy (BE) was calibrated by acquiring the Au4f$_{7/2}$ (84.0 eV of BE) core-level after each measurement. The measurements were done after annealing the samples at 320 °C for 1 h, at 110 °C for 15 h and 320 °C for another hour in UHV.

Electrochemical measurements were done in a three-electrode cell. Our material, which had an apparent area of 1.3 cm^2, was placed as the working electrode (WE), a platinum sheet (9 cm^2) as the counter electrode (CE), and, as the reference electrode (RE), an Ag/AgCl electrode filled with 1M KNO_3 (XR440 from Radiometer Analytical) was used. Its electrode potential was 484 mV vs. a normal hydrogen electrode (NHE). These three electrodes were immersed in 0.1 M (pH = 13.0) and 1.0 M (pH = 13.7) KOH aqueous electrolyte. The potentials are converted to the reversible hydrogen electrode (RHE) using Equations (1) and (2):

$$E_{NHE} = E_{Ag/AgCl} + E^0_{Ag/AgCl} \qquad (1)$$

$$E_{RHE} = E_{NHE} + 0.059 \cdot pH \qquad (2)$$

where E_{NHE} is the electrode potential in the NHE scale, $E_{Ag/AgCl}$ is the experimental electrode potential measured vs. Ag/AgCl reference electrode, $E^0_{Ag/AgCl}$ is the electrode potential vs. the NHE, E_{RHE} is the electrode potential in the RHE scale, and pH is the pH of the solution used.

The electrochemical and photoelectrochemical (PEC) measurements were done using a potentiostat–galvanostat PGTAT302N (Autolab) provided with an integrated impedance module, FRAII. The WE was illuminated with a Xe lamp (Jobin Yvon) of 75 W in the visible-UV range. The maximum light intensity reaching the sample was 140 mW. During electrochemical measurements, argon was bubbled at a constant flow of 20.0 sccm. To characterize the electrodes under both dark and illumination conditions, linear sweep voltammetry (LSV), cyclic voltammetry (CV) and current density measurements at a fixed potential were employed. In addition, electrochemical impedance spectroscopy (EIS) was used to characterize the interface electrolyte-semiconductor, by using a sinusoidal AC voltage signal with an amplitude of 10 mV and a variable frequency between 100–1000 Hz.

More details about the photoelectrochemical experimental system can be found in the Supplementary Materials (Figure S2).

3. Results and Discussion

3.1. Morphological and Structural Characterization

The oxidation of the TiS_3 nanoribbons was performed to change the atomic structure and the composition of the material, without changing the morphology of the sample, in order to obtain TiO_2 nanoribbons. Figure 1a reports the morphology of TiS_3 before and after the oxidation. It is clear from the SEM measurements, that there was no significant change in the morphology, apart from a slight increase of the roughness on the TiO_2

surface compared to TiS$_3$. We also acquired optical microscopy images to have a deeper understanding of the change on the surface of these nanoribbons (Figure 1b), which allowed us to determine that the material became more transparent to the optical microscope, in good agreement with the results reported by Ghasemi et al. [17] and with our optical characterizations shown below (see Figure 4).

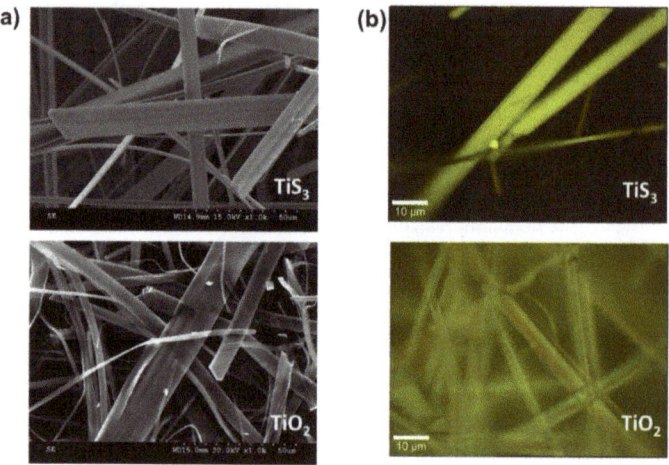

Figure 1. (a) Scanning electron micrographs of TiS$_3$ (top) and TiO$_2$ (bottom). (b) Optical microscopy image of TiS$_3$ (top) and TiO$_2$ (bottom).

Structural changes were also monitored via XRD after the oxidation and after growing the BCN on top of this oxidized material (Figure 2). There were clear changes in the structure due to the thermal annealing, as there was a transition from monoclinic TiS$_3$ to tetragonal TiO$_2$-anatase with some peaks of tetragonal TiO$_2$-rutile.

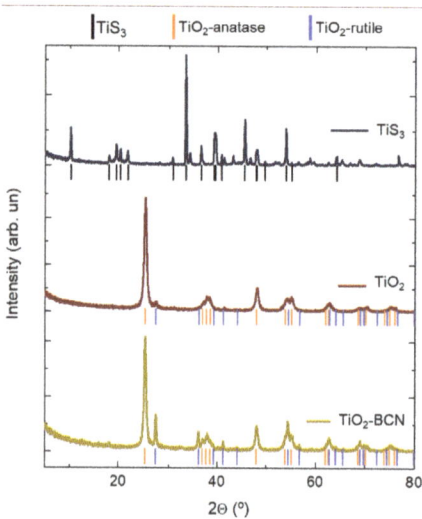

Figure 2. X-ray diffraction patterns for TiS$_3$, TiO$_2$ and TiO$_2$-BCN samples. TiS$_3$ peaks correspond to JDPDF 00-015-0783, TiO$_2$-anatase peaks correspond to JDPDF 01-071-1167, and TiO$_2$-rutile peaks correspond to JDPDF 01-073-2224.

After the BCN synthesis, the intensity of the rutile peaks increased. This was ascribed to the high temperature that the sample achieved during exposure to the plasma, which modified the structure of the bulk material by crystallizing the sample from anatase TiO_2 to rutile TiO_2.

Raman spectra of TiS_3 and TiO_2 nanoribbons, reported in Figure 3, show that there is a complete change in their structure from TiS_3 (175 cm^{-1}, 303 cm^{-1}, 373 cm^{-1}, 562 cm^{-1}) [22] to TiO_2-anatase (145 cm^{-1}, 396 cm^{-1}, 518 cm^{-1}, 643 cm^{-1}) [10,23,24]. In the case of TiO_2-BCN, there was no change in the Raman spectra of the nanoribbons, thus they maintained their anatase composition. The growth of the BCN in this sample was confirmed by the presence of D (1375 cm^{-1}) and G (1590 cm^{-1}) Raman bands, ascribed to the BCN layers graphene-like sp2 structure [15,16].

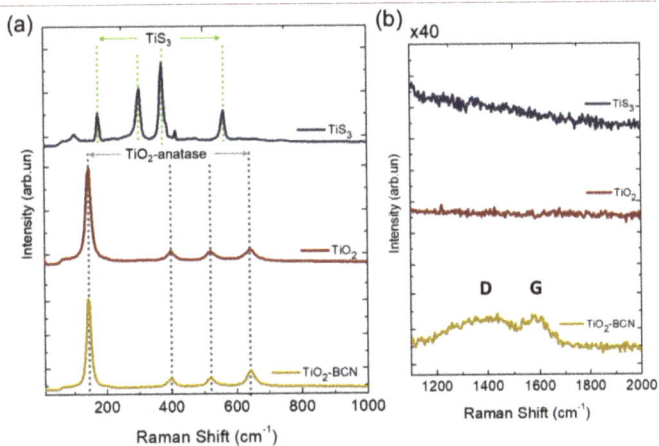

Figure 3. Raman spectra for TiS_3, TiO_2 and TiO_2-BCN samples in the (a) 100–1000 cm^{-1} Raman shift range and (b) 1100–2000 cm^{-1} Raman shift range with a ×40 zoom in the intensity.

It must be mentioned that in the TiO_2-BCN sample, some TiO_2 rutile nanoribbons [25,26] were also found in the Raman spectra (Figure S3). This result is in good agreement with the increase in the rutile peaks in the XRD diffractogram for this sample. The nanoribbons were found at the borders of the samples and did not contribute to the electrochemistry as they were not exposed to the electrolyte due to the shape of the electrode holder.

3.2. Optical Characterization

Diffuse reflectance measurements were done using an integrating sphere to characterize the change in the energy bandgap. The reflectance (R) data were converted into a Kubelka–Munk function, F(R) [27] (Equation (3)), which was proportional to the optical density in the optical absorption measurements.

$$F(R) = \frac{(1-R)^2}{2R} \quad (3)$$

Results as a function of the incident light energy can be seen in Figure 4a. At first glance, the presence of 1.23 eV, 1.73 eV and 2.1 eV absorption bands, characteristic of the TiS_3 samples [19] can be observed, but they disappear after the oxidation process due to the transformation of TiS_3 into TiO_2.

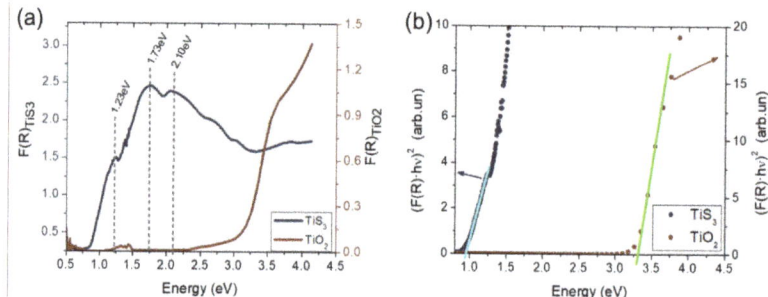

Figure 4. (a) Kubelka–Munk function obtained from diffuse reflectance measurements of TiS$_3$ (left axis) and TiO$_2$ (right axis). (b) Tauc plots of Kubelka–Munk function with the corresponding linear fit for TiS$_3$ (left axis) and TiO$_2$ (right axis).

To obtain the band gap energy of TiS$_3$ and TiO$_2$ samples, Tauc fitting equation for direct band gaps (($F(R) \cdot h\nu)^2$) has been applied to the F(R) function (Figure 4b). A clear change can be observed from 0.94 ± 0.04 eV in TiS$_3$, to 3.3 ± 0.4 eV in TiO$_2$. These results are in good agreement with the values previously obtained for TiS$_3$ [19] and TiO$_2$ [28].

3.3. Surface Chemical Composition and Element Distribution

For further characterization of the chemical bonding state in the surface of TiO$_2$ and TiO$_2$-BCN samples, XPS measurements were carried out. The Ti2p level for the TiO$_2$ sample is shown in Figure 5a, and the B1s, C1s and Ti2p core levels for TiO$_2$-BCN can be seen in Figure 5b–d. A fitting analysis using Voigt line shapes (combination of Gaussian and Lorentzian curves) was performed to take into account the overall experimental uncertainty and the intrinsic linewidth, respectively. A Shirley background was introduced as a fitting parameter for all the analyzed peaks. All the fitting parameters for the graphs in Figure 5 are reported in Table S1.

From this result, it is evident that the Ti2p levels show the presence of Ti^{4+} in both samples, which can be attributed to the formation of TiO$_2$ [29]. However, in the case of TiO$_2$-BCN, a small additional peak at 457.55 eV of BE appears, which is ascribed to the presence of Ti^{3+} bonds [30,31], probably due to small areas with defects close to the edge of the sample. Nevertheless, this component represents the 0.5% of the relative intensity, meaning that the plasma did not significantly affect the surface of the substrate for the growth times used.

In the TiO$_2$-BCN sample, XPS measurements revealed that the BCN layer was composed of C and h-BN domains with high mutual doping levels of B and N in C, and of C in BN, respectively. The B1s and C1s peaks appear with a large width and are composed of more than one peak, indicating the presence of different components associated with the mutual bonding among the elements.

The B1s core level in TiO$_2$-BCN shows three peaks due to the B–C (190.51 eV), B–N (191.41 eV) [15,16,21,32–34] and B–O (192.96 eV) bonds [15,16,20,21,35]. The C1s core level exhibit four main components due to the C–C bonding at 284.6 eV, signature of the sp2-bonded carbon, C–N bonds (285.48 eV), C–B bonds (282.65 eV), and multiple components due to different bonds of C to O [15,16,21,32–34], which are unresolved and appear as a single broad band at 286.43 eV. We were not able to decompose the N1s peak in different Voigt functions, as we did with C1s and B1s peaks, since the signal of N1s is overlapped with a small signal of the Ta4p$_{3/2}$ peak, ascribed to the tantalum clips used to fix the sample to the sample holder. The spectrum in an energy range close to the N1s peak (385–413 eV) is shown in Figure S4.

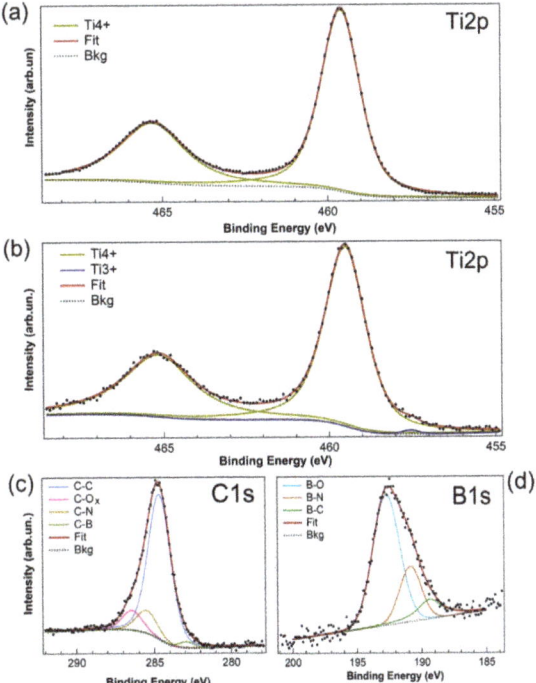

Figure 5. Ti2p XPS spectra for (**a**) TiO$_2$ sample and (**b**) TiO$_2$-BCN sample. C1s (**c**) and B1s (**d**) spectra for TiO$_2$-BCN sample. Experimental data (dots), Shirley background is represented by a grey dotted line, the fitting curve for each measurement is in red, and single fitting components, as described in the legend.

Thus, the B and C 1s core levels present the typical features associated with the mutual chemical bonds in sp2-hybridized compounds. This is a clear signature of the formation of a ternary BCN planar layer [15,16]. The amount of BCN in the sample was estimated by XPS quantification, taking into account the cross-section, obtaining a ratio between Ti and B of 7:1, which suggests that the layer of BCN was ultrathin.

3.4. Photoelectrocatalytic Activity for the OER

The effect of BCN on the electrocatalytic activity of TiO$_2$ was investigated at first via linear sweep voltammetry (LSV) measurements under intermittent illumination conditions comparing the results from TiO$_2$-BCN heterostructure and these from bare TiO$_2$ one (Figure 6a). Regarding the dark condition, bare TiO$_2$ presented low currents and poor catalytic activity for the OER compared to those of the TiO$_2$-BCN sample, which points out the electrocatalytic effect of BCN. In fact, at the maximum applied potential (1.9 V vs. RHE), the dark current was 12-fold higher than that of the bare TiO$_2$. Photocurrents were measured at different applied stationary potentials (Figure 6b), and, as can be seen, it increases from around 1 µA (bare TiO$_2$) to around 40 µA (TiO$_2$-BCN) at 1.23 V vs. RHE. The photocurrents were stable (Figure S5) and have a positive value, in good agreement with the nature of TiO$_2$ as an n-type semiconductor [36]. It can be concluded that the BCN layer does not only improves the electrocatalytic properties of the TiO$_2$ at dark conditions, but also under illumination by increasing the photocurrents. The LSV curves, as well as CV curves, were measured for different scan rates, which made it possible to determine that the current that appears in the TiO$_2$-BCN sample before the OER in the LSV curve is due to the scan rate and not to another reaction (Figure S6).

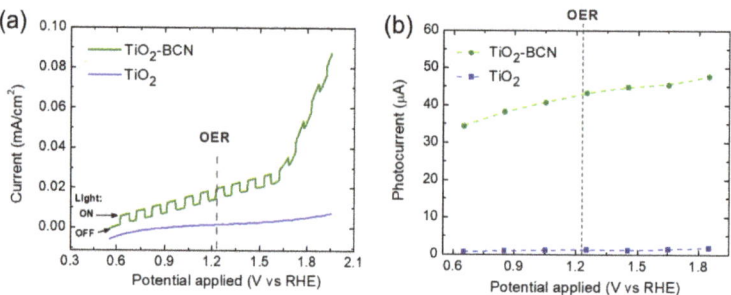

Figure 6. (a) Linear sweep voltammetry curves for bare TiO$_2$ and TiO$_2$-BCN under intermittent illumination conditions. (b) Photocurrents obtained for different constant applied potentials.

The stability has been monitored by doing series of 50 cycles at 0.05 V/s, proving the high stability of our material, as can be seen in Figure S7. Raman measurements were also made before and after the photoelectrochemical measurements to prove that there was no degradation of the sample and that the BCN layer was still there (Figure S8).

In order to characterize the interface between the electrodes (TiO$_2$ and TiO$_2$-BCN) and the electrolyte (KOH), EIS measurements were done. By measuring the impedance at different frequencies, the capacitance in the spatial charge region (C_{SC}) could be acquired, and with the Mott–Schottky equation (Equation (4)), the value of the flat band potential (V_{fb}) could be obtained.

$$\frac{1}{C_{SC}^2} = \left(\frac{2}{\varepsilon \cdot A^2 \cdot e \cdot N_D}\right) \cdot \left(V_{bias} - V_{fb} - \frac{k_B \cdot T}{e}\right) \quad (4)$$

where ε is the dielectric constant of the electrode, ε_0 is the vacuum permittivity, A is the area of the electrode, e is the charge of the electron, N_D is the donor density, V_{bias} is the applied potential, k_B is the Boltzmann's constant and T is the temperature. The term $\frac{k_B \cdot T}{e}$ can be neglected due to its low value when compared to V_{bias} and V_{fb}.

The value of the flat band potential is highly significant in the electrochemical characterization of a semiconductor–electrolyte interface, as it is related to the bottom energy of the conduction band. Together with the bandgap energy, it is used to describe the energy levels position at the electrode–electrolyte interface, determining the adequacy of a material to carry out or not any reaction, in this case, water splitting. Figure 7a,b shows the Mott–Schottky plot at three representative frequencies for the bare TiO$_2$ and TiO$_2$-BCN samples. The value obtained for the flat band potential is 0.2 ± 0.1 V vs. RHE for bare TiO$_2$, and 0.2 ± 0.1 V vs. RHE for TiO$_2$-BCN. These results are in good agreement with values of the flat band potential previously obtained for TiO$_2$ samples [9,10], and with a previous report about the good electrocatalytic activity of BCN for OER, in which it was shown that the BCN layer does not affect the flat band potential of the underlying material [15]. Table S2 summarizes the main differences between TiO$_2$ and TiO$_2$-BCN.

The slope of the Mott–Schottky fitting gives information about the dielectric constant of the material and its donor density. By using the data of the dielectric constant of TiO$_2$ at 20 °C reported by Wypych et al. [37] and the slope of bare TiO$_2$, its donor density was determined to be independent of the frequency and has a value of $(1.53 \pm 0.02) 10^{19}$ cm^{-3} (Figure S9). This is in good agreement with the fact that the donor density is a characteristic parameter of the material and should not change with the frequency. However, as observed in [37], the dielectric constant is a function of the frequency.

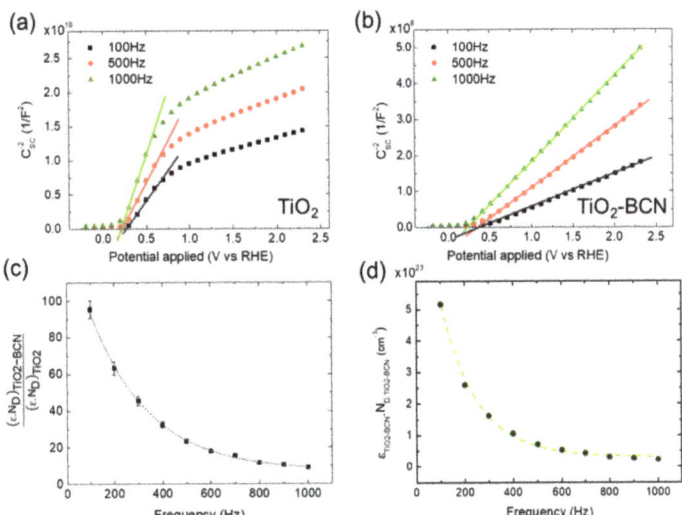

Figure 7. Mott–Schottky plot at three different frequencies for (a) sample TiO$_2$ and (b) sample TiO$_2$-BCN. (c) Relationship between the dielectric constant and the density of donors ($\varepsilon \cdot N_D$ factor) of the sample TiO$_2$-BCN over that of the bare TiO$_2$, as a function of the frequency. (d) $\varepsilon \cdot N_D$ factor for TiO$_2$-BCN sample vs. the frequency. Fitting parameters of the exponential decay fittings represented by a grey dotted line in (c) and green dashed line in (d) are shown in Table S3.

By comparing the slopes of TiO$_2$ (m_{TiO_2}) and TiO$_2$-BCN (m_{TiO_2-BCN}) at each frequency (Equation (5)) the relationship between the product of ε and N_D for both samples can be determined (Figure 7c).

$$\frac{m_{TiO_2}}{m_{TiO_2-BCN}} = \frac{\varepsilon_{TiO_2-BCN} \cdot N_{D\,TiO_2-BCN}}{\varepsilon_{TiO_2} \cdot N_{D\,TiO_2}} \tag{5}$$

It can be concluded that the TiO$_2$-BCN heterostructure has a greater value of factor $\varepsilon \cdot N_D$ than the bare TiO$_2$. This can be partially ascribed to a higher N_D in the TiO$_2$-BCN sample, induced by the plasma treatment during BCN growth. The higher donor density in the TiO$_2$-BCN heterostructure is also in good agreement with the increase in the photocurrents in this sample in comparison with TiO$_2$ (Figure 6b). Moreover, the dielectric constant of the TiO$_2$-BCN heterostructure was also modified as the factor $\frac{\varepsilon_{TiO_2-BCN} \cdot N_{D\,TiO_2-BCN}}{\varepsilon_{TiO_2} \cdot N_{D\,TiO_2}}$ is not constant with the frequency. With the value of $\varepsilon_{TiO_2} \cdot N_{D\,TiO_2}$, the product $\varepsilon_{TiO_2-BCN} \cdot N_{D\,TiO_2-BCN}$ vs. the frequency can be obtained, as shown in Figure 7d. This factor also follows an exponential decay tendency, similar to that obtained for TiO$_2$ [37] in the region of 100–1000 Hz, ascribed to the $\varepsilon_{TiO_2\,BCN}$ as the $N_{D\,TiO_2-BCN}$ is expected to be independent of the frequency.

To summarize our results, the photo-electrochemical response characterizations suggest that the BCN layer acts as an efficient electrocatalyst, improving electron transfer between the electrolyte and the underlying TiO$_2$ electrode. It has been reported before that these good electrocatalytic properties are related to the heterogeneity of these compounds, formed by highly doped C-rich and h-BN domains. Substitutional doping and grain boundaries defects act as electrocatalytic sites for OER [15]. The improvement of the electrocatalytic properties of the samples is also confirmed by EIS measurements, that show an increase in the product of the donor density and the dielectric constant induced by BCN growth. On the other hand, it must be noticed that the BCN layer is very thin and absorbs mainly in the UV region [16], so it is not expected that it contributes to the creation of electron–hole pairs due to the appearance of new energy levels

4. Conclusions

We have successfully changed the composition and structure of TiS_3 by thermal annealing at 300 °C to obtain TiO_2 without changing the nanoribbon morphology of the samples. The XRD and Raman measurements reveal a change from TiS_3 monoclinic to anatase tetragonal TiO_2, and the diffuse reflectance measurements a change in the gap from 0.94 ± 0.04 eV to 3.3 ± 0.4 eV. On top of the TiO_2 nanoribbons, we have successfully grown a BCN layer by plasma-enhanced CVD without affecting the TiO_2 structure and morphology as it has been proved by XRD, Raman and XPS techniques. The analysis of the chemical composition and bonding scheme of the BCN layer revealed that our layer is composed of C and h-BN nanodomains with high mutual doping levels of B and N in C, and of C in BN, respectively.

The heterostructure TiO_2-BCN has been measured in a photoelectrochemical cell, corroborating the good photoelectrocatalytic properties of the BCN to carry out the OER when compared with the bare TiO_2 substrate. Cycling stability tests and Raman measurements after the experiments demonstrated that the BCN layer remained on the sample and that the structure of the sample has not changed, showing the high stability of our samples. Finally, with EIS measurements, flat band potentials for TiO_2 and TiO_2-BCN have been determined to have the same value (0.2 ± 0.1 V vs RHE). By analyzing the slopes of the Mott–Schottky plots, it has been determined that the factor $\varepsilon \cdot N_D$ is higher in the case of the TiO_2-BCN sample which is in good agreement with the increase observed in the photocurrents. In conclusion, the present results point out the excellent photoelectrocatalytic properties of the BCN as a metal-free material to be used in water splitting devices.

Supplementary Materials: The following are available online at https://www.mdpi.com/article/10.3390/ma14195490/s1, Figure S1: Photograph of a sample before and after the oxidation, Figure S2: Scheme of the photoelectrochemical cell used in this work, Figure S3: Raman spectra of one of the samples with BCN. The spectra have been measured in an external nanoribbon (border) and one in the center of the sample, Figure S4: XPS measurement between 384eV and 413 eV. N1s is overlapped with a small signal of the $Ta4p_{3/2}$ peak, ascribed to the tantalum clips used to fix the sample to the sample holder, Figure S5: Photocurrent of TiO_2-BCN heterostructure under Xe lamp illumination at 0.6 V vs Ag/AgCl in 0.1 M KOH aqueous solution, Figure S6: (a) LSW at different scan rates (b) CV at different scan rates (c) Difference between the anodic and cathodic current of the CV measurements, Figure S7: To prove the stability of our electrode, series of cyclic voltammetry at 0.05mV/s have been done. This figure shows the current at the maximum potential applied (1.8V vs RHE) for each one of the cycles, for the TiO_2-BCN sample in 1.0M KOH aqueous electrolyte, Figure S8: Raman spectra of TiO_2 and TiO_2-BCN before the photoelectrochemical measurements (PEC) and TiO_2-BCN after the PEC, Figure S9: Donor density of the TiO_2 sample, Table S1: Position (binding energy, BE), full peak width at a half maximum (FWHM), and relative intensities (with and without oxygen contribution) for the TiO_2-BCN and bare TiO_2 samples, Table S2: Comparison between TiO_2 and TiO_2-BCN of the current density in dark condition at 1.85V vs RHE (I_{dark}), photocurrent at 1.85V vs RHE (I_{ph}) and flat band potential, Table S3: Fitting parameters of graphs Figure 7c,d, to an exponential decay: $y = y_0 + A \cdot \exp\left(-\frac{f}{t}\right)$ where y is the parameter of the Y-axis, and f is the frequency.

Author Contributions: Conceptualization: N.J.-A., F.L., I.J.F., E.F. and J.R.A., morphological and structural characterization: N.J.-A., E.F., F.L., I.J.F. and J.R.A. optical characterization: N.J.-A., E.F., F.L., I.J.F. and J.R.A. photoelectrochemical measurements: N.J.-A., I.J.F. and F.L. surface chemical composition and element distribution: N.J.-A., A.G., M.S., C.M. and M.G.B. writing—original draft preparation: N.J.-A. writing—review and editing: I.J.F., F.L., A.G., C.M., M.S., E.F., J.R.A. and M.G.B. All authors have read and agreed to the published version of the manuscript.

Funding: This research was funded by Spanish MICINN under RTI2018-099794-B-I00 grant.

Institutional Review Board Statement: Not applicable.

Informed Consent Statement: Not applicable.

Data Availability Statement: Not applicable.

Acknowledgments: Authors wish to thank technical assistance from F. Moreno, SIdI and Segainvex Facilities at Universidad Autónoma de Madrid. Jiménez-Arévalo, N. acknowledges Comunidad de Madrid and European Social Fund for the PEJD-2019-PRE/IND-16301 predoctoral contract, and Universidad Autónoma de Madrid and Banco Santander for the mobility grant *Ayudas UAM-Santander para la movilidad de jóvenes investigadores* to carry out the XPS measurements. Giampietri, A. and Mariani, C. thank the support by PRIN FERMAT (No, 2017KFY7XF) from Italian Ministry MIUR and by Sapienza Ateneo Funds.

Conflicts of Interest: The authors declare no conflict of interest.

References

1. Lewis, N.S.; Nocera, D.G. Powering the planet: Chemical challenges in solar energy utilization. *Proc. Natl. Acad. Sci. USA* **2006**, *103*, 15729–15735. [CrossRef]
2. Osterloh, F.E.; Parkinson, B.A. Recent developments in solar water-splitting photocatalysis. *MRS Bull.* **2011**, *36*, 17–22. [CrossRef]
3. Van der Krol, R.; Grätzel, M. *Photoelectrochemical Hydrogen Production*; Springer: New York, NY, USA, 2012.
4. Lewis, N.S. Toward cost-effective solar energy use. *Science* **2007**, *315*, 798–801. [CrossRef]
5. Fujishima, A.; Honda, K. Electrochemical Photolysis of Water at a Semiconductor Electrode. *Nature* **1972**, *238*, 37–38. [CrossRef] [PubMed]
6. Walter, M.G.; Warren, E.L.; McKone, J.R.; Boettcher, S.W.; Mi, Q.; Santori, E.A.; Lewis, N.S. Solar water splitting cells. *Chem. Rev.* **2010**, *110*, 6446–6473. [CrossRef] [PubMed]
7. Chen, X.; Shen, S.; Guo, L.; Mao, S.S. Semiconductor-based photocatalytic hydrogen generation. *Chem. Rev.* **2010**, *110*, 6503–6570. [CrossRef] [PubMed]
8. Jamesh, M.I.; Xiaoming, S. Recent progess on earth abundant electrocatalysts for oxygen evolution reaction (OER) in alkaline medium to achieve efficient water splitting- A review. *J. Power Sources* **2018**, *400*, 31–68. [CrossRef]
9. Berger, T.; Monllor-Satoca, D.; Jankulovska, M.; Lana-Villareal, T.; Gómez, R. The electrochemistry of nanostructured titanium dioxide electrodes. *ChemPhysChem* **2012**, *13*, 2824–2875. [CrossRef]
10. Cao, F.; Xiong, J.; Wu, F.; Liu, Q.; Shi, Z.; Yu, Y.; Wang, X.; Li, L. Enhanced photoelectrochemical performance from rationally designed anatase/rutile TiO_2 heterostructures. *ACS Appl. Mater. Interfaces* **2016**, *8*, 12239–12245. [CrossRef]
11. Hu, S.; Shaner, M.R.; Beardslee, J.A.; Lichterman, M.; Brunschwig, B.S.; Lewis, N.S. Amorphous TiO_2 coatings stabilize Si, GaAs, and GaP photoanodes for efficient water oxidation. *Science* **2014**, *344*, 1005–1009. [CrossRef]
12. Li, Z.; Chen, Y.; Ma, T.; Jiang, Y.; Chen, J.; Pan, H.; Sun, W. 2D Metal-Free Nanomaterials Beyond Graphene and its Analogues toward Electrocatalysis Applications. *Adv. Energy Mater.* **2021**, *11*, 2101202. [CrossRef]
13. Chhetri, M.; Maitra, S.; Chakraborty, H.; Waghmare, U.V.; Rao, C.N.R. Superior performance of borocarbonitrides, BxCyNz, as stable, low-cost metal-free electrocatalysts for the hydrogen evolution reaction. *Energy Environ. Sci.* **2016**, *9*, 95–101. [CrossRef]
14. Rao, C.N.R.; Chhetri, M. Borocarbonitrides as Metal-Free Catalysts for the Hydrogen Evolution Reaction. *Adv. Mater.* **2019**, *31*, 1–13. [CrossRef]
15. Jiménez-Arévalo, N.; Leardini, F.; Ferrer, I.J.; Ares, J.R.; Sánchez, C.; Abdelnabi, M.M.S.; Betti, M.G.; Mariani, C. Ultrathin Transparent B-C-N Layers Grown on Titanium Substrates with Excellent Electrocatalytic Activity for the Oxygen Evolution Reaction. *ACS Appl. Energy Mater.* **2020**, *3*, 1922–1932. [CrossRef]
16. Leardini, F.; Jiménez-Arévalo, N.; Ferrer, I.J.; Ares, J.R.; Molina, P.; Gómez Navarro, C.; Manzanares, Y.; Granados, D.; Urbanos, F.J.; García-García, F.J.; et al. A fast synthesis route of boron-carbon-nitrogen ultrathin layers towards highly mixed ternary B-C-N phases. *2D Mater.* **2019**, *6*, 035015. [CrossRef]
17. Ghasemi, F.; Frisenda, R.; Flores, E.; Papadopoulos, N.; Biele, R.; Perez de Lara, D.; Van der Zant, H.S.J.; Watanabe, K.; Taniguchi, T.; D'Agosta, R.; et al. Tunable Photodetectors via In Situ Thermal Conversion of TiS_3 to TiO_2. *Nanomaterials* **2020**, *10*, 711. [CrossRef] [PubMed]
18. Ferrer, I.J.; Maciá, M.D.; Carcelén, V.; Ares, J.R.; Sánchez, C. On the photoelectrochemical properties of TiS_3 films. *Energy Procedia* **2012**, *22*, 48–52. [CrossRef]
19. Ferrer, I.J.; Ares, J.R.; Clamagirand, J.M.; Barawi, M.; Sánchez, C. Optical properties of titanium trisulphide (TiS_3) thin films. *Thin Solid Films* **2013**, *535*, 398–401. [CrossRef]
20. Massimi, L.; Betti, M.G.; Caramazza, S.; Postorino, P.; Mariani, C.; Latini, A.; Leardini, F. In-Vacuum Thermolysis of ethane 1,2-diamineborane for the synthesis of ternary borocarbonitrides. *Nanotechnology* **2016**, *27*, 435601. [CrossRef]
21. Leardini, F.; Flores, E.; Ferrer, I.J.; Ares, J.R.; Sánchez, C.; Molina, P.; van der Meulen, H.P.; Navarro, C.G.; Polin, G.L.; Urbanos, F.J.; et al. Chemical vapor deposition growth of boron-carbon-nitrogen layers from methylamine borane thermolysis products. *Nanotechnology* **2018**, *29*, 025603. [CrossRef]
22. Pawbaje, A.S.; Island, J.O.; Flores, E.; Ares, J.R.; Sánchez, C.; Ferrer, I.J.; Jadkar, S.R.; van der Zant, H.S.J.; Castellanos-Gomez, A.; Late, D.J. Temperature-Dependent Raman Spectroscopy of Titanium Trisulfide (TiS_3) Nanoribbons and Nanosheets. *ACS Appl. Mater. Interfaces* **2015**, *43*, 24185–24190. [CrossRef] [PubMed]
23. Ohsaka, T. Temperature Dependence of the Raman Spectrum in Anatase TiO_2. *J. Phys. Soc. Jpn.* **1980**, *48*, 1661–1668. [CrossRef]

24. Zhang, Y.; Wu, W.; Zhang, K.; Liu, C.; Yu, A.; Peng, M.; Zhai, J. Raman Study of 2D anatase TiO$_2$ nanosheets. *Phys. Chem. Chem. Phys.* **2016**, *18*, 32178–32184. [CrossRef] [PubMed]
25. Melendres, C.A.; Narayanasamy, A.; Maroni, V.A.; Siegel, R.W. Raman Spectroscopy of nanophase TiO$_2$. *J. Mater. Res.* **1989**, *4*, 1246–1250. [CrossRef]
26. Mammone, J.F.; Sharma, S.K.; Nicol, M. Raman study of rutile (TiO$_2$) at high pressures. *Solid State Commun.* **1980**, *34*, 799–802. [CrossRef]
27. Murphy, A.B. Modified Kubelka-Munk model for calculations of the reflectance of coatings with optically-rough surfaces. *J. Phys. D Appl. Phys.* **2006**, *39*, 3571. [CrossRef]
28. Xu, B.; Sohn, H.Y.; Mohassab, Y.; Lan, Y. Structures, preparation and applications of titanium suboxides. *RSC Adv.* **2016**, *6*, 79706–79722. [CrossRef]
29. Diebold, U.; Mandey, T.E. TiO$_2$ by XPS. *Surf. Sci. Spectra* **1996**, *4*, 227–231. [CrossRef]
30. Göpel, W.; Anderson, J.A.; Frankel, D.; Jaehnig, M.; Philips, K.; Schäfer, J.A.; Rocker, G. Surface defects of TiO$_2$ (110): A combined XPS, XAES and ELS study. *Surf. Sci.* **1984**, *139*, 333–346. [CrossRef]
31. Guillemot, F.; Porté, M.C.; Labrugère, C.; Baquey, C. Ti4+ to Ti3+ conversion of TiO$_2$ uppermost layer by low-temperature vacuum annealing: Interest for titanium biomedical applications. *J. Colloid Interface Sci.* **2002**, *255*, 75–78. [CrossRef]
32. Nappini, S.; Bondino, F.; Píš, I.; Chelleri, R.; Greco, S.L.; Lazzarino, M.; Magnano, E. Chemical composition and interaction strength of two-dimensional boron-nitrogen-carbon heterostructures driven by polycrystalline metallic surfaces. *Appl. Surf. Sci.* **2019**, *479*, 903–913. [CrossRef]
33. Attri, R.; Sreedhara, M.B.; Rao, C.N. R Compositional tuning of electrical and optical properties of PLD-generated thin films of 2D borocarbonitrides (BN)1-x (C) x. *ACS Appl. Electron. Mater.* **2019**, *1*, 569–576. [CrossRef]
34. Ci, L.; Song, L.; Jin, C.; Jariwala, D.; Wu, D.; Li, Y.; Srivastava, A.; Wang, Z.F.; Storr, K.; Balicas, L.; et al. Atomic layers of hybridized boron nitride and graphene domains. *Nat. Mater.* **2010**, *9*, 430–435. [CrossRef] [PubMed]
35. Ong, C.W.; Huang, H.; Zheng, B.; Kwok, R.W.M.; Hui, Y.Y.; Lau, W.M. X-ray photoemission spectroscopy of nonmetallic materials: Electronic structures of boron and B$_x$O$_y$. *J. Appl. Phys.* **2004**, *95*, 3527. [CrossRef]
36. Schneider, J.; Matsuoka, M.; Takeuchi, M.; Zhang, J.; Horiuchi, Y.; Anpo, M.; Bahnemann, D.W. Understanding TiO$_2$ Photocatalysis: Mechanisms and materials. *Chem. Rev.* **2014**, *114*, 9919–9986. [CrossRef] [PubMed]
37. Wypych, A.; Bobowska, I.; Tracz, M.; Opasinska, A.; Kadlubowski, S.; Krzywania-Kaliszewska, A.; Grobelny, J.; Wojciechowski, P. Dielectric Properties and Characterisation of Titanium Dioxide Obtained by Different Chemistry Methods. *J. Nanomater.* **2014**, *2014*, 1–9. [CrossRef]

Article

LDH-Co-Fe-Acetate: A New Efficient Sorbent for Azoic Dye Removal and Elaboration by Hydrolysis in Polyol, Characterization, Adsorption, and Anionic Exchange of Direct Red 2 as a Model Anionic Dye

Nawal Drici-Setti [1,2,*], Paolo Lelli [3] and Noureddine Jouini [1,3,*]

1. Laboratoire des Sciences des Procédés et des Matériaux (LSPM), Centre National de Recherche Scientifique (CNRS), Université Sorbonne Paris Nord, LSPM-CNRS-UPR 3407, 99 Avenue Jean-Baptiste Clément, 93430 Villetaneuse, France
2. Laboratoire de Physico-Chimie des Matériaux, Département de Génie des Matériaux, Faculté de Chimie, Université des Sciences et de Technologie-Mohamed Boudiaf d'Oran (USTO-MB), M'Nouar 1505, Oran 31000, Algeria
3. Département Hygiène, Sécurité, Environnement, Institut Universitaire de Technologie, Université Sorbonne Paris Nord, 8 Place du 8 mai 1945, 93200 Saint-Denis, France; labochimie@yahoo.fr
* Correspondence: drici_nawel@yahoo.fr (N.D.-S.); jouini@univ-paris13.fr (N.J.); Tel.: +213-41627176 (N.D.-S.); +331-49403435 (N.J.)

Received: 31 May 2020; Accepted: 14 July 2020; Published: 16 July 2020

Abstract: A new, double hydroxide based on Co and Fe was elaborated on by forced hydrolysis in a polyol medium. Complementary characterization techniques show that this new phase belongs to the layered double hydroxide family (LDH) with Co^{2+} and Fe^{3+} ions located in the octahedral sites of the bucite-like structure. The acetate anions occupy interlayer space with an interlamellar distance of 12.70 Å. This large distance likely facilitates the exchange reaction. Acetates were exchanged by carbonates. The as-obtained compound Co-Fe-Ac$_{Ex}$ shows an interlamellar distance of 7.67 Å. The adsorption of direct red 2 by Co-Fe-Ac-LDH has been examined in order to measure the capability of this new LDH to eliminate highly toxic azoic anionic dyes from waste water and was compared with that of Co-Fe-Ac$_{Ex}$ and Co-Fe-CO$_3$/A (synthesized in an aqueous medium). The adsorption capacity was found to depend on contact time, pH, initial dye concentration, and heating temperature. Concerning CoFeAc-LDH, the dye uptake reaches a high level (588 mg/g) due to the occurrence of both adsorption processes: physisorption on the external surface and chemical sorption due to the intercalation of dye by exchange with an acetate anion. The study enables us to quantify the uptake amount of each effect in which the intercalation has the most important amount (418 mg/g).

Keywords: layered double hydroxide; nanomaterials; forced hydrolysis; polyol; dye removal; adsorption; anionic exchange; intercalation

1. Introduction

Layered double hydroxides (LDHs) also called hydrotalcite-like compounds are brucite-like layered materials, which consist of positively charged hydroxide layers and interlayered anions with a general formula $[M^{II}_{1-x}M^{III}_x(OH)_2]^{x+}·(A^{n-})_{x/n}·mH_2O$, where M^{II} and M^{III} are divalent and trivalent cations, respectively. A^{n-} is an exchangeable anion located as water molecules in the interlayer space, and x is the molar ratio, $M^{III}/(M^{II}+M^{III})$, which determines layer charge density [1]. These compounds have been the subject of great attention for several decades because of their use in a large number of fields ranging from the delivery of drugs to the protection against corrosion of metallurgical parts [2–5].

These performances have their origin in the very diverse chemical composition of these materials. The cationic layers accommodate a large number of metal cations while the interlayer space can accommodate a wide variety of inorganic or organic anions or even macromolecules. Besides this richness in chemical composition, LDHs exhibit interesting microstructural and structural characteristics. They present a relatively open structure due to its bi-dimensional character. Thus, the interlayer space serves as a micro-reactor where targeted anion exchange reactions can be carried out depending on the desired application (exchange reaction, exfoliation, lamination, and reconstruction.). Additionally, they can present a highly specific surface depending on the synthesis method [6–9].

In this context and thanks to their specific morphological properties and their high potential ability as ion exchangers with organic and inorganic anions, layered double hydroxides (LDHs) find their applications as sorbent agents for depolluting waste water and especially for eliminating anionic dyes, which are recognized to be a threat toward humans and ecosystems because of their high potential toxicity [10].

A review of the literature shows that the most widely used method to synthesise these materials is the coprecipitation in an aqueous medium of the hydroxides at room temperature by varying the pH of the solution. In the majority of cases, this coprecipitation is carried out in air and in the presence of sodium carbonate. In this case and whatever the nature of the starting salts, the obtained phase is a layered double hydroxide with carbonate anion intercalated between the sheets (LDH-CO_3). This preferential intercalation is due to the great affinity of this anion for positively charged hydroxide sheets [11]. These carbonated LDHs display weak adsorption capacities. The carbonate ion is difficult to displace by an exchange reaction and, therefore, the adsorption remains limited to the surface [12].

To overcome this difficulty, one proceeds to the calcination of these materials at a moderate temperature between 400–500 °C in air. The oxides obtained are finely divided and show a memory effect. In fact, when put back into an aqueous solution, these oxides transform into LDH by intercalating the anions present in the solution. This intercalation in addition to the surface adsorption of these anions significantly improves the adsorption capacities compared to those of LDH-carbonates [13].

Another strategy consists of the synthesis of layered double hydroxides by coprecipitation in an aqueous medium where more easily exchangeable anions are intercalated such as chloride [14], nitrate [15,16], chlorate [17], dodecylsulfate [18], and sulfate [19]. To avoid the presence of CO_2, reactions must be carried out under nitrogen, and distilled deionized water must be freshly decarbonated by vigorously boiling prior to use.

Since the nature of the intercalated anion plays an important role in adsorption capacity, we focus in this work on LDH with acetate (Ac) as an intercalated anion. To the best of our knowledge, this anion has never been used for this purpose. Furthermore, this new layered double hydroxide containing Co^{2+} and Fe^{3+} cations in the brucite-like layers (CoFe-Ac/$_p$) was produced by forced hydrolysis in a polyol medium. The advantages of this synthesis method will be discussed in light of the obtained results. The choice of the Co and Fe elements falls within the framework of our project to synthesize multifunctional LDHs based on the third transition element (Ni, Co, Fe). Like their counterparts based on Mg-Al, they present anion exchange properties, which are the basis of the work presented in this case. In addition, the presence of paramagnetic elements with several valences (Ni, Co, Fe) opens up other various application fields. These LDH are of great interest in the energy field thanks to their electrocatalytic activity for water oxidation [20]. They also serve as precursors for obtaining finely divided particles of magnetic oxides or metal alloys sought for several applications: air depollution [21], drug delivery [22], and magnetic recording [23].

To enrich and supplement this work, we investigated the anionic exchange properties of this newly synthesised LDH, along with two LDHs where carbonate anion is intercalated. The first was derived from the precursor CoFe-Ac/$_p$ by exchanging acetate with carbonate anions (CoFe-Ac/$_{Ex}$) and the second (CoFe-CO_3/$_A$) synthesised by a standard coprecipitation in an aqueous medium.

Direct red 2 was chosen as a model for the toxic azoic anionic dyes. Its adsorptive capability by sorbents is seldom reported in the literature [24,25]. The adsorption onto CoFe-Ac/$_p$, CoFe-Ac/$_{Ex}$, and CoFe-CO_3/$_A$ was studied as a function of various factors such as contact time, pH, initial dye concentration, and temperature. The results were discussed and compared to those previously reported for the adsorption of direct red 2 and similar dyes on LDH materials.

2. Materials and Methods

2.1. Materials

For all preparative procedures, Co(CH$_3$COO)$_2$·4H$_2$O, CoCl$_2$·4H$_2$O, Fe(CH$_3$COO)$_2$, FeCl$_3$, NaOH, Na$_2$CO$_3$, and diethylene glycol (DEG) were purchased from Acros and used without any further purification. Direct red 2 dye (purity > 99%) was provided by sigma-Aldrich and used as received. Its molecular weight is 724.73 g/mol. Figure 1 represents the estimated dimensions of the molecule obtained using Avogadro software [26].

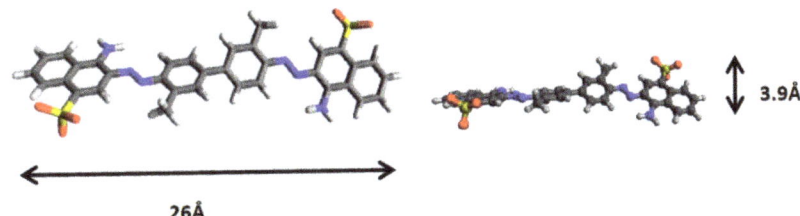

Figure 1. Estimated dimensions of direct red 2 anionic dye.

The initial pH value of direct red 2 solution is about 6.6. A series of direct red 2 solutions of desired pH values was obtained by adjusting with dilute HCl or NaOH solutions.

Synthesis of LDH Samples

(a) CoFe-Ac/$_p$

The CoFe-Ac LDH where acetate anion is intercalated was synthesized with a molar ratio (Co/Fe) of the three following, previously described methods based on a forced hydrolysis reaction in a polyol medium [27]. Accordingly, a mixture of acetate salts dissolved in DEG with a total molar concentration of 0.1 mol/L that is heated at 130 °C under continuous stirring for 6 h. The corresponding LDH precipitated when the hydrolysis and alkalinity ratios h and b were fixed at 100 and 2, respectively, where h = nH$_2$O/n (Co+Fe) and b = nNaOH/n (Co + Fe). The solid formed is separated by centrifugation. Then it is washed several times with ethanol, dried under air at 60 °C, and named CoFe-Ac/$_p$. As it will be shown below by Mossbauer spectroscopy, the Fe^{2+} present in the precursor has been oxidized to Fe^{3+} in the polyol medium despite the reducing nature of this solvent. This oxidation is due to the presence of a large amount of water, which inhibits reduction, promotes oxidation, and the formation of hydroxides or oxides in addition to the easy oxidation of ferrous ions. The valence of Co^{2+} is preserved in these conditions [28,29].

(b) CoFe-Ac/$_{Ex}$

The anion exchange properties of CoFe-Ac/$_p$ with carbonate anions were investigated by mixing 1 g of the synthesized CoFe-Ac/$_p$ LDH with 100 mL of a 2 M Na$_2$CO$_3$ solution. Despite the fact that the Co^{2+} in the cationic layers appears stable with respect to oxidation (see UV-Vis-NIR analysis), the exchange has been carried out as a precaution in an inert atmosphere. After equilibrating for 24 h at room temperature, the solid was separated by centrifugation, washed several times with ethanol, dried under air at 60 °C, and then named CoFe-Ac/$_{Ex}$.

(c) CoFe-CO$_3$/$_A$

The LDH intercalated with carbonate anions (CoFe-CO$_3$) was prepared by coprecipitation in an aqueous medium [30]. An acid solution of CoCl$_2$·4H$_2$O and FeCl$_3$, with a Co^{2+}/Fe^{3+} molar ratio R = 3 and a total concentration of metallic cations of 0.75 mol/L, was added drop-by-drop to a vigorously stirred alkaline solution of NaOH (1 M) and Na$_2$CO$_3$ (2 M) in an inert atmosphere in order to avoid the oxidation of Co^{2+} into Co^{3+}. The pH of the reaction mixture was adjusted to 10. The resulting slurry was aged at 70 °C for 24 h, separated by centrifugation, and washed extensively using distilled water until the supernatant was chloride-free, as indicated by the AgNO$_3$ test. The product was dried at 60 °C under air and ground in an agate mortar. The obtained material is called CoFe-CO$_3$/$_A$.

2.2. Methods

2.2.1. Characterization

The radio crystallographic characterizations and the identification of the phases were achieved using a diffractometer (INEL, Artenay, France) with Co-Kα1 radiation (λCoα1 = 1.7889 Å). The microstructure was studied by means of observations carried out in scanning electron microscopy (SEM) using LEICA STEREOSCAN 440 instrument (Cambridge, UK) and Transmission electron microscopy (TEM) performed on a JEOL-100 CX II microscope (Tokyo, Japan). The infrared spectroscopy study was carried out by transmission on a PERKIN ELMER 1750 spectrometer (Watham, MA, USA) on pressed KBr pellets with 4 cm^{-1} resolution between 400 and 4000 cm^{-1}. The stoichiometry of the final product was determined by inductively coupled plasma analysis (ICP, Agilent, Santa Clara, CA, USA) at central analysis service of the national centre for scientific research (Solaize, France). The thermal stability has been specified using a Setaram TG 92-12 thermal analyser (Caluire, France). Few milligrams are introduced in an alumina crucible and submitted to thermal analysis at a heating rate of 1 °C/min under argon. A specific surface area was measured in a Micromeritics Tristar 3000 (Norcoss, GA, USA) by nitrogen adsorption N$_2$ (77 K) after degassing the sample in vacuum by flowing nitrogen overnight at 100 °C.

UV-Vis-NIR (Ultra-Violet-Visible-Near Infrared) spectrum was recorded between 200 and 1800 nm with a Cary 5/Varian spectrometer (Agilent, Santa Clara, CA, USA). Polytetrafluoroethylene (PTFE) was used as a reference.

The ^{57}Fe Mössbauer spectroscopic study was carried out in a transmission mode, at room temperature, using a ^{57}Co/Rh γ-ray source and a conventional Mössbauer spectrometer. The spectrum was fitted by the least-squares method with a lorentzian function. The isomer shift (δ) was measured at 300 K relative α-Fe shift used as a reference. The sample (area: 3 cm^2) is constituted of 40 mg of the studied compound dispersed in a specific resin.

2.2.2. Adsorption Experiments

The adsorption experiments were carried out by a batch method in an open medium and at room temperature. A well-known mass of adsorbent (CoFe-Ac/$_p$, CoFe-Ac/$_{Ex}$, and CoFe-CO$_3$/$_A$ LDHs) is added in a 100 mL conical beaker to a volume of 50 mL of dye solution with a variable concentration whose temperature and pH are measured beforehand. The initial pH values and temperature were not adjusted except when the effect of these two parameters on adsorption was investigated. The adsorbent was left in contact with the dye solution for several durations. In each case, the adsorbent was then separated by centrifugation. An UV-Visible spectrophotometer (SAFAS, Monaco) was used to monitor the dye removal by measuring its remaining concentration in the solution. The measurements were made at a wavelength of λ_{max} = 500 nm, which corresponds to the maximum absorbance of the direct red 2.

The adsorption capacity q_e (mg/g), which represents the amount of adsorbed dye per amount of dry adsorbent, was calculated using the following equation.

$$q_e = (C_i - C_e) \times \frac{V}{m} \qquad (1)$$

where C_i and C_e are the initial and equilibrium concentrations (mg/L) of dye, respectively, and m is the mass of adsorbent (g) and V is the solution volume (L).

2.2.3. Theory and Modelling

Kinetic Study

In order to investigate the mechanism of adsorption and to fit the kinetics experimental data, three kinetics models were used and analysed including pseudo-first-order Equation (2) [31], pseudo-second-order Equation (3) [32], and Weber's intraparticle diffusion Equation (4) [33].

$$\text{Log}(q_e - q_t) = \text{Log} q_e - \frac{k_1 \times t}{2.303} \qquad (2)$$

$$\frac{t}{q_t} = \frac{1}{k_2 \times q_e^2} + \left(\frac{1}{q_e}\right) \times t \qquad (3)$$

$$q_t = k_p \times t^{1/2} + C \qquad (4)$$

where q_e and q_t are, respectively, the amount of dye adsorbed (mg/g) at equilibrium and at any time t, k_1 is the rate constant of pseudo-first-order adsorption (min^{-1}), values of k_1 and q_e are determined from the plot of Log $(q_e - q_t) = f(t)$, k_2 is the rate constant of pseudo-second-order adsorption (g/mg.min). The equilibrium adsorption capacity (q_e) and the pseudo-second-order constant k_2 are determined from the slope and intercept of plot of $t/q_t = f(t)$, k_p (mg·g^{-1}mn$^{-0.5}$) is the intraparticle diffusion rate constant, and C (mg·g^{-1}) represents the effect of boundary layer thickness, k_p and C can be evaluated, respectively, from the slope and the intercept of the linear plot $q_t = f(t^{1/2})$.

Isotherm Study

To investigate the nature of the interaction of direct red 2 anions and the LDHs adsorbents, two models were selected and used in order to match the experimental data, namely Langmuir and Freundlich isotherms.

(a) Langmuir Isotherm

In the Langmuir isotherm, it is assumed that the maximum adsorption is limited to a monolayer of molecules distributed homogeneously over the entire surface and without interactions between them [34]. It is given by the following linear equation.

$$\frac{C_e}{q_e} = \frac{1}{Q_{max} \times K_L} + \frac{C_e}{Q_{max}} \qquad (5)$$

where K_L is the equilibrium adsorption coefficient (L mg^{-1}), Q_{max} is the maximum adsorption capacity (mg/g), C_e is the equilibrium concentration (mg L^{-1}), and q_e is the adsorbed amount at equilibrium (mg/g). K_L and Q_{max} values were calculated from the slope and intercept of the plot of $C_e/q_e = f(C_e)$.

(b) Freundlich Isotherm

The Freundlich model is based on an empirical equation, which considers that the sorption occurred on a surface where the active sites have heterogeneous energetic distribution. Additionally, it supposes multilayer adsorption with interactions between the adsorbed molecules [34,35]. It is represented by the following linear equation.

$$\ln q_e = \ln K_f + \frac{1}{n} \times \ln C_e \quad (6)$$

C_e is the equilibrium concentration (mg·L^{-1}), q_e is the adsorbed amount at equilibrium (mg·g^{-1}), and K_f and $1/n$ are the Freundlich constants. The constant n is related to the energy and the intensity of adsorption and K_f indicates the adsorption capacity (mg g^{-1}). K_f and $1/n$ values were inferred from the slope and intercept of the plot of Ln q_e = f (ln C_e).

Thermodynamic Parameters

To better understand the temperature effect on the adsorption, the thermodynamic parameters such as Gibbs-free energy change $\Delta G°$, standard enthalpy $\Delta H°$, and standard entropy $\Delta S°$ were studied. They were obtained from experiments at various temperatures using the following equations [36].

$$\Delta G° = -R \times T \times \ln k_L \quad (7)$$

$$\ln k_L = \left(-\frac{\Delta H°}{R}\right) \times \frac{1}{T} + \left(\frac{\Delta S°}{R}\right) \quad (8)$$

$$\Delta G° = \Delta H° - T \times \Delta S° \quad (9)$$

where k_L is the equilibrium adsorption constant, R is the molar gas constant, and T is the absolute temperature. $\Delta H°$ and $\Delta S°$ thermodynamic parameters were calculated from the values of the slopes and the intercepts of Van't Hoff plots of $\ln k_L$ = f (1/T).

3. Results

3.1. Characterization of Adsorbents

3.1.1. X-ray Diffraction

The X-ray diffraction patterns of all the as-prepared samples are displayed in Figure 2. They exhibit the typical signature of a crystalline layered double hydroxide belonging to the space group R-3m (JCPDS card 25-0521) [37], and display the principal characteristic reflections of LDHs by the presence of three sharp symmetrical peaks (00l) at low 2θ angle and a weak asymmetrical peaks (hk0) at high 2θ angle, which indicates a turbo-static disorder of the layers' stacking [38].

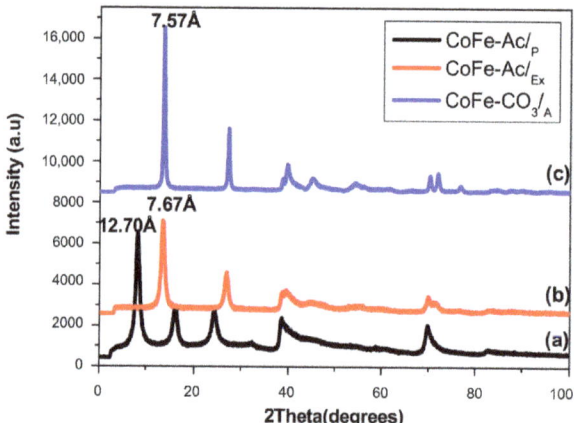

Figure 2. X-ray diffractogram of (a) CoFe-Ac/p, (b) CoFe-Ac/Ex, and (c) CoFe-CO3/A LDHs.

The acetate anion was successfully intercalated into the interlayer gallery of CoFe-Ac/p (Figure 2a), which gave an interlayer spacing of 12.70 Å. This value is very close to that of NiFe-Ac [39].

According to Figure 2b, it is clear that, after anion exchange by carbonate anions, the material preserved its lamellar structure and it showed the disappearance of the peak due to the acetate anions (d_{003} = 12.70 Å) and the appearance of another peak at high 2θ value attributable to the presence of intercalated carbonate anions (d_{003} = 7.67 Å). This value is in keeping with the results for the reference phase CoFe-CO3/A (d_{003} = 7.57 Å) (Figure 2c) and those reported in the literature [1].

According to the lattice parameters of CoFe-Ac/p, CoFe-Ac/Ex, and CoFe-CO3/A LDHs summarized in Table 1, we can say that the interlayer distance depends mainly only on the nature of the inserted anion without any change in the value of the parameter a = $2d_{110}$, which depends on the size of the metal cations.

Table 1. Evolution of the lattice parameters for CoFe-Ac/Ex compared to CoFe-Ac/p and CoFe-CO3/A (reference) LDHs.

Compound	d_{003} (Å)	a Parameter (Å)
CoFe-Ac/p	12.70	3.11
CoFe-Ac/Ex	7.67	3.12
CoFe-CO3/A (reference)	7.57	3.12

3.1.2. Morphology

The samples CoFe-Ac/p and CoFe-Ac/Ex have similar morphologies. They appear formed of an aggregation of very fine platelets (Figures S1 and S2) (in Supplementary Materials). TEM images reveal that these platelets approach a hexagonal shape characteristic of LDH compounds and have a diameter around 50 nm. The sample CoFe-CO3/A prepared by coprecipitation in an aqueous medium has a different morphology, which is also observed in the case of LDH (Figures S1c and S2c) [40]. The particles are in rounded form with a micrometric size. Aggregations of these particles are also observed. TEM observations show that these particles are microporous and made up of fine particles in the nanometer size.

3.1.3. Spectroscopy Studies

The infrared spectra for all synthesized compounds (Figure 3) show the characteristic absorption bands of layered double hydroxide compounds [41]. Particularly, the large band located at high frequency (3400–3500 cm^{-1}) can be assigned to the OH stretching of water molecules and hydroxyl groups of the brucitic layers. The absorption band around 1640 cm^{-1} is assigned to δH_2O vibration of the water molecules and bands at a lower wave level (ν < 800 cm^{-1}) are due to vibrations implying M-O, M-O-M, and O-M-O bonds in the layer.

As shown in Figure 3a, the intercalation of acetate anions in CoFe-Ac/p LDH is confirmed by the presence of two large and intense bands in 1600–1000 cm^{-1} domain, assigned to ν_{as}(-COO-) = 1561 cm^{-1} and ν_s (-COO-) = 1410 cm^{-1} respectively. Additionally, it is confirmed by the band with low intensity appearing at 1344 cm^{-1} that is ascribed to the vibration of the CH3 group (δCH_3) of this anion. The overall spectrum in the range 1990–1300 cm^{-1} is very similar to that reported for the layered hydroxy-acetate $Ni_{1-x}Zn_{2x}(OH)_2(CH_3COO)_{2x} \cdot nH_2O$ [42]. The value of $\Delta\nu$ = ($\nu_{as}-\nu_s$) was in the range 175 ≤ $\Delta\nu$ ≤ 151 cm^{-1}, which indicates that the acetate anions are intercalated as free species in between the layers [43]. A careful examination of the CoFe-Ac/p LDH spectrum (Figure 3a) shows that the band close to 1360 cm^{-1} is absent, which suggests that the carbonate anion is not intercalated even if the contamination by this anion coming from air cannot be excluded.

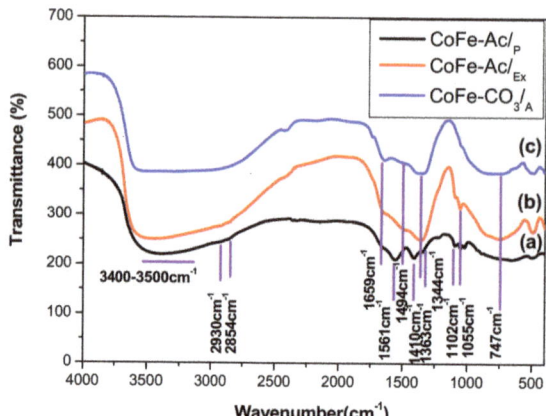

Figure 3. FT-IR spectra for a CoFe-Ac/p, (b) CoFe-Ac/Ex, (c) CoFe-CO3/A LDHs.

In the same way, the comparison of the infrared spectrum for CoFe-Ac/Ex (Figure 3b) with the spectra for CoFe-Ac/p (Figure 3a) and CoFe-CO3/A LDHs (Figure 3c) shows the complete exchange of the acetate by the carbonate anions, which confirms the results from the XRD. The two characteristic bands of the acetate ion disappear, leaving two new bands at 1494 cm^{-1} and 1363 cm^{-1} characteristic of the carbonate ion [42].

Lastly, for both CoFe-Ac/p and CoFe-Ac/Ex compounds, the presence of absorption bands at low intensity located at approximately 2930 and 2854 cm^{-1} along with two weak absorption bands are located between 1300 and 1000 cm^{-1}. They are attributed to the presence of the DEG solvent adsorbed on the surface of LDHs [39].

The UV-Vis-NIR spectrum of CoFe-Ac/p is shown in Figure S3. We note the presence of the absorption bands located at around 1200 to 500 nm, which are attributed to a high spin transition of octahedral coordinated Co^{2+} [44]:

$$\nu_1\ (Co^{2+}{}_{Oh}){:}^4T_{1g}(F) \rightarrow {}^4T_{2g}(F)$$

$$\nu_2\ (Co^{2+}{}_{Oh}){:}^4T_{1g}(F) \rightarrow {}^4A_{2g}(F)$$

$$\text{and } \nu_3\ (Co^{2+}{}_{Oh}){:}^4T_{1g}(F) \rightarrow {}^4T_{1g}(P)$$

The absorption band at around 700 nm is absent, which corresponds to Co^{3+} ions in octahedral sites. Similarly, the results show the absence of Co^{2+} at tetrahedral sites since the characteristic bands located around 1567 nm for d-d transitions are absent:

$$\nu_2\ (Co^{2+}{}_{Td}){:}\ {}^4A_2(F) \rightarrow {}^4T_1(F),\ \nu_2\ (Co^{2+}{}_{Td}){:}\ {}^4A_2(F) \rightarrow {}^4T_1(F)$$

and, at around 592 and 641 nm, for the transition [45]: $\nu_3\ (Co^{2+}{}_{Td}{:}\ {}^4A_2(F)) \rightarrow {}^4T_1(P)$

The UV-Vis-NIR study was conducted six months after the synthesis of the compound was kept for this period in air. It clearly shows that the compound is stable with respect to oxidation, as shown by the absence of Co^{3+}.

Figure S4 shows the Mössbauer spectrum for the CoFe-Ac/p sample at room temperature in order to probe the local magnetic environment around the Fe sites and to determine the oxidation state of iron in the LDH matrix. The spectrum shows only one doublet with an isomer shift $\delta(Fe^{3+})$ = 0.339 mm/s and quadrupole splitting $\Delta(Fe^{3+})$ = 0.44 mm/s, which indicates an Fe^{3+} nature of the Fe atom, and no evidence of the Fe^{2+} signal was found. The small value of quadrupole splitting (Δ) corresponds to high spin Fe^{3+} ions in octahedral sites [46].

3.1.4. Thermal Analysis

The TG/DTA (Thermogravimetry/Differential Thermal Analysis) for CoFe-Ac/$_p$ LDH is presented in Figure 4. The TG diagram is characteristic of LDH thermal behaviour showing three weight losses [47]. Upon heating, we observe a weight loss of 16% due to the departure of adsorbed and interlayer water molecules. This departure is characterized by two endothermic peaks at 97 °C and 168 °C. The de-hydroxylation of the brucite-like layers is characterized by the endothermic peak at 262 °C and corresponds to the second weight loss of 11.5%. Lastly, the third step occurred in a large range of temperature (342–700 °C). The corresponding total loss (28.5%) actually corresponds to the loss of acetate ions (10%) along with that adsorbed DEG molecules (7%) and also corresponds to the release of oxygen (11%) due to the reduction of cobalt ions. X-ray diffraction analysis shows that the residue is a mixture of iron oxide and metallic cobalt.

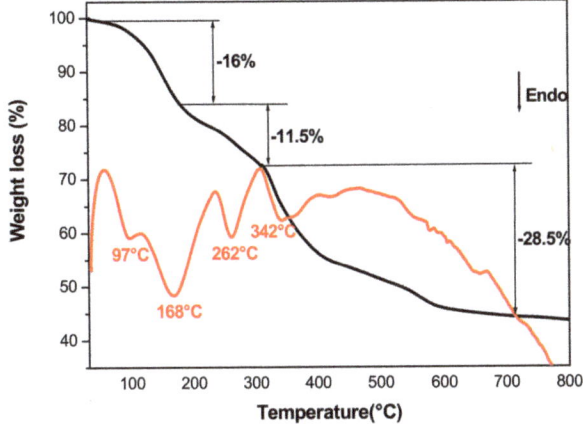

Figure 4. TG/DTA curves of CoFe-Ac/$_p$ LDH.

3.1.5. Chemical Analysis

Table 2 summarizes the results of elemental chemical analysis of the CoFe-Ac/$_p$ compound and its corresponding formula. The calculated Co^{II}/Fe^{III} ratio aligns with the experimental data. The observed carbon content is slightly higher than that corresponding to the acetate anions. This confirms the results of the infrared study showing the presence of DEG molecules likely adsorbed on the surface of the particles. This appears as a common feature of inorganic compounds prepared in a polyol medium [48]. Thus, we have taken into account the amount of adsorbed polyol in the chemical formula in order to calculate the theoretical weight loss, which matches that observed by TG analysis (Table 2).

Table 2. Elemental chemical analysis data of CoFe-Ac/$_p$ LDH.

Compound	Mass Fraction (%)				Molar Ratio Co^{II}/Fe^{III}		X = Fe^{III}/Co^{II} + Fe^{III}	
	Co	Fe	C	H	Solution	Solid	Solution	Solid
CoFe-Ac/$_p$	28.68	8.96	6.80	4.34	3.0	3.0	0.25	0.25

Chemical formula	DEG	H$_2$O% Exp.	Total weight loss % Exp.	Cal.
Co$_{0.75}$Fe$_{0.25}$·(OH)$_2$·Ac$_{0.25}$,1.50H$_2$O, 0.094DEG	0.094	16.0	56.0	54.4

3.1.6. Surface Area Measurements

The nitrogen adsorption-desorption isotherms for the CoFe-Ac/$_p$ and CoFe-Ac/ex phases are shown in Figure S5 and the results of the surface area measurement are included in Table S1. The two phases present almost similar textural characteristics. Their surface areas are very close at 48 and 50 m²/g, respectively, and their isotherms match those recorded for a layered double hydroxide type [49]. The shape of the isotherms is type IV, which indicates a predominant mesoporous character with pores measuring between 2 and 50 nm [50].

3.2. Adsorption Study

3.2.1. Effect of Contact Time

The effect of contact time on the removal of direct red 2 by all LDHs adsorbents illustrated in Figure 5 was studied at room temperature at a pH equal to 6.6 and with initial dye concentration of 1 g/L. It is clear that the fastest kinetics is observed for CoFe-Ac/$_p$ LDH where the shape of the curve indicates fast diffusion. In this case, the highest quantity of dye (94%) is already fixed at approximately 5 min, which is followed by a slower stage until reaching the total removal. This corresponds to an equilibrium time of 15 min. For the two other materials, the adsorption was slower and their equilibrium time was reached at about 180 and 240 min with regard to CoFe-Ac/$_{Ex}$ and CoFe-CO$_3$/$_A$, respectively.

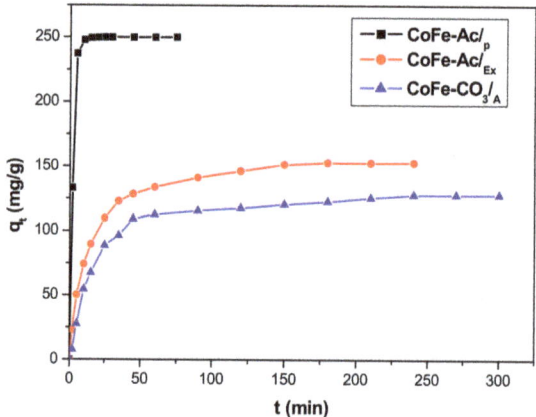

Figure 5. Adsorption kinetics of direct red 2 on CoFe-Ac/$_p$, CoFe-Ac/$_{Ex}$, and CoFe-CO$_3$/$_A$ LDHs. (C = 1 g/L, pH = 6.6, T = 20 °C).

3.2.2. Kinetic Modelling

The calculated kinetic parameters for direct red 2 adsorbed by CoFe-Ac/$_p$, CoFe-Ac/$_{Ex}$, and CoFe-CO$_3$/$_A$ are reported in Table 3. Experimental results fit the first order kinetic model with regression coefficient values of 0.9599, 0.9652, and 0.9128 for CoFe-Ac/$_p$, CoFe-Ac/$_{Ex}$, and CoFe-CO$_3$/$_A$, respectively. Moreover, large differences between experimental and calculated values of the equilibrium adsorption capacities are observed. On the contrary, the second-order-kinetic model curves (Figure S6) show much better correlation coefficients (value higher than 0.99) in all cases and the calculated q_e is much closer to the experimental values. The second-order-kinetic model seems to be the more appropriate one to describe the adsorption of anionic dye on LDHs [51].

As shown in Table 3, k_2 (pseudo-second-order constant) is inversely related to the equilibrium time. A high value of k_2 implies a shorter kinetic for the same concentration and that the limiting adsorption step is a chemical interaction adsorbate-adsorbent [51,52]. This is the case of adsorption of direct red 2 on CoFe-Ac/$_p$. None of the two kinetic models could identify the diffusion mechanism. Thus, the intra-particle diffusion model was further investigated.

Table 3. Kinetic parameters for the adsorption of direct red 2 onto LDH samples.

	Method/Adsorbent			
Pseudo-First-Order	q_{exp} (mg/g)	k_1 (min^{-1})	q_e (mg/g)	R^2
CoFe-Ac/$_P$	250	0.494	234.54	0.9599
CoFe-Ac/$_{Ex}$	153	0.026	101.24	0.9652
CoFe-CO$_3$/$_A$	128	0.016	70.24	0.9128
Pseudo-Second-Order	q_e (mg/g)	$k_2 \cdot 10^3$ (g/mg·min)	R^2	-
CoFe-Ac/$_P$	250	8.42	0.9996	-
CoFe-Ac/$_{Ex}$	161.3	0.529	0.9999	-
CoFe-CO$_3$/$_A$	136.98	0.426	0.9978	-
Intraparticule Diffusion	k_p (mg/g·mn$^{1/2}$)	C (mg/g)	R^2	-
CoFe-Ac/$_P$	5.4882	227.25	0.8108	-
CoFe-Ac/$_{Ex}$	3.2649	107.65	0.9271	-
CoFe-CO$_3$/$_A$	1.7811	98.875	0.9768	-

Figure 6 shows the variation of q_t versus $t^{1/2}$. Even if the values R^2 vary between 0.8108 and 0.9768 for all systems and remain lower than those found for the pseudo-second order model (Table 3), this model brings interesting insight into the adsorption mechanism occurring.

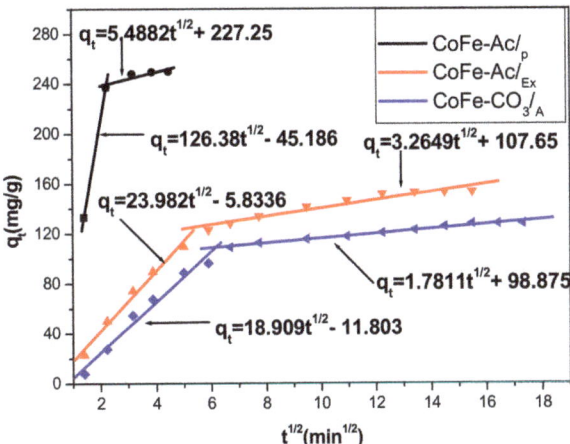

Figure 6. Intraparticle diffusion plot for adsorption of direct red 2 on CoFeAc/$_P$, CoFe-Ac/$_{Ex}$, and CoFe-CO$_3$/$_A$ LDHs.

The curves are composed of two linear segments indicating the presence of two steps and confirming that the intraparticle diffusion is not the only process implied for these systems since a simple straight line would be observed if this was the case.

According to several studies, the first segment corresponds to the instantaneous step that controls the diffusion at the external surface. When the external surface saturates, the dye diffuses inside the particle and allows the ion exchange to occur in between the layers as will be observed in the discussion section. This progressive intra-particle diffusion process is presented by the second segment [19,52].

From the same curves, it is clear that the lines do not pass through the origin. That can be due to the difference in the rate of mass transfer during the initial and final stages of adsorption. It also states that the intraparticle diffusion is not the only stage limiting adsorption and that a surface adsorption can occur simultaneously [53].

The k_p values indicate (Table 3) that the intraparticle diffusion is affected by the type of material used for the adsorption. We notice that the diffusion in the case of CoFe-Ac/$_P$ is most significant

(k_p = 5.4882 mg/g·mn$^{1/2}$). That is likely due to the appearance of a significant porous surface for this material along with the ease of access to the interlamellar space, as will be shown in the discussion section. The high values of C can be due to a great contribution of surface adsorption on the phenomenon of the intraparticle diffusion [54].

3.2.3. Effect of pH

The initial pH of the dye solution is considered an important parameter, which controls the adsorption at water-adsorbent interfaces, it affects the surface charge of adsorbent as well as the degree of ionization of pollutants. Therefore, the adsorption of direct red 2 on the different LDHs adsorbents was examined at different concentrations and different pH values ranging from 7 to 10 at 20 °C. As shown in Figure 7, it was observed that the adsorption capacity of direct red 2 on CoFe-Ac/$_p$ is practically constant and it was nearly independent of pH in the pH range of 7–9 (Figure 7a). Similar results were obtained for the adsorption of the indigo carmine dye by calcined hydrotalcite MgAlCO$_3$ [55].

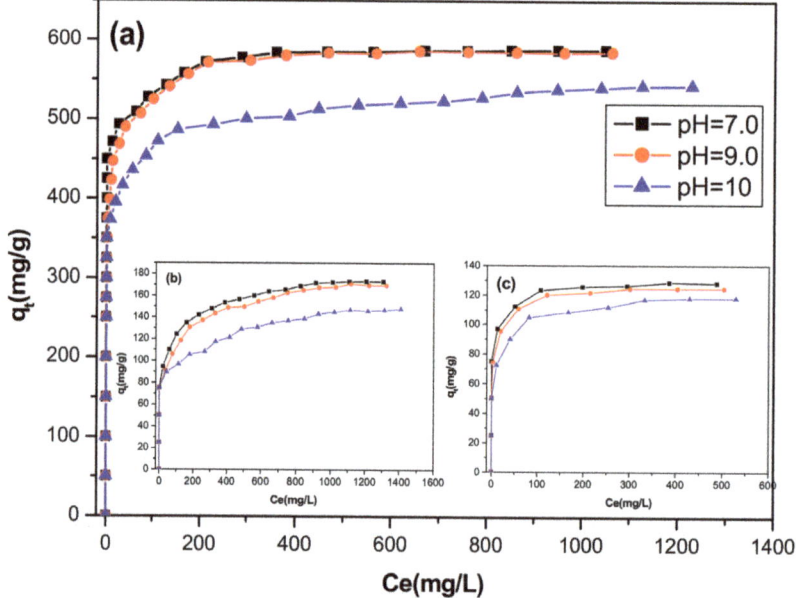

Figure 7. Effect of initial pH on adsorption isotherms of direct red 2 on: (**a**) CoFe-Ac/$_p$, (**b**) CoFe-Ac/$_{Ex}$, and (**c**) CoFe-CO$_3$/$_A$ LDHs.

However, a remarkable diminution in dye adsorption on all materials occurred when the pH was greater than 9. This diminution was greater at a high concentration of dye solutions.

For CoFe-Ac /$_{Ex}$ and CoFe-CO$_3$/$_A$ LDHs (Figure 7b,c), a decrease in the adsorption capacity is observed when the pH increases from 7 to 10. This may be due to the presence of competitive adsorption of OH$^-$ and CO$_3$$^{2-}$ with coloured ions. These anions have a very great affinity for LDHs [56]. Similar results were found for adsorption of fluoride on calcined hydrotalcite [57] and for adsorption of selenite on calcined, layered double hydroxide MgFe-CO$_3$ [51].

3.2.4. Effect of Temperature and Thermodynamic Study

The temperature is a very significant parameter for the adsorption process. For this reason, the adsorption of direct red 2 on CoFe-Ac/$_P$, CoFe-Ac/$_{Ex}$, and CoFe-CO$_3$/$_A$ LDHs was studied by carrying out a series of isotherms at 283, 293, and 323 K.

As shown in Figure 8, it is clear that sorption capacities of dye onto all adsorbents increased with an increasing temperature from 283 to 323 K at a high concentration of solution dye, which implies that adsorption of direct red 2 onto all materials is endothermic in nature.

This result may be attributed to the enhanced mobility of direct red 2 ions at high temperature due to greater vibrational energies of the molecules, which facilitates the penetration of dye into the internal structure of LDHs and increases its diffusion in the pores [58]. These results fit well with those described for treating methyl orange by calcined, layered double hydroxide ZnAl-CO$_3$ [59].

$\Delta G°$, $\Delta H°$, and $\Delta S°$ values are reported in Table 4. For all adsorbents, the positive values of $\Delta H°$ ($\Delta H° > 0$) confirm that the adsorption process is endothermic [60]. The decrease in the negative $\Delta G°$ values with an increase in temperature show the spontaneous nature of adsorption, and indicates that the adsorption process becomes more favourable at a higher temperature [61].

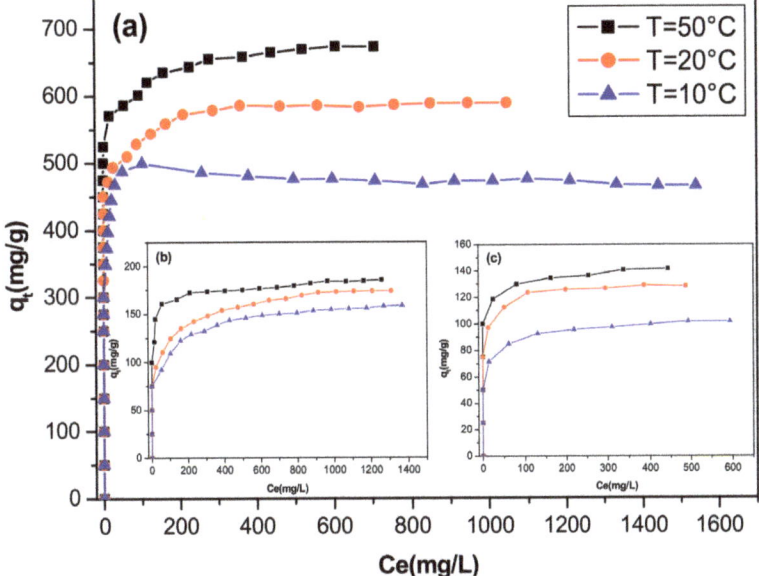

Figure 8. Effect of temperature on adsorption isotherms of direct red 2 on (a) CoFe-Ac/$_P$, (b) CoFe-Ac/$_{Ex}$, and (c) CoFe-CO$_3$/$_A$ LDHs.

The positive values of standard entropy $\Delta S°$ reflect the affinity of all adsorbent LDHs for direct red 2 dye in an aqueous medium and suggest that some structural changes occur on the adsorbent with the increase of randomness at the solid/liquid interface in the adsorption system [62].

Table 4. Thermodynamic parameters for the adsorption of direct red 2 onto LDH samples.

Compound	$\Delta H°$ (kJ/mol)	$\Delta S°$ (J/mol·K)	$\Delta G°$ (kJ/mol)		
			283K	293K	323K
CoFe-Ac/$_P$	41.31	202.54	−16.00	−19.04	−24.11
CoFe-Ac/$_{Ex}$	8.75	78.89	−13.57	−14.76	−16.73
CoFe-CO$_3$/$_A$	11.48	84.88	−12.54	−13.81	−15.93

3.2.5. Adsorption Isotherms

Adsorption data are often presented as an adsorption isotherm, which is important for understanding the interactions between adsorbent and adsorbate. Figure 9 shows the adsorption isotherms of direct red 2 on the three studied LDHs. They all display the typical L shape Sub-group 2 (Langmuir monolayer) according to Giles's classification [63], which indicates a great affinity between the adsorbate and the adsorbent and corresponds to the formation of a monolayer of dye.

According to the same figure, we can also observe the decrease in adsorption capacities (Q_{max}) following the series: CoFe-Ac/$_p$ ($\approx$588 mg/g) > CoFe-Ac/$_{Ex}$ ($\approx$170 mg/g) > CoFe-CO$_3$/$_A$ ($\approx$127 mg/g).

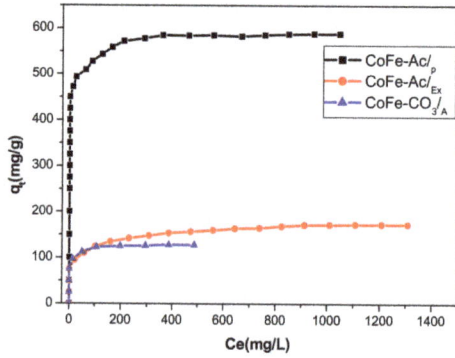

Figure 9. Adsorption isotherms of direct red 2 on CoFe-Ac/$_p$, CoFe-Ac/$_{Ex}$, and CoFe-CO$_3$/$_A$ LDHs. (pH = 6.6, T = 20 °C).

According to the curves of the tested models (Langmuir, Freundlich), Figure S7 and, following the higher correlation coefficients R^2 values close to 1 (Table 5), we can note that the Langmuir model is the best to describe the phenomenon of adsorption. The Toth model confirms that the Langmuir isotherm better describes this adsorption [64] (Figure S8 and Table S2).

Table 5. Isotherm parameters for the adsorption of direct red 2 onto LDH samples.

	Method/Adsorbent			
Langmuir Isothem	Q_{max} (mg/g)	K_L (l/mg)	R^2	Q_{max} (exp) (mg/g)
CoFe-Ac/$_p$	588.23	0.404	0.9999	588
CoFe-Ac/$_{Ex}$	175.44	0.034	0.9998	170
CoFe-CO$_3$/$_A$	128.205	0.426	0.9998	127
Freundlich Isotherm	K_f (mg/g)	n	R^2	
CoFe-Ac/$_p$	433.93	21.64	0.9914	-
CoFe-Ac/$_{Ex}$	1.82	5.12	0.9848	-
CoFe-CO$_3$/$_A$	81.95	12.93	0.9252	-

The Langmuir parameters Q_{max} and K_L were calculated and listed in Table 5. As shown, the K_L parameter reflects the affinity of LDHs for direct red 2 [65] and the calculated maximum adsorption capacities are very close to those obtained in experiments. However, the best adsorption capacity is observed for the CoFe-Ac/$_p$ ($Q_{max} \approx$ 588 mg/g), which corresponds to an uptake capacity of 3.4 times greater than that of CoFe-Ac/$_{Ex}$ and 4.6 times greater than that of CoFe-CO$_3$/$_A$. CoFe-Ac/$_p$ LDH prepared here in polyol medium presents high adsorption capacity of direct red 2 than CTAB-bentonite ($\approx$109.89 mg/g) [24] and calcined hydrotalcite ($\approx$417mg/g) [25]. Furthermore, this new adsorbent shows removal efficiency of 100% until 1.5 g/L of dye solution.

3.3. X-ray and IR Characterizations of the LDHs after Adsorption

Infrared spectra clearly confirm the adsorption of direct Red 2 on the three LDHs: CoFe-Ac/$_{Ex}$, CoFe-Ac/$_p$ (Figure 10), and CoFe-CO$_3$/A (Figure S9). The main characteristic bands of this dye are present. The absorption bands located between 1250 and 1000 cm^{-1} are assigned for the sulfonic SO$_3{}^-$ vibrations group. The absorption band at approximately 1600 cm^{-1} is attributed to the absorption of C-C aromatic stretch and the band at around 1500 cm^{-1} is ascribed to the azoic group [14].

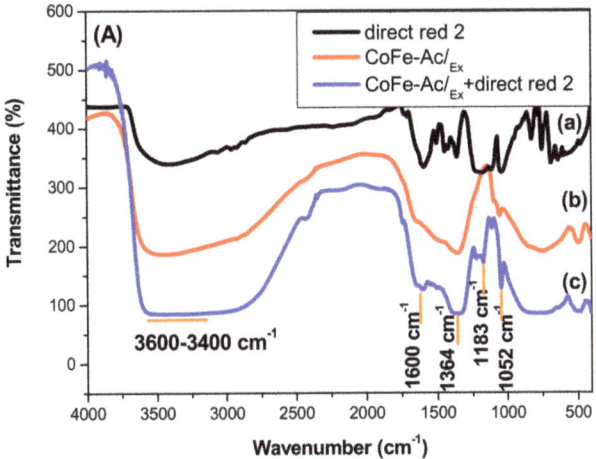

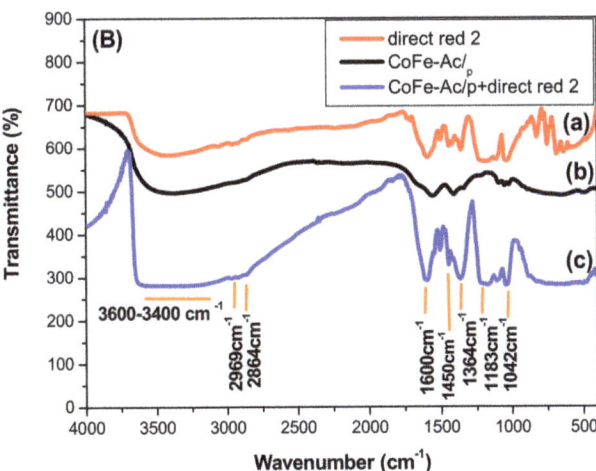

Figure 10. FT-IR spectra before and after adsorption of dye for (**A**) CoFe-Ac/$_{Ex}$, (**B**) CoFe-Ac/$_p$ LDHs.

However, the IR spectra of CoFe-CO$_3$/A and CoFe-Ac/$_{Ex}$ after adsorption still present the same asymmetrical shape in 1600–1300 cm^{-1} region as that before adsorption. In this region, we notice the intense band at 1364 cm^{-1} showing the predominance of the carbonate ion in the interlamellar space. Conversely, the spectrum of CoFe-Ac/$_p$ after adsorption has a more symmetrical shape in the same region. We observe the disappearance of the characteristic bands of acetate ion and the appearance of four bands in which two are intense and coincide with those of the direct red 2 anion, which shows that the acetate has been exchanged by this anion. However, the concomitant intercalation of the carbonate ion coming from air cannot be excluded since spectra of the two species overlap in the 1600–1300 cm^{-1} domain.

X-ray diffraction analysis sheds additional light. It shows that CoFe-CO$_3$/A (Figure S10) and CoFe-Ac/Ex (Figure 11A) maintain, before and after adsorption, the same interlayer distance revealing that the carbonate has not been exchanged by the direct red 2 anion and that adsorption has occurred only on the surface of the particle.

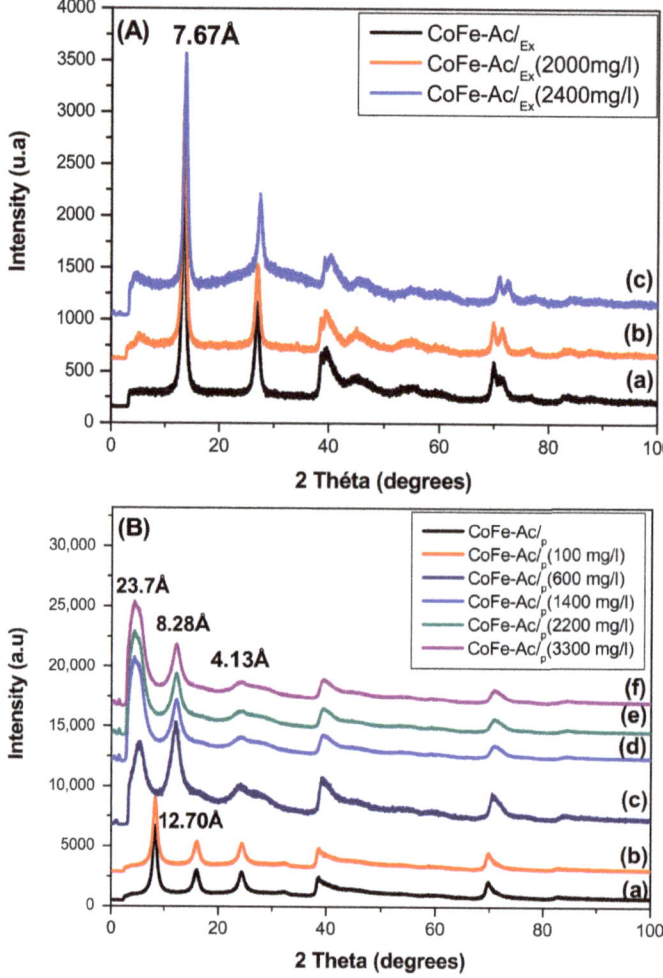

Figure 11. X-ray diffractogram before and after adsorption of dye for (**A**) CoFe-Ac/Ex, (**B**) CoFe-Ac/p LDHs.

Conversely, in the case of CoFe-Ac/p (Figure 11B), the characteristic interlamellar distance of the acetate ion (12.70 Å) disappears after adsorption, which results in two new distances (23.7 Å and 8.28 Å). This reveals that the direct red 2 anion was intercalated in the interlayer space.

4. Discussion: Mechanism of Direct Red 2 Removal and Comparison with Previous Works

Table 6 compares the results obtained during this work with the adsorption performance of azoic dyes belonging to the same family (with at least two sulfonate groups) by LDHs prepared via the common synthetic routes discussed in the introduction.

Table 6. Adsorption performances of azoic dyes belonging to the same family by LDHs prepared by different synthetic routes.

LDH	Ratio M^{2+}/M^{3+}	Dye	Initial d_{003} (Å)	Final d_{003} (Å)	Q_{max} (mg/g)	Reference
MgFe-CO$_3$	3/1	Acid Brown 14	≈7.8	≈7.8	41.7	[66]
C(MgFe-CO$_3$)			-	≈7.8	370.0	
MgAl-CO$_3$	2/1	Congo Red	7.58	7.88	129.9	[67]
C(MgAl-CO$_3$)			-	7.92	143.27	
MgAl-CO$_3$	2/1	Brilliant Blue R	7.43	7.55	54.59	[68]
C(MgAl-CO$_3$)			-	7.76	613.6	
MgAl-CO$_3$	2/1	Direct Red 2	7.57	7.77	153.88	[25]
C(MgAl-CO$_3$)			-	23.77	417.3	
MgAl-CO$_3$	3/1	Acid Green 68:1	7.6	7.6	99.1	[51]
C(MgAl-CO$_3$)			-	7.3	154.8	
ZnAl-CO$_3$	2/1	Congo Red	7.6	Not given	Not given	[69]
C(ZnAl-CO$_3$)			-	30.0	1540	
ZnAl-CO$_3$	2/1	RR (X-3B)	7.6	Not given	Not given	
C(ZnAl-CO$_3$)			-	7.6–8.0	390	
MgAl-SO$_4$	3/1	Remazol Brilliant Red 3FB	8.12	7.9	85	[19]
ZnAl-NO$_3$	2/1	Direct Red 16	8.84	11.78	69.85	[15]
MgAl-NO$_3$	2/1	Reactive blue 19	8.79	8.41	281	[16]
MgAl-SDS	2/1	Direct Blue G-RB	25.5	Not given	707.76	[18]
ZnAl-Cl	2/1	Evan Blue (EB)	7.73	20.6	0.512 mmol/g	[14]
CoFe-Ac/$_P$		Direct Red 2	12.70	23.77/8.28	588 (0.812 mmol/g)	This work
CoFe-Ac/$_{Ex}$	3/1		7.67	7.67	170 (0.235 mmol/g)	
CoFe-CO$_3$/$_A$			7.57	7.57	127 (0.175 mmol/g)	

C: Calcined.

As can be seen, CoFe-CO$_3$/$_A$ and CoFe-Ac/$_{Ex}$ behave similarly to LDHs prepared by coprecipitation in an air atmosphere. Their final interlamellar distances remain unchanged after adsorption and their removal capacities are low (128 and 175 mg/g). This is due to the great affinity of the carbonate anion with the positively charged LDH layers. Thus, the anionic exchange between the carbonate and the anionic dye is energetically disadvantaged and, exclusively, surface adsorption takes place (physisorption). This physisorption is confirmed by the small positive values of $\Delta H°$ ($\Delta H° < 40$ kJ/mol) obtained for these two compounds (Table 4) and by the low k_2 constant of the second order kinetic model (Table 3).

As shown in Table 6, LDHs with carbonate anion heated at about 500 °C present significantly higher adsorption capacities due to the high surface properties and to the memory effect resulting after calcination. In this case, besides physisorption on the external surface, the intercalation of the anionic dye appears to be the main sorption process during the reconstruction of the layer structure. The intercalation may lead to anionic dye standing perpendicularly to the lamellar sheets and giving a high basal spacing. The anionic dye may also be intercalated in a flat position, which results in a limited interlamellar distance [69]. Even if the calcination-reconstruction process clearly improves the LDH adsorption capacities, the intercalation of the anions remains below the theoretical value allowed by the M^{2+}/M^{3+} ratio. The charge compensation is fulfilled by the simultaneous intercalation of the carbonate anion present in the effluent as clearly shown by IR studies. Among the examples cited in Table 6, only the intercalation of Congo Red into calcined Zn-Al-CO$_3$ reaches the theoretical value (1540 mg/g). The performance of LDHs also prepared by coprecipitation and containing other intercalated anions depends on the size of these anions. LDH intercalated with small anions: nitrate, chloride, and sulfate have interlamellar distances very close to the LDH-CO$_3$ distance. Their adsorption performance is variable as shown in Table 6. For sulfate and nitrate, the adsorption capacity is medium or even low. Significantly larger anions such as dodecylsulfate (SDS) lead to the formation of LDH with a significant interlamellar distance (25.5 Å). This weakens cohesion interlayers and, therefore, facilitates the exchange. Such conditions are at the origin of the high adsorption capacity observed in this case (707.76 mg/g).

The sorbent proposed in the present work (LDH-CoFe-Ac/$_p$) is based on a new approach. This material was prepared by forced hydrolysis in a polyol medium. This route, described for the first time for the preparation of the Layered Hydroxide Salts (LHS) [70], has been extended to the synthesis of LDH based on the 3D transition elements [27,39,71]. This synthesis method presents two main advantages: (i) the reaction is conducted without atmosphere control since polyol avoids the carbonate anions' contamination and (ii) the pH is mainly controlled by the amount of acetate anion present in the precursor salts [70]. In this new LDH, the acetate anion was chosen as the intercalate anion. This leads to a relatively large inter-layer distance (12.70 Å). As dodecylsulphate, this likely weakens the cohesion of sheets and, thus, will favor the exchange reaction. Such factors enhance the adsorption capacity leading to the high adsorption efficiency observed for this compound (588 mg/g). The high $\Delta H°$ value obtained $\Delta H° > 40$ kJ/mole means that the adsorption is characterized by a physisorption effect followed by a chemical effect (intercalation reaction). This is in keeping with results reported on the sorption of EB and similar dyes on ZnAl-Cl [14] and the sorption of direct red 2 on calcined hydrotalcite [25]. The occurrence of both physisorption and chemical sorption via intercalation is also confirmed by the high constant K_2 of the second order-kinetic model (Table 3).

X-ray diffraction analyses clearly confirm these results. They show that the adsorption processes are mainly controlled by the dye concentration. When the concentration is low (100 mg/L), only physisorption occurred (adsorption on the external surface). No change in basal spacing (12.70 Å) was observed. XRD patterns remain identical before and after adsorption (Figure 11B(a,b)). At a higher concentration, the chemisorption phenomenon (exchange/intercalation) begins to act in addition to the surface adsorption. The basal spacing 12.70 Å disappeared giving way to the appearance of two new distances (23.7 Å and 8.28 Å) (Figure 11B(c–f)).

These results allowed us to deduce that the intercalation of direct red 2 gave rise to the formation of two LDH phases with the same brucite-like layers but with two different interlamellar distances. The first phase corresponding to the highest interlamellar distance (23.7 Å) results from the intercalation of direct red 2 in an inclined orientation toward the brucite-like layers. Taking into account the long axis length (26 Å) (Figure 1) and the thickness of the brucite-like layer (4.8 Å), the inclination angle calculated is about 46°. Similar interlamellar distance was obtained for the adsorption of the same dye on calcined hydrotalcite [25].

The second phase is characterized by two peaks: an intense, symmetrical peak at 8.28 Å and its harmonic at 4.13 Å with dissymmetrical shape and low intensity. This lamellar phase with an interlamellar distance of 8.28 Å can result from the intercalation of direct red 2 in an almost flat position parallel to the layers, as was previously observed for RR-3FB intercalated into MgAl-SO$_4$ [19] and for RR-X-3B intercalated into calcined ZnAl-CO$_3$ [69]. The short axis length of direct red 2 (3.9 Å) (Figure 1) is very close to that of RR-3FB (3.7 Å) and to that of RR-X-3B (4 Å). Taking into account the thickness of the brucite-like layer, a predicted interlamellar distance of 8.70 Å would be expected. This value corresponds to what was observed. This orientation corresponds to a slight inclination (7°) of the direct red 2 long axis relative to the layers.

There exists a balance between the two phases depending on the dye concentration in the solution. Relatively low concentrations favour the lamellar phase corresponding to the direct red 2 anion intercalated almost parallel to the layers (600 mg /L, Figure 11B(c)) (Interlamellar distance 8.28 Å). Higher concentrations facilitate the organization of anions in an inclined position relative to the layers (Interlamellar distance 23.7 Å) (Figure 11B(d–f)).

Lastly, the comparative study of both CoFe-Ac/$_P$ and CoFe-Ac/$_{Ex}$ brings an interesting insight into the relative importance of physisorption (external surface adsorption) and chemical sorption (intercalation). CoFe-Ac/$_{Ex}$, containing carbonate as an intercalate anion, was obtained by a topotactic anion exchange reaction from the CoFe-Ac/p compound. It follows that its microstructure likely remains identical (porosity and specific surface). Thus, we can suppose that the external surface adsorption for CoFe-Ac/p is very close to that of CoFe-Ac/Ex (i.e., 170 mg/g). This makes it possible to estimate the contribution of intercalaion around 418 mg/g showing that chemical sorption is predominant. It corresponds to 2/3 of the theoretical exchange capacity. This difference observed with respect to the theoretical intercalation rate appears to be a common characteristic of the LDHs used as adsorbents, regardless of their synthesis method (see references cited in Table 6 and Reference [17]). In the present case, the compensating interlayer anions are mostly acetate. However, the concomitant intercalation of the carbonate ion coming from air cannot be excluded since the spectra of the two species overlap in the 1600–1300 cm^{-1} domain. Such a phenomenon was also observed regarding other organic dyes such as Methyl Orange adsorbed on Ni and ZnAl-LDH where compensating anions are nitrate and/or carbonate [72] and in the case of fluorescein on ZnAl-Hydrotalcite with perchlorate as a compensating anion [17].

5. Conclusions

In the present study, a new CoFe-Ac/p LDH was prepared by forced hydrolysis in a polyol medium, and its structural properties were confirmed by complementary characterization techniques.

The synthesised CoFe-Ac/p LDH has shown anion exchange properties. Indeed, acetate interlayer species were successfully exchanged by carbonate anions with a topotactic reaction.

The adsorption capacities of dye on CoFe-Ac/p were significantly affected by the initial contact time, initial dye concentration, pH, temperature, and by the structural properties (the nature of the initial interlayer anion and morphology).

In comparison with CoFe-Ac/Ex and CoFe-CO$_3$/A LDHs, it is demonstrated that the CoFe-Ac/p compound displays unique adsorption properties for direct red 2 dye. The adsorption was induced by physisorption at a low concentration of dye. High concentrations of dye favour the intercalation of dye anions via the exchange reaction with acetate anions. In addition to physisorption, this intercalation, which is considered as a chemical sorption, confers to this material an efficient uptake capacity ($\approx$588 mg/g).

The intercalation of the azo dye anion leads to the formation of two LDH phases differing in the orientation of this anion relative to the layers. The first phase corresponds to an intercalation of the anion almost parallel to the layers. In the second phase, the azoic anions are inclined about 46° relative to the layers. Higher concentrations favour the latter orientation.

Lastly, when taking into account the obtained results, the forced hydrolysis in a polyol medium constitutes an original way and easy method to elaborate layered double hydroxides exempt of contamination by carbonates with controlled morphology, nanometric size, a more significant surface area, and, consequently, a good dispersion of particles. In addition, the acetate anion has been chosen as the intercalated specie since it leads to high interlamellar distance facilitating the exchange reaction. Altogether, these characteristics are at the origin of the efficient adsorbent capacity of this material and, accordingly, confer on it a major interest in the retention of the pollutants contained in the industrial effluents.

Supplementary Materials: The following are available online at http://www.mdpi.com/1996-1944/13/14/3183/s1. Figure S1: SEM images of (a) CoFe-Ac/p, (b) CoFe-Ac/Ex, (c) CoFe-CO$_3$/A LDHs. Figure S2: TEM images of (a) CoFe-Ac/p. (b) CoFe-Ac/Ex, (c) CoFe-CO$_3$/A LDHs. Figure S3: UV-Vis-NIR of CoFe-Ac/p LDH. Figure S4: Mössbauer spectrum of CoFe-Ac/p LDH. Figure S5: N$_2$ adsorption–desorption isotherms of CoFe-Ac/p and CoFe-Ac/Ex LDHs. Table S1: Textural properties of CoFe-Ac/p and CoFe-Ac/Ex LDHs. Figure S6: (a) Pseudo-first-order and (b) pseudo-second-order kinetics for adsorption of direct red 2 on CoFeAc/p, CoFe-Ac/Ex, and CoFe-CO$_3$/A LDHs. Figure S7: (a) Freundlich and (b) Langmuir isotherms for adsorption of direct red 2 on CoFe-Ac/p, CoFe-Ac/Ex, and CoFe-CO$_3$/A LDHs. Figure S8: Toth isotherm for adsorption of direct red 2 on CoFe-Ac/p, CoFe-Ac/Ex, and CoFe-CO$_3$/A LDHs. Table S2: Isotherm parameters for the adsorption of direct red 2 onto LDH samples. Figure S9: FT-IR spectra, before and after adsorption, for CoFe-CO$_3$/A. Figure S10: X-ray diffractogram before and after adsorption for CoFe-CO$_3$/A.

Author Contributions: N.D.-S. and N.J. conceived and designed the experiments. N.D.-S. and P.L. performed the experiments. N.D.-S. analyzed the data. N.D.-S. wrote the original draft. N.D.-S. and N.J. wrote, edited and revised the paper. N.J. supervised the work. All authors have read and agreed to the published version of the manuscript.

Funding: This research received no external funding.

Acknowledgments: Thanks to J.Y. Piquemal for UV measurements, J.M. Grenèche for Mössbauer Measurements and for Jennifer Morrice for her help in proofreading and improving the English style and expressions in the manuscript.

Conflicts of Interest: The authors declare no conflict of interest.

References

1. Cavani, F.; Trifiro, F.; Vaccari, A. Hydrotalcite-Type Anionic Clays: Preparation, properties and applications. *Catal. Today* **1991**, *11*, 173–301. [CrossRef]
2. Rives, V.; del Arco, M.; Martín, C. Intercalation of drugs in layered double hydroxides and their controlled release: A review. *Appl. Clay Sci.* **2014**, *88–89*, 239–269. [CrossRef]
3. Richetta, M. Characteristics, Preparation Routes and Metallurgical Applications of LDHs: An Overview. *J. Mater. Sci. Eng.* **2017**, *6*, 397. [CrossRef]
4. Mishra, G.; Dash, B.; Pandey, S. Layered double hydroxides: A brief review from fundamentals to application as evolving biomaterials. *Appl. Clay Sci.* **2018**, *153*, 172–186. [CrossRef]
5. Duan, X.; Evans, D.G. Layered Double Hydroxydes. *Struct. Bond.* **2006**, *119*, 1–234.
6. Del Arco, M.; Trujillano, R.; Rives, V. Cobalt–iron hydroxycarbonates and their evolution to mixed oxides with spinel structure. *J. Mater. Chem.* **1998**, *8*, 761–767. [CrossRef]
7. Faour, A.; Prévot, V.; Taviot-Gueho, C. Microstructural study of different LDH morphologies obtained via different synthesis routes. *J. Phys.Chem. Solids* **2010**, *71*, 487–490. [CrossRef]
8. Qiao, C.; Zhang, Y.; Zhu, Y.; Cao, C.; Bao, X.; Xu, J. One-step synthesis of zinc–cobalt layered double hydroxide (Zn-Co-LDH) nanosheets for high-efficiency oxygen evolution reaction. *J. Mater. Chem. A* **2015**, *3*, 6878–6883. [CrossRef]
9. Zaghouane-Boudiaf, H.; Boutahala, M.; Arab, L. Removal of methyl orange from aqueous solution by uncalcined and calcined MgNiAl layered double hydroxides (LDHs). *J. Chem. Eng.* **2012**, *187*, 142–149. [CrossRef]
10. Mohapatra, L.; Parida, K.M. Zn-Cr layered double hydroxide: Visible light responsive photocatalyst for photocatalytic degradation of organic pollutants. *Separ. Purif. Tech.* **2012**, *91*, 73–80. [CrossRef]
11. Reichle, W.-T. Synthesis of anionic clay minerals (Mixed Metal Hydroxides, Hydrotalcite). *Solid State Ionics* **1986**, *22*, 135–141. [CrossRef]
12. Ulibarri, M.A.; Pavlovic, I.; Barriga, C.; Hermosín, M.C.; Cornejo, J. Adsorption of anionic species on hydrotalcite-like compounds: Effect of interlayer anion and crystallinity. *Appl. Clay Sci.* **2001**, *18*, 17–27. [CrossRef]
13. Chibwe, K.; Jones, W. Intercalation of organic and inorganic anions into layered double hydroxides. *J. Chem. Soc. Chem. Commun.* **1989**, *14*, 926–927. [CrossRef]
14. Marangoni, R.; Bouhent, M.; Taviot-Guého, C.; Wypych, F.; Leroux, F. Zn_2Al layered double hydroxides intercalated and adsorbed with anionic blue dyes: A physico-chemical characterization. *J. Colloid Interf. Sci.* **2009**, *333*, 120–127. [CrossRef]
15. Abdolmohammad-Zadeh, H.; Nejati, K.; Ghorbani, E. Synthesis, Characterization, and Application of Zn–Al Layered Double Hydroxide as a Nano-Sorbent for the Removal of Direct Red 16 from Industrial Wastewater Effluents. *Chem. Eng. Commun.* **2015**, *202*, 1349–1359. [CrossRef]
16. Alexandrica, M.C.; Silion, M.; Hritcu, D.; Popa, M.I. Layered Double Hyrdoxides as adsorbents for anionic dye removal from aqueous solutions. *J. Environ. Eng. Manag.* **2015**, *14*, 381–388.
17. Costantino, U.; Coletti, N.; Nocchetti, M.; Aloisi, G.G.; Elisei, F.; Latterini, L. Surface Uptake and Intercalation of Fluorescein Anions into Zn–Al–Hydrotalcite. Photophysical Characterization of Materials Obtained. *Langmuir* **2000**, *16*, 10351–10358. [CrossRef]

18. Wu, P.; Wu, T.; He, W.; Sun, L.; Li, Y.; Sun, D. Adsorption properties of dodecylsulfate-intercalated layered double hydroxide for various dyes in water. *Colloids Surf. A Physicochem. Eng. Aspects* **2013**, *436*, 726–731. [CrossRef]
19. El Hassani, K.; Jabkhiro, H.; Kalnina, D.; Beakou, B.H.; Anouar, A. Effect of drying step on layered double hydroxides properties: Application in reactive dye intercalation. *Appl. Clay Sci.* **2019**, *182*, 105246. [CrossRef]
20. Anantharaj, S.; Karthick, K.; Kundu, S. Evolution of layered double hydroxides (LDH) as high performance water oxidation electrocatalysts: A review with insights on structure, activity and mechanism. *Mater. Today Energy* **2017**, *6*, 1–26. [CrossRef]
21. Wang, R.; Wu, X.; Zou, C.; Li, X.; Du, Y. NOx Removal by Selective Catalytic Reduction with Ammonia over a Hydrotalcite-Derived NiFe Mixed Oxide. *Catalysts* **2018**, *8*, 384. [CrossRef]
22. Carja, G.; Chiriac, H.; Lupu, N. New magnetic organic–inorganic composites based on hydrotalcite-like anionic clays for drug delivery. *J. Magn. Magn. Mater.* **2007**, *311*, 26–30. [CrossRef]
23. Wang, L.; Liu, J.; Zhou, Y.; Song, Y.; He, J.; Evans, D.G. Synthesis of CoFe alloy nanoparticles embedded in a MgO crystal matrix using a single-source inorganic precursor. *R. Soc. Chem. Chem. Commun.* **2010**, *46*, 3911–3913. [CrossRef] [PubMed]
24. Bouberka, Z.; Khenifi, A.; Sekrane, F.; Bettahar, N.; Derriche, Z. Adsorption of Direct Red 2 on bentonite modified by cetyltrimethylammonium bromide. *J. Chem. Engr.* **2008**, *136*, 295–305.
25. Setti, N.D.; Jouini, N.; Derriche, Z. Sorption study of an anionic dye—Benzopurpurine 4B—On calcined and uncalcined Mg-Al layered double hydroxides. *J. Phys. Chem. Solids* **2010**, *71*, 556–559. [CrossRef]
26. Hanwell, M.D.; Curtis, D.E.; Lonie, D.C.; Vandermeersch, T.; Zurek, E.; Hutchison, G.R. Avogadro: An advanced semantic chemical editor, visualization, and analysis platform. *J. Cheminform.* **2012**, *4*. [CrossRef]
27. Drici Setti, N.; Derriche, Z.; Jouini, N. New acetate-intercalated CoNiFe, ZnNiFe and CoZnFe layered double hydroxides: Synthesis by forced hydrolysis in polyol medium and characterization. *J. Adv. Chem.* **2014**, *10*, 93097–93107.
28. Poul, L.; Ammar, S.; Jouini, N.; Fievet, F.; Villain, F. Synthesis of Inorganic Compounds (Metal, Oxide and Hydroxide) in Polyol Medium: A Versatile Route Related to the Sol-Gel Process. *J. Sol-Gel Sci. Technol.* **2003**, *26*, 261–265. [CrossRef]
29. Beji, Z.; Ben Chaabane, A.; Smiri, L.S.; Ammar, S.; Fiévet, F.; Jouini, N.; Greneche, J.M. Synthesis of nickel-zinc ferrite nanoparticles in polyol: Morphological, structural and magnetic properties. *Phys. Stat. Solid A* **2005**, *203*, 504–512. [CrossRef]
30. Aramendia, M.A.; Borau, V.; Jimé, N.C.; Maria, M.J.; José, R.F.; Rafael, R.J. Synthesis, Characterization, and ^{1}H and ^{71}Ga MAS NMR Spectroscopy of a Novel Mg/Ga Double Layered Hydroxide. *J. Solid State Chem.* **1997**, *131*, 78–83. [CrossRef]
31. Lagergren, S. Zur theorie der sogenannten adsorption gelosterstoffe. Kungliga Svenska Vetenskapsakademiem. *Handlingar* **1898**, *24*, 1–39.
32. Namasivayam, C.; Sumithra, C. Adsorptive removal of catechol on waste Fe (III)/Cr (III) hydroxide: Equilibrium and kinetics study. *Ind. Eng. Chem. Res.* **2004**, *43*, 7581–7587. [CrossRef]
33. Weber, J.R.; Morris, J.C. Kinetics of adsorption on carbon from solution. *J. Sanit. Eng. Div. Am. Soc. Civ. Eng.* **1963**, *89*, 31–59.
34. Chatterjee, S.; Lee, D.S.; Lee, M.W.; Wooa, S.H. Enhanced adsorption of congo red from aqueous solutions by chitosan hydrogel beads impregnated with cetyl trimethyl ammonium bromide. *Bioresour. Technol.* **2009**, *100*, 2803–2809. [CrossRef]
35. Freundlich, H.M.F. über die adsorption in lösungen. *Z. Phys. Chem.* **1906**, *57*, 385–470. [CrossRef]
36. Deliyanni, E.A.; Peleka, E.N.; Lazaridis, N.K. Comparative study of phosphates removal from aqueous solutions by nanocrystalline akaganéite and hybrid surfactant-akaganéite. *Sep. Purif. Tech.* **2007**, *52*, 478–486. [CrossRef]
37. Triantafyllidis, K.S.; Peleka, E.N.; Komvokis, V.G.; Mavros, P.P. Iron-modified hydrotalcite-like materials as highly efficient phosphate sorbents. *J. Colloid Interf. Sci.* **2010**, *342*, 427–436. [CrossRef]
38. Trujillano, R.; Holgado, M.J.; Pigazo, F.; Rives, V. Preparation, physicochemical characterisation and magnetic properties of Cu-Al layered double hydroxides with CO_3^{2-} and anionic surfactants with different alkyl chains in the interlayer. *Phys. B Condenser. Mater.* **2006**, *373*, 267–273. [CrossRef]

39. Taibi, M.; Ammar, S.; Schoenstein, F.; Jouini, N.; Fiévet, F.; Chauveau, T.; Greneche, J.-M. Powder and film of nickel and iron-layered double hydroxide: Elaboration in polyol medium and characterization. *J. Phys. Chem. Solids.* **2008**, *69*, 1052–1055. [CrossRef]
40. Oriakhi, O.; Farr, I.V.; Lerner, M.M. Incorporation of poly (acrylic acid), poly (vinylsulfonate) and poly (styrenesulfonate) within layered double hydroxides. *J. Mater. Chem.* **1996**, *6*, 103. [CrossRef]
41. Bruna, F.; Celis, R.; Pavlovic, I.; Barriga, C.; Cornejo, J.; Ulibarri, M.A. Layered double hydroxides as adsorbents and carriers of the herbicide (4-chloro-2-methylphenoxy)acetic acid (MCPA): Systems Mg-Al, Mg-Fe and Mg-Al-Fe. *J. Hazard Mater.* **2009**, *168*, 1476–1481. [CrossRef]
42. Choy, J.-H.; Kwon, Y.-M.; Han, K.-S.; Song, S.-W.; Chang, S.H. Intra- and inter-layer structures of layered hydroxy double salts, $Ni_{1-x}Zn_{2x}(OH)_2(CH_3CO_2)_{2x} \cdot nH2O$. *Mater. Lett.* **1998**, *34*, 356–363. [CrossRef]
43. Nakamoto, K. *Infrared and Raman Spectra of Inorganic and Coordination Compounds*, 4th ed.; Wiley: New York, NY, USA, 1986.
44. Verberckmoes, A.A.; Weckhuysen, M.; Schoonheydt, R.A. Spectroscopy and coordination chemistry of cobalt in molecular sieves. *Microporous Mesoporous Mater.* **1998**, *22*, 16. [CrossRef]
45. Noronha, F.B.; Perez, C.A.; Schmal, M.; Férty, R. Determination of cobalt species in niobia supported catalysts. *Phys. Chem. Chem. Phys.* **1999**, *1*, 2861. [CrossRef]
46. Sharma, P.-K.; Dutta, R.-K.; Pandey, A.-C.; Layek, S.; Verma, H.C. Effect of iron doping concentration on magnetic properties of ZnO nanoparticles. *J. Magn. Magn. Mater.* **2009**, *321*, 2587–2591. [CrossRef]
47. Morpurgo, S.; Lo Jacono, M.; Porta, P. Copper-zinc-cobalt-chromium hydroxycarbonates and oxides. *J. Solid State Chem.* **1995**, *119*, 246–253. [CrossRef]
48. Taibi, M.; Ammar, S.; Jouini, N.; Fievet, F. Layered nickel-cobalt hydroxyacetates and hydroxycarbonates: Chimie douce synthesis and structural features. *J. Phys. Chem. Solids.* **2006**, *67*, 932–937. [CrossRef]
49. Carja, G.; Nakamura, R.; Aida, T.; Niiyama, H. Textural properties of layered double hydroxides: Effect of magnesium substitution by copper or iron. *Microporous Mesoporous Mater.* **2001**, *47*, 275–284. [CrossRef]
50. Aramendia, M.A.; Borau, V.; Jimenez, C.; Marinas, J.M.; Ruiz, J.R.; Urbano, F.J. Comparative Study of Mg/M(III) (M = Al, Ga, In) Layered Double Hydroxides Obtained by Coprecipitation and the Sol–Gel Method. *J. Solid State Chem.* **2002**, *168*, 156–161. [CrossRef]
51. Dos Santos, R.M.M.; Gonçalves, R.G.L.; Constantino, V.R.L.; Costa, L.M.D.; Silva, L.H.M.D.; Tronto, J.; Pinto, F.G. Removal of Acid Green 68:1 from aqueous solutions by calcined and uncalcined layered double hydroxides. *Appl. Clay Sci.* **2013**, *80–81*, 189–195. [CrossRef]
52. Li, Y.; Gao, B.; Wu, T.; Chen, W.; Li, X.; Wang, B. Adsorption kinetics for removal of thiocyanate from aqueous solution by calcined hydrotalcite. *Colloids Surf. A Physicochem. Eng. Aspects* **2009**, *325*, 38–43. [CrossRef]
53. Nandi, B.K.; Goswami, A.; Purkait, M.K. Adsorption characteristics of brilliant green dye on kaolin. *J. Hazard Mater.* **2009**, *161*, 387–395. [CrossRef]
54. Kannan, K.; Sundaram, M.M. Kinetics and mechanism of removal of methylene blue by adsorption on various carbons—A comparative study. *Dyes Pigments* **2001**, *51*, 25. [CrossRef]
55. El Gaini, L.; Lakraimi, M.; Sebbar, E.; Meghea, A.; Bakasse, M. Removal of indigo carmine dye from water to Mg-Al-CO_3-calcined layered double hydroxides: Review. *J. Hazard Mater.* **2009**, *61*, 627–632. [CrossRef]
56. Miyata, S. Anion-exchange properties of hydrotalcite-like compounds. *Clays Clay Miner.* **1983**, *31*, 305–311. [CrossRef]
57. Lv, L.; He, J.; Evans, D.G.; Duan, X. Factors influencing the removal of fluoride from aqueous solution by calcined Mg-Al-CO3 layered double hydroxides. *J. Hazard Mater. B* **2006**, *133*, 119–128. [CrossRef] [PubMed]
58. Venkat, S.M.; Indra, D.M.; Vimal, C.S. Use of bagasse fly ash as an adsorbent for the removal of brilliant green dye from aqueous solution. *Dyes. Pigments* **2007**, *73*, 269–278.
59. Ni, Z.-M.; Xia, S.-J.; Wang, L.-G.; Xing, F.-F.; Pan, G.-X. Treatment of methyl orange by calcined layered double hydroxides in aqueous solution: Adsorption property and kinetic studies. *J. Colloid Interf. Sci.* **2007**, *316*, 284–291. [CrossRef]
60. Lian, L.; Guo, L.; Guo, C. Adsorption of Congo red from aqueous solutions onto Ca-bentonite. *J. Hazard Mater.* **2009**, *161*, 126–131. [CrossRef]
61. Zaki, A.B.; El-Sheikh, M.Y.; Evans, J.; El-Safty, S.A. Kinetics and Mechanism of the Sorption of Some Aromatic Amines onto Amberlite IRA-904 Anion-Exchange Resin. *J. Colloid Interf. Sci.* **2000**, *221*, 58–63. [CrossRef]

62. Gupta, V.K. Equilibrium Uptake, Sorption Dynamics, Process Development, and Column Operations for the Removal of Copper and Nickel from Aqueous Solution and Wastewater Using Activated Slag, a Low-Cost Adsorbent. *Ind. Eng. Chem. Res.* **1998**, *37*, 192–202. [CrossRef]
63. Giles, C.H.; MacEwan, T.H.; Nakhwa, S.N.; Smith, D. Studies in adsorption. Part XI. A system of classification of solution adsorption isotherms, and its use in diagnosis of adsorption mechanisms and in measurement of specific surface areas of solids. *J. Chem. Soc.* **1960**, *3*, 3973–3993. [CrossRef]
64. Toth, J. Gas-(Dampf-)Adsorption auf festen Oberflecken inhomogener Aktivität. *Acta Chim. Acad. Sci. Hung.* **1962**, *32*, 39.
65. Auxilio, A.R.; Andrews, P.C.; Junk, P.C.; Spiccia, L.; Neumann, D.; Raverty, W.; Vanderhoek, N. Adsorption and intercalation of Acid Blue 9 on Mg–Al layered double hydroxides of variable metal composition. *Polyhedron* **2007**, *26*, 3479–3490. [CrossRef]
66. Guo, Y.; Zhu, Z.; Qiu, Y.; Zhao, J. Enhanced adsorption of acid brown 14 dye on calcined Mg/Fe layered Enhanced adsorption of acid brown 14 dye on calcined Mg/Fe layered double hydroxide with memory effect. *J. Chem. Eng.* **2013**, *219*, 69–77. [CrossRef]
67. Li, B.; Zhang, Y.; Zhou, X.M.; Liu, Z.; Li, X. Different dye removal mechanisms between monodispersed and Different dye removal mechanisms between monodispersed and uniform hexagonal thin plate-like MgAlCO$_3^{2-}$ LDH and its calcined product in efficient removal of Congo red from water. *J. Alloy. Compd.* **2016**, *673*, 265–271. [CrossRef]
68. Zhu, M.; Li, H.; Xie, M.; Xin, H. Sorption of anionic dye by uncalcined and calcined Layered Double Hydroxides. *J. Hazard Mater.* **2005**, *120*, 163–171. [CrossRef]
69. Huang, G.; Sun, Y.; Zhao, C.; Zhao, Y.; Song, Z.; Chen, S.; Ma, J.; Du, J.; Yin, Z. Water—n-BuOH solvothermal synthesis of ZnAl-LDHs with different morphologies and its calcined product in efficient dyes removal. *J. Colloid Interf. Sci.* **2017**, *494*, 215–222. [CrossRef]
70. Poul, L.; Jouini, N.; Fievet, F. Layered hydroxide metal acetates (metal = zinc, cobalt, and nickel): Elaboration via hydrolysis in polyol medium and comparative study. *Chem. Mater.* **2000**, *12*, 3123–3132. [CrossRef]
71. Prevot, V.; Forano, C.; Besse, J.P. Hydrolysis in polyol: New route for hybrid-layered double hydroxides preparation. *Chem. Mater.* **2005**, *17*, 6695–6701. [CrossRef]
72. Mandal, S.; Lerner, D.A.; Marcotte, N.; Tichit, D. Structural characterization of azoic dye hosted layered double hydroxides. *Z. Krist.* **2009**, *224*, 282–286. [CrossRef]

© 2020 by the authors. Licensee MDPI, Basel, Switzerland. This article is an open access article distributed under the terms and conditions of the Creative Commons Attribution (CC BY) license (http://creativecommons.org/licenses/by/4.0/).

MDPI
St. Alban-Anlage 66
4052 Basel
Switzerland
Tel. +41 61 683 77 34
Fax +41 61 302 89 18
www.mdpi.com

Materials Editorial Office
E-mail: materials@mdpi.com
www.mdpi.com/journal/materials

www.ingramcontent.com/pod-product-compliance
Lightning Source LLC
LaVergne TN
LVHW070652100526
838202LV00013B/942